CHAPMAN & HALL/CRC APPLIED MATHEMATICS
AND NONLINEAR SCIENCE SERIES

Stochastic Partial
Differential Equations

T0200517

Published Titles

Computing with hp-ADAPTIVE FINITE ELEMENTS, Volume 1 One and Two Dimensional Elliptic and Maxwell Problems, Leszek Demkowicz

CRC Standard Curves and Surfaces with Mathematica®*: Second Edition,* David H. von Seggern

Exact Solutions and Invariant Subspaces of Nonlinear Partial Differential Equations in Mechanics and Physics, Victor A. Galaktionov and Sergey R. Svirshchevskii

Geometric Sturmian Theory of Nonlinear Parabolic Equations and Applications, Victor A. Galaktionov

Introduction to Fuzzy Systems, Guanrong Chen and Trung Tat Pham

Introduction to non-Kerr Law Optical Solitons, Anjan Biswas and Swapan Konar

Introduction to Partial Differential Equations with MATLAB®, Matthew P. Coleman

Mathematical Methods in Physics and Engineering with Mathematica, Ferdinand F. Cap

Optimal Estimation of Dynamic Systems, John L. Crassidis and John L. Junkins

Quantum Computing Devices: Principles, Designs, and Analysis, Goong Chen, David A. Church, Berthold-Georg Englert, Carsten Henkel, Bernd Rohwedder, Marlan O. Scully, and M. Suhail Zubairy

Stochastic Partial Differential Equations, Pao-Liu Chow

Forthcoming Titles

Computing with hp-ADAPTIVE FINITE ELEMENTS, Volume II Frontiers: Three Dimensional Elliptic and Maxwell Problems with Applications, Leszek Demkowicz, Jason Kurtz, David Pardo, Maciej Paszynski, Waldemar Rachowicz, and Adam Zdunek

Mathematical Theory of Quantum Computation, Goong Chen and Zijian Diao

Mixed Boundary Value Problems, Dean G. Duffy

Multi-Resolution Methods for Modeling and Control of Dynamical Systems, John L. Junkins and Puneet Singla

CHAPMAN & HALL/CRC APPLIED MATHEMATICS
AND NONLINEAR SCIENCE SERIES

Stochastic Partial Differential Equations

Pao-Liu Chow

Wayne State University
Detroit, Michigan, U.S.A.

CRC Press
Taylor & Francis Group
Boca Raton London New York

CRC Press is an imprint of the
Taylor & Francis Group, an **informa** business

A CHAPMAN & HALL BOOK

First published 2007 by Chapman & Hall

Published 2019 by CRC Press
Taylor & Francis Group
6000 Broken Sound Parkway NW, Suite 300
Boca Raton, FL 33487-2742

© 2007 by Taylor & Francis Group, LLC
CRC Press is an imprint of Taylor & Francis Group, an Informa business

First issued in paperback 2019

No claim to original U.S. Government works

ISBN-13: 978-0-367-45312-1 (pbk)
ISBN-13: 978-1-58488-443-9 (hbk)

Visit the Taylor & Francis Web site at
http://www.taylorandfrancis.com

and the CRC Press Web site at
http://www.crcpress.com

Library of Congress Cataloging-in-Publication Data

Chow, P. L. (Pao Liu), 1936-
  Stochastic partial differential equations / Pao-Liu Chow.
    p. cm. -- (Chapman & Hall/CRC applied mathematics & nonlinear science)
  Includes bibliographical references and index.
  ISBN-13: 978-1-58488-443-9 (alk. paper)
  ISBN-10: 1-58488-443-6 (alk. paper)
  1. Stochastic partial differential equations. I. Title. II. Series.

QA274.25.C48 2007
519.2--dc22   2006101018

Preface

This is an introductory book on Stochastic Partial Differential Equations, by which here we mean partial differential equations (PDEs) of evolutional type with coefficients (including the inhomogeneous terms) being random functions of space and time. In particular such random functions or random fields may be generalized random fields, for instance, spatially dependent white noises. In the case of ordinary differential equations (ODEs), this type of stochastic equations was made precise by the theory of Itô's stochastic integral equations. Before 1970, there was no general framework for the study of stochastic PDEs. Soon afterward, by recasting stochastic PDEs as stochastic evolution equations or stochastic ODEs in Hilbert or Banach spaces, a more coherent theory of stochastic PDEs, under the cover of stochastic evolution equations, began to develop steadily. Since then the stochastic PDEs are, more or less, synonymous with stochastic evolution equations. In contrast with the deterministic PDE theory, it began with the study of concrete model equations in mathematical physics and evolved gradually into a branch of modern analysis, including the theory of evolution equations in function spaces. As a relatively new area in mathematics, the subject is still in its tender age and has not yet received much attention in the mathematical community.

So far there are very few books on stochastic PDEs. The most well-known and comprehensive book on this subject is *Stochastic Equations in Infinite Dimensions* (Cambridge University Press, 1992) by G. Da Prato and J. Zabczyk. This book is concerned with stochastic evolution equations, mainly in Hilbert spaces, equipped with the prerequisite material, such as probability measures and stochastic integration in function spaces. When I decided to write a book on this subject, the initial plan was to expand and update a set of my lecture notes on stochastic PDEs and to turned it into a book. If I proceeded with the original plan, I envisioned the end product to be an expository book, with a somewhat different style, on a subject treated so well by the aforementioned authors. In the meantime, there was a lack of an introductory book on this subject, even today, without a prior knowledge of infinite-dimensional stochastic analysis. On second thought, I changed my plan and decided to write an introductory book that required only a graduate level course in stochastic processes, in particular, the ordinary Itô's differential equations, without assuming specific knowledge of partial differential equations. The main objective of the book is to introduce the readers conversant with basic probability theory to the subject and to highlight some of the computational and analytical techniques involved. The route to be taken is to first bring the subject back to its root in the classical concrete problems and then to proceed to a unified theory in stochastic evolution equations with some applications. At the end we will point out the connection of stochastic PDEs to infinite-dimensional stochastic analysis. Therefore the subject not only is of practical interest, but it also provides many challenging problems in stochastic analysis.

As an introductory book, an attempt is made to provide an overview of some relevant topics involved, but it is by no means exhaustive. For instance, we will not cover the martingale solutions, Poisson's type of white noises and the stochastic flow problems. The same thing can be said about the references given therein. Some of them may not be original but are more accessible or familiar to the author. Since, in various parts of the book, the material is taken from my published works, it is natural to cite my own papers more often. In order to come to the subject quickly without having to work through the highly technical stuff, the prerequisite material, such as probability measures and stochastic integration in function spaces, is not given systematically at the beginning. Instead, the necessary information will be provided, somewhat informally, wherever the need arises. In fact, as a personal experience, many students in probability theory tend to shy away from the subject due to its heavy prerequisite material. Hopefully, this "less technical" book has a redeemable value and will serve the purpose of enticing some interested readers to pursue the subject further. Through many concrete examples, the book may also be of interest to those who use stochastic PDEs as models in applied sciences.

Writing this book has been a long journey for me. Since the fall of 2003, I have worked on the book intermittently. In the process many people have been extremely helpful to me. First of all, I must thank my wife for her unfailing support and encouragement as well as her valuable assistance in technical typing. I am greatly indebted to my colleagues: Jose-Luis Menaldi, who read the first draft of the manuscript and pointed out some errors and omissions for correction and improvement; George Yin, who was very helpful in resolving many of the LaTeX problems, and to my long-time friend Hui-Hsiung Kuo, who encouraged me to write the book and provided his moral support. In my professional life, I owe my early career to my mentor Professor Joseph B. Keller of Stanford University, who initiated my interest in stochastic PDEs through applications and provided me a most stimulating environment in which I worked, for several years, at the Courant Institute of Mathematical Sciences, New York University. There I had the good fortune to learn a great deal from the late Professor Monroe D. Donsker whose inspiring lectures on integration in function spaces got me interested in the subject of stochastic analysis. To both of them I wish to express my deepest gratitude.

This book would not have been written if it were not prompted by a campus visit in the fall of 2002 from Robert B. Stern, Senior Editor of Chapman Hall/CRC Press. During the visit I showed him a copy of my lecture notes on stochastic PDEs. He then encouraged me to expand it into a book and later kindly offered me a contract. Afterward, for the reason given before, I decided to write quite a different type of book instead. That has caused a long delay in completing the book. To Mr. Stern I would like to give my hearty thanks for his continuing support and infinite patience during the preparation of my manuscript. Also I appreciate the able assistance by Jessica Vakili, the Project Coordinator, and Ari Silver, the Project Editor, in expediting the book's publication.

Pao-Liu Chow

Detroit, Michigan

Contents

Chapter 1

Preliminaries

1.1 Introduction

The theory of stochastic ordinary differential equations has been well developed since K. Itô introduced the stochastic integral and the stochastic integral equation in the mid 1940s [40, 41]. Therefore such an equation is also known as an Itô equation in honor of its originator. Due to its diverse applications ranging from biology and physics to finance, the subject has become increasingly important and popular. For an introduction to the theory and many references, one is referred to the books [2, 32, 39, 55, 61] among many others.

Up to the early 1960s, most works on stochastic differential equations had been confined to ordinary differential equations. Since then, spurred by the demand from modern applications, partial differential equations with random parameters, such as the coefficients or the forcing term, have begun to attract the attention of many researchers. Most of them were motivated by applications to physical and biological problems. Notable examples are turbulent flow in fluid dynamics, diffusion and waves in random media [5, 8]. In general the random parameters or random fields involved need not be of white-noise type, but, in many applications, models with white noises provide reasonable approximations. Besides, as a mathematical subject, they pose many interesting and challenging problems in stochastic analysis. By a generalization of the Itô equations in $\mathbf{R}^d$, it seems natural to consider stochastic partial differential equations of Itô type as a stochastic evolution equation in some Hilbert or Banach space. This book will be exclusively devoted to stochastic partial differential equations of Itô type.

The study of stochastic partial differential equations in a Hilbert space goes back to Baklan [3]. He proved the existence theorem for a stochastic parabolic equation or parabolic Itô equation by recasting it as an integral equation with the aid of the associated Green's function. This is the precursor to what is now known as the semigroup method and the solution in the Itô sense is called a mild solution [19]. Alternatively, by writing such an equation as an Itô integral equation with respect to the time variable only, the corresponding solution in a variational formulation is known as a strong solution. The existence and uniqueness of strong solutions were first discussed by Bensoussan and Temam [4, 5] and further developed by Pardoux [67], Krylov and Rozovskii [49], among many others. The aforementioned solutions are both weak solutions in the sense of partial differential equations. However there are other notions of weak solutions, such as distribution-valued solutions and the martingale solutions. The latter solutions will not be considered in this book in favor of more regular versions of solutions that are amenable to analytical techniques from stochastic analysis and partial differential equations.

In the deterministic case, partial differential equations originated from mathematical models for physical problems, such as heat conduction and wave propagation in continuous media. They gradually developed into an important mathematical subject both in theory and application. In contrast, the theoretical development of stochastic partial equations leapt from concrete problems to the general theory quickly. Most stochastic PDEs were treated as a stochastic evolution equation in a Hilbert or Banach space without close connection to specific equations arising from applications. As an introductory book, it seems wise to start by studying some prototypes of concrete equations before taking up a unified theory of stochastic evolution equations. With this in mind we shall first study the stochastic transport equation and then proceed to the stochastic heat and wave equations. To analyze these basic equations constructively, we will systematically employ the familiar tools in partial differential equations, such as the methods of eigenfunction expansions, the Green's functions and Fourier transforms, together with the conventional techniques in stochastic analysis. Then we are going to show how these concrete results lead to the investigation of stochastic evolution equations in a Hilbert space. The abstract theorems on existence, uniqueness and regularity of solutions will be proved and applied later to study the asymptotic behavior of solutions and other types of stochastic partial differential equations that have not been treated previously in the book.

To be more specific, the book can be divided into two parts. In the first part we begin with some well-known examples and a brief review of some basic results in stochastic processes and the ordinary Itô's equations. Then, in view of seemingly a disconnect of the general theory of stochastic evolution equations to concrete stochastic PDEs, the book will proceed to study some archetype of stochastic PDEs, such as the transport equation, the heat and wave equations. Following the tradition in PDEs, we will start with a class of first-order scalar equations in Chapter Two. For simplicity the random coefficients are assumed to be spatially dependent white noises in finite dimensions. They are relatively simple and their path-wise solutions can be obtained by the method of stochastic characteristics. In Chapter Three and Chapter Four, the stochastic parabolic equations, in bounded domain and the whole space, are treated, respectively. They will be analyzed by the Fourier methods: the method of eigenfunctions expansion in a bounded domain and the method of Fourier transform in the whole space. Combining the Fourier methods with some techniques for stochastic ODEs and the convergence of random functions, it is possible to prove, constructively, the existence, uniqueness and regularity properties of solutions. By adopting a similar approach, stochastic hyperbolic equations are analyzed in Chapter Five. So far, based on the Fourier methods and concrete calculations, without resorting to some heavy machinery in stochastic analysis, we are able to study the solutions of typical stochastic PDEs in explicit form. By the way, the methods of performing such calculations are also of interest and well worth knowing. After gaining some feelings about the solution properties of concrete stochastic PDEs, we move on to the second part of the book: the stochastic evolution equations we alluded to earlier. In Chapter Six, the existence and uniqueness theorems for linear and nonlinear stochastic evolution equations are proved for two kinds of solutions: the mild solution and the strong solution. The mild solution is usually associated with the semigroup approach, which is treated extensively in the book by Da Prato and Zabczyk [19], while the strong solution, based on the variational formulation, is not covered in that book. Under some stronger assumptions, such as the conditions of coercivity and monotone nonlinearity, the strong solutions have more regularity. As a consequence, the Itô formula holds for the corresponding stochastic evolution equations. This allows a deeper study of solutions, such as the asymptotic behavior of a large time or for small noises. In particular, the boundedness, stability and the existence of invariant measures for the solutions as well as small perturbation problems will be discussed in Chapter Seven. To show that the theorems given in Chapter Six has a wider range of applications, several more examples arising from physical models in turbulence are provided in Chapter Eight. Finally, in Chapter Nine, we will give a brief exposition on the connection between the stochastic PDEs and diffusion equations in infinite dimensions. In particular the associated Kolmogorov and Hopf equations are studied in some Gauss-Sobolev spaces.

This chapter serves as a general introduction to the subject. To motivate the study of stochastic PDEs, we shall first give several examples of stochastic partial differential equations arising from applied sciences. Then we briefly review some basic facts about stochastic processes and stochastic differential equations that will be needed in the subsequent chapters.

1.2   Some Examples

(Example 1) Nonlinear Filtering

In the nonlinear filtering theory, one is interested in estimating the state of a partially observable dynamical system by computing the relevant conditional probability density function. It was shown by Zakai [87] that an un-normalized conditional probability density function $u(x,t)$ satisfies the stochastic parabolic equation:

$$\frac{\partial u}{\partial t} = \frac{1}{2}\sum_{j,k=1}^{d}\frac{\partial}{\partial x_j}[a_{jk}(x)\frac{\partial u}{\partial x_k}] + \sum_{k=1}^{d}\frac{\partial}{\partial x_k}[g_k(x)u]$$

$$+[\sum_{k=1}^{d}h_j(x)\dot{w}_j(t)]u, \quad x \in \mathbf{R}^d, \, t > 0,$$

$$u(x,0) = u_0(x).$$

(1.1)

Here the coefficients a_{jk}, g_k and the initial state u_0 are given functions of x, and $\dot{V} = \frac{\partial}{\partial t}V$, where

$$V(x,t) = \sum_{k=1}^{d}h_j(x)w_j(t)$$

with known coefficients h_j and $w(t) = (w_1, \cdots, w_d)(t)$ is the standard Brownian motion or a Wiener process in $\mathbf{R}^d$. The formal derivative $\dot{w}(t)$ of the Brownian motion $w(t)$ is known as a white noise. Due to its important applications to systems science, this equation has attracted the attention of many workers in engineering and mathematics.

(Example 2) Turbulent Transport

In a turbulent flow, let $u(x,t)$ denote the concentration of a passive substance, such as the smoke particles or pollutants, which undergoes the molecular diffusion and turbulent transport. Let ν be the diffusion coefficient and let $v(x,t,\omega) = (v_1, v_2, v_3)(x,t,\omega)$ be the turbulent velocity field. The turbulent mixing of the passive substance contained in a domain $D \subset \mathbf{R}^3$ is governed by the following initial-boundary value problem:

$$\frac{\partial u}{\partial t} = \nu\Delta u - \sum_{k=1}^{3}v_k\frac{\partial u}{\partial x_k} + q(x,t),$$

$$\frac{\partial u}{\partial n}|_{\partial D} = 0, \quad u(x,0) = u_0(x),$$

(1.2)

where Δ is the Laplacian operator, $u_0(x)$ and $q(x,t)$ are the initial and the source distribution, respectively, and $\frac{\partial}{\partial n}$ denotes the normal derivative to the boundary ∂D. When the random velocity field $v(x,t)$ fluctuates rapidly, it may be approximated by a Gaussian white noise $\dot{W}(x,t)$ with a spatial parameter to be introduced later. Then the concentration $u(x,t)$ satisfies the parabolic Itô equation (1.2).

(Example 3) Random Schrödinger Equation

As is well known, the Schrödinger equation arises in quantum mechanics. In the case of a random potential, it takes the form

$$i\frac{\partial u}{\partial t} = \Delta u + \dot{V}(x,t)u, \quad x \in \mathbf{R}^d, t \in (0,T),$$

$$u(x,0) = u_0(x),$$

(1.3)

where $i = \sqrt{-1}$, u_0 is the initial state and $\dot{V}$ is a random potential. For $d = 2$, the equation (1.3) arises from time-harmonic random wave propagation problem under a forward scattering approximation (see Section 8.3), where t is the third space variable in the direction of wave propagation. At high frequencies, $\dot{V}(x,t)$ is taken to be a spatially dependent Gaussian white noise.

(Example 4) Stochastic Sine-Gordon Equation

The Sine-Gordon equation is used to describe the dynamics of coupled Josephson junctions driven by a fluctuating current source. As a continuous model, the time rate of change in voltage $u(x,t)$ at x satisfies the nonlinear stochastic wave equation

$$\frac{\partial^2 u}{\partial t^2} = \alpha \Delta u + \beta \sin u + \dot{V}(x,t), \quad x \in \mathbf{R}^d, t > 0,$$

$$u(x,0) = u_0(x), \quad \frac{\partial u}{\partial t}(x,0) = u_1(x),$$

(1.4)

where α, β are some positive parameters, u_0, u_1 are the initial states, and the current source $\dot{V}(x,t)$ is a spatially dependent Gaussian white noise. The equation (1.4) is an example of the hyperbolic Itô equations.

(Example 5) Stochastic Burgers Equation

The Burgers equation is a nonlinear parabolic equation which was introduced as a simplified model for the Navier-Stokes equations in fluid mechanics. In the statistical theory of turbulence, a random force method was proposed in the hope of finding a physically meaningful stationary distribution of turbulence (see Section 8.6). In the miniature version, one considers the randomly forced Burgers' equation in $D \subset \mathbf{R}^3$. The velocity field $u = (u_1, u_2, u_3)$ satisfies

$$\frac{\partial u}{\partial t} + (u \cdot \nabla)u = \nu \Delta u + +\dot{V}(x,t)$$

$$u|_{\partial D} = 0, \quad u(x,0) = u_0(x),$$

(1.5)

where

$$(u \cdot \nabla) = \sum_{k=1}^{3} u_k \frac{\partial}{\partial x_k},$$

$\nu > 0$ is the diffusion coefficient, $\dot{V}(x,t)$ is a spatially dependent white noise in $\mathbf{R}^3$, and $u_0(x)$ is the initial state which may be random. This equation will be considered later in Section 8.2.

1.3   Brownian Motions and Martingales

Let $(\Omega, \mathcal{F}, P)$ be a probability space, where Ω is a set with elements ω; $\mathcal{F}$ denotes the Borel σ−field of subsets of Ω, and P is a probability measure. A measurable function $X : \Omega \to \mathbf{R}^d$ is called a *random variable* with values in a d−dimensional Euclidean space. Let $\mathbf{T} = \mathbf{R}^+$ or be an interval in $\mathbf{R}^+ = [0, \infty)$. A family of random variables $X(t)$, or written as $X_t, t \in \mathbf{T}$, is known as a *stochastic process* with values in $\mathbf{R}^d$. It is said to be a *continuous stochastic process* if its sample function $X(t, \omega)$ or $X_t(\omega)$ is a continuous function of $t \in \mathbf{T}$ for almost every $\omega \in \Omega$. Continuity in probability or in the mean can be defined similarly as in the convergence of random variables. A stochastic process Y_t is said to be a *modification* or a *version* of X_t if $P\{\omega : X_t(\omega) = Y_t(\omega)\} = 1$ for each $t \in \mathbf{T}$. To check continuity, the following theorem is well known.

Theorem 2.1 (Kolmogorov's Continuity Criterion)   Let $X_t, t \in \mathbf{T}$, be a stochastic process in $\mathbf{R}^d$. Suppose that there exist positive constants α, β, C such that

$$E|X_t - X_s|^\alpha \leq C|t - s|^{1+\beta}, \tag{1.6}$$

for any $t, s \in \mathbf{T}$. Then X has a continuous version. Moreover, the process X_t is Hölder-continuous with exponent $\gamma < \alpha/\beta$.   □

Remark: In fact the above theorem holds for X_t being a stochastic process in a Banach space B with the Euclidean norm $|\cdot|$ in (1.6) replaced by the B−norm $\|\cdot\|$ (§1.4, [51]).

A stochastic process $w_t = (w_t^1, \cdots, w_t^d)$, $t \geq 0$, is called a (standard) *Brownian motion* or a *Wiener process* in $\mathbf{R}^d$ if it is a continuous Gaussian process with independent increments such that $w_0 = 0$ a.s. (almost surely), the mean $Ew_t = 0$ and the covariances $Cov\{w_t^i, w_s^j\} = \delta_{ij}(t \wedge s)$ for any $s, t > 0, i, j = 1, \cdots, d$. Here δ_{ij} denotes the Kronecker delta symbol with $\delta_{ij} = 0$ for $i \neq j$ and $\delta_{ii} = 1$, and $(t \wedge s) = \min\{t, s\}$.

Let $\{\mathcal{F}_t : t \in \mathbf{T}\}$, or simply $\{\mathcal{F}_t\}$, be a family of sub σ-fields of $\mathcal{F}$. It is called a *filtration* of the sub σ-fields if $\{\mathcal{F}_t\}$ is right continuous, increasing,

$\mathcal{F}_s \subset \mathcal{F}_t$ for $s < t$, and $\mathcal{F}_t$ contains all P-null sets for each $t \in \mathbf{T}$. A process $X_t, t \in \mathbf{T}$ is said to be $\mathcal{F}_t$-*adapted* if X_t is $\mathcal{F}_t$-measurable for each $t \in \mathbf{T}$. Given a stochastic process $X_t, t \in \mathbf{T}$, let $\mathcal{F}_t$ be the smallest σ-field generated by $X_s, s \le t$. Then X_t is $\mathcal{F}_t$-adapted.

Given a filtration $\{\mathcal{F}_t\}$, a $\mathcal{F}_t$-adapted stochastic process X_t is said to be a $\mathcal{F}_t$-*martingale* (or simply *martingale*) if it is integrable such that the following holds

$$E\{X_t|\mathcal{F}_s\} = X_s, \quad a.s. \text{ for any } t > s.$$

If X_t is a real-valued process, it is called a *submartingale (supermartingale)* if it satisfies

$$E\{X_t|\mathcal{F}_s\} \ge (\le) \ X_s, \quad a.s. \text{ for any } t > s.$$

If X_t is a $\mathbf{R}^d$-valued martingale with $E|X_t|^p < \infty, t \in \mathbf{T}$, for some $p \ge 1$, then it is called a L^p-*martingale*. It is easy to check that $|X_t|^p$ is a submartingale. The most well-known example of a continuous martingale is the Brownian motion $w_t, t \ge 0$. For $\mathbf{T} = [0, T]$ or $\mathbf{R}^+$, an extended real-valued random variable τ is called a *stopping time* if $\{\tau \le t\} \in \mathcal{F}_t$ holds for any $t \in \mathbf{T}$. Given a $\mathbf{R}^d$-valued process X_t, it is said to be a *local martingale* if there exists an increasing sequence of stopping times $\tau_n \uparrow \tau$ a.s. such that the stopped process $X_{t \wedge \tau_n}$ is a $\mathcal{F}_t$- martingale for each n. *Local submartingale* and *local supermartingale* are defined similarly. A continuous $\mathcal{F}_t$- adapted process is said to be a *semimartingale* if it can be written as the sum of a local martingale and a process of bounded variation. In particular, if b_t is a continuous $\mathcal{F}_t$- adapted process and M_t is a continuous martingale with quadratic variation process Q_t, then X_t defined by

$$X_t = \int_0^t b_s \, ds + M_t,$$

is a continuous semimartingale. If Q_t has a density q_t such that

$$Q_t = [M]_t = \int_0^t q_s \, ds,$$

then the pair (q_t, b_t) is called the *local characteristic* of the semimartingale X_t.

The following Doob's submartingale inequalities are well known and will be used often later on.

Theorem 2.2 (Doob's Inequalities)   Let $\xi_t, t \in \mathbf{T}$, be a positive continuous L^p-submartingale. Then for any $p \ge 1$ and $\lambda > 0$,

$$\lambda^p P\{\sup_{s \le t} \xi_s^p\} \le E\{\xi_t^p; \sup_{s \le t} \xi_s \ge \lambda\}, \tag{1.7}$$

and, for $p > 1$, the following holds

$$E\{\sup_{s \le t} \xi_s^p\} \le q^p E\{\xi_t^p\}, \tag{1.8}$$

for any $t \in \mathbf{T}$, where $q = p/(p-1)$.   □

Let $X_t, t \in [0, T]$, be a continuous real-valued $\mathcal{F}_t$- adapted stochastic process. Let $\pi_T = \{0 = t_0 < t_1 < \cdots < t_m = T\}$ be a partition of $[0, T]$ with $|\pi_T| = \max_{1 \le k \le m} (t_k - t_{k-1})$. Define $\xi_t(\pi_T)$ as

$$\xi_t(\pi_T) = \sum_{k=1}^{m} (X_{t_k \wedge t} - X_{t_{k-1} \wedge t})^2.$$

If, for any sequence of partitions π_T^n, $\xi_t(\pi_T^n)$ converges in probability to a limit $[X]_t$, or $[X_t]$, as $|\pi_T^n| \to 0$ for $t \in [0, T]$, then $[X]_t$ is called the *quadratic variation* of X_t. For instance, if $X_t = w_t$ is a Brownian motion, then $[w]_t = t$ a.s. If X_t is a process of bounded variation, then $[X]_t = 0$ a.s. Similarly, let $X_t, Y_t, 0 \le t \le T$ be two continuous real-valued, $\mathcal{F}_t$-adapted processes. Define

$$\eta_t(\pi_T) = \sum_{k=1}^{m} (X_{t_k \wedge t} - X_{t_{k-1} \wedge t})(Y_{t_k \wedge t} - Y_{t_{k-1} \wedge t}).$$

Then the *mutual variation*, or the *covariation* of X_t and Y_t, denoted by $\langle X, Y \rangle_t$ or $\langle X_t, Y_t \rangle$ is defined as the limit of $\eta_t(\pi_T^n)$ in probability as $|\pi_T^n| \to 0$. Alternatively the mutual variation can be expressed in terms of the quadratic variations as follows

$$\langle X, Y \rangle_t = \frac{1}{4} \{ [X + Y]_t - [X - Y]_t \}. \tag{1.9}$$

1.4   Stochastic Integrals

The stochastic integral was first introduced by K. Itô based on a standard Brownian motion. It was later generalized to that of a local martingale and semimartingale. Here we confine our exposition to the special case of continuous L^2-martingale. Let M_t be a continuous, real-valued L^2-martingale and let $f(t)$ be a continuous adapted process in $\mathbf{R}$ for $0 \le t \le T$. For any partition $\triangle_T^n = \{0 = t_0 < t_1 < \cdots < t_n = T\}$, define

$$I_t^n = \sum_{k=1}^{n} f_{t_{k-1} \wedge t}(M_{t_k \wedge t} - M_{t_{k-1} \wedge t}). \tag{1.10}$$

Then I_t^n is a continuous martingale with the quadratic variation

$$[I^n]_t = \int_0^t f_s^n d[M]_s,$$

where $f_t^n = f_{t_{k-1}}$ for $t_{k-1} \le t < t_k$.

Suppose that

$$\int_0^T |f_t|^2 \, d[M]_t < \infty, \quad \text{a.s.} \tag{1.11}$$

Then the sequence I_t^n will converge uniformly in probability as $|\triangle_T^n| \to 0$ to a limit

$$I_t = \int_0^t f_s \, dM_t, \tag{1.12}$$

which is independent of the choice of the partition. The limit I_t is called the *Itô integral* of f_t with respect to the martingale M_t. Instead of I_t^n given by (1.10), define

$$J_t^n = \sum_{k=1}^n \frac{1}{2} [f_{t_{k-1} \wedge t} + f_{t_k \wedge t}] (M_{t_k \wedge t} - M_{t_{k-1} \wedge t}). \tag{1.13}$$

The corresponding limit J_t of J_t^n written as

$$J_t = \int_0^t f_s \circ dM_t, \tag{1.14}$$

is known as the *Stratonovich integral* of f_t with respect to M_t. Similar to the Itô integral, the Stratonovich integral (1.14) is a generalization from the case when $M_t = w_t$ is a Brownian motion.

Remarks: For simplicity, the integrand f_t was assumed to be a continuous adapted process. In fact it is known that the stochastic integrals introduced above can be defined for a more general class of integrands, known as predictable processes. Technically, a *predictable σ-field* $\mathcal{P}$ is the smallest σ-field on the product set $[0, T] \times \Omega$ generated by subsets of the form: $(s, t] \times B$ with $(s, t] \subset [0, T]$, $B \in \mathcal{F}_s$, and $\{0\} \times B_0$ with $B_0 \in \mathcal{F}_0$. A stochastic process f_t is said to be *predictable* if the function: $(t, \omega) \to f_t(\omega)$ is $\mathcal{P}$-measurable on $[0, T] \times \Omega$. In particular, a left-continuous adapted process is predictable.

Theorem 3.1 Let M_t and f_t be given as above. Then the Itô integral $I_t = \int_0^t f_s \, dM_t$ is a continuous local martingale satisfying $E\, I_t = 0$ and

$$[I]_t = \int_0^t |f_s|^2 \, d[M]_s, \quad \text{a.s.} \tag{1.15}$$

Moreover the Stratonovich integral is related to the Itô integral as follows

$$\int_0^t f_s \circ dM_t = \int_0^t f_s \, dM_t + \frac{1}{2} \langle f, M \rangle_t. \qquad \square \tag{1.16}$$

Now let $Z_t = (Z_t^1, \cdots, Z_t^d)$ be a continuous martingale and let $b_t = (b_t^1, \cdots, b_t^d)$ be an adapted integrable process over $[0, T]$ in $\mathbf{R}^d$. Let $X_t = (X_t^1, \cdots, X_t^d)$ be a continuous semimartingale defined by

$$X_t^i = \int_0^t b_s^k \, ds + Z_t^i, \quad i = 1, \cdots, d. \tag{1.17}$$

In what follows we will quote the famous Itô's formula.

Theorem 3.2 (Itô's Formula)   Let X_t be a continuous semimartingale given by (1.17). Suppose that $\Phi : \mathbf{R}^d \times [0, T] \to \mathbf{R}$ is a continuous function such that $\Phi(x, t)$ is continuously differentiable twice in x and once in t. Then the following formula holds

$$\begin{aligned}
\Phi(X_t, t) = \ & \Phi(X_0, 0) + \int_0^t \frac{\partial \Phi}{\partial s}(X_s, s) \, ds + \sum_{i=1}^d \int_0^t \frac{\partial \Phi}{\partial x_i}(X_s, s) \, dZ_s^i \\
& + \frac{1}{2} \sum_{i,j=1}^d \int_0^t \frac{\partial^2 \Phi}{\partial x_i \partial x_j}(X_s, s) \, d\langle Z^i, Z^j \rangle_s.
\end{aligned} \tag{1.18}$$

If, in addition, $\Phi(x, t)$ is three-time differentiable in x, then the above formula can be written simply as

$$\Phi(X_t, t) = \Phi(X_0, 0) + \int_0^t \frac{\partial \Phi}{\partial s}(X_s, s) \, ds + \sum_{i=1}^d \int_0^t \frac{\partial \Phi}{\partial x_i}(X_s, s) \circ dZ_s^i. \ \Box \tag{1.19}$$

In particular, let Z_t be a $\mathbf{R}^d$-valued Itô's integral with respect to the standard Brownian motion w_t in $\mathbf{R}^m$ defined by

$$Z_t^i = \sum_{j=1}^m \int_0^t \sigma_{ij}(s) \, dw_s^j, \quad i = 1, \cdots, d, \tag{1.20}$$

where $\sigma_{ij}(t), i = 1, \cdots, d, j = 1, \cdots, m$, are predictable processes so that

$$\int_0^t \sigma_{ij}^2(s) ds < \infty, \quad a.s. \quad \text{for } i = 1, \cdots, d, \ j = 1, \cdots, m.$$

Then the corresponding Itô integrals exist and are local martingales, and the equation (1.18) yields the conventional Itô's formula

$$\begin{aligned}
\Phi(X_t, t) = \ & \Phi(X_0, 0) + \int_0^t \frac{\partial \Phi}{\partial s}(X_s, s) \, ds + \sum_{i=1}^d \int_0^t \frac{\partial \Phi}{\partial x_i}(X_s, s) b_s^i \, ds \\
& + \sum_{i=1}^d \sum_{j=1}^m \int_0^t \frac{\partial \Phi}{\partial x_i}(X_s, s) \sigma_{ij}(s) \, dw_s^j \\
& + \frac{1}{2} \sum_{i,j=1}^d \sum_{k=1}^m \int_0^t \sigma_{ik}(s) \sigma_{jk}(s) \frac{\partial^2 \Phi}{\partial x_i \partial x_j}(X_s, s) \, ds.
\end{aligned} \tag{1.21}$$

For $x, y \in \mathbf{R}^d$, let (x, y) denote the inner product of x and y, and let $D\phi(x)$ be the gradient vector of ϕ. In the matrix notation, a vector will be regarded as a column matrix, and a $m \times n$ -matrix A with entries a_{ij} will be written as $[a_{ij}]_{m \times n}$ or simply $[a_{ij}]$. Denote by $D^2\phi(x) = [\frac{\partial^2 \phi(x)}{\partial x_i \partial x_j}]_{(d \times d)}$ the Hessian matrix of ϕ and $[Z]_t = [\langle Z^i, Z^j \rangle_t]_{(d \times d)}$ being the quadratic variation matrix of Z. Then the Itô formulas (1.18) and (1.19) can be written, respectively, as

$$
\begin{aligned}
\Phi(X_t, t) = {} & \Phi(X_0, 0) + \int_0^t \frac{\partial \Phi}{\partial s}(X_s, s)\, ds + \int_0^t (D\Phi(X_s, s), b_s)\, ds \\
& + \int_0^t (D\Phi(X_s, s), dZ_s) + \frac{1}{2} \int_0^t Tr\{D^2\Phi(X_s, s)\, d\,[Z]_s\},
\end{aligned}
\tag{1.22}
$$

$$
\begin{aligned}
\Phi(X_t, t) = {} & \Phi(X_0, 0) + \int_0^t \frac{\partial \Phi}{\partial s}(X_s, s)\, ds + \int_0^t (D\Phi(X_s, s), b_s)\, ds \\
& + \int_0^t (D\Phi(X_s, s), \circ dZ_s),
\end{aligned}
\tag{1.23}
$$

where, for a matrix $A = [a_{ij}]_{d \times d}$, $Tr\, A = \sum_{i=1}^d a_{ii}$ denotes the trace of A. Similarly, let $\sigma = [\sigma_{ij}]_{d \times m}$ be a diffusion matrix with its transpose denoted by σ^*. One can also rewrite (1.21) as

$$
\begin{aligned}
\Phi(X_t, t) = {} & \Phi(X_0, 0) + \int_0^t \frac{\partial \Phi}{\partial s}(X_s, s)\, ds + \int_0^t (D\Phi(X_s, s), b_s)\, ds \\
& + \int_0^t (D\Phi(X_s, s), \sigma(s)\, dw_s) + \frac{1}{2} \int_0^t Tr\{D^2\Phi(X_s, s)\sigma(s)\sigma^*(s)\}\, ds.
\end{aligned}
\tag{1.24}
$$

By means of Itô's formula and a Doob's inequality, it can be shown that the following well-known Burkholder-Davis-Gundy (B-D-G) inequality holds true.

Theorem 3.3 (B-D-G Inequality)   Let $M_t, t \in [0, T]$ be any continuous real-valued martingale with $M_0 = 0$ and $E|M_T|^p < \infty$. Then for any $p > 0$, there exist two positive constants c_p and C_p such that

$$
c_p E[M]_T^{p/2} \leq E\{\sup_{t \leq T} |M_t|^p\} \leq C_p E[M]_T^{p/2}. \qquad \square
\tag{1.25}
$$

As a consequence, if Z_t is the Itô integral given by (1.20), then we have

Corollary 3.4   For any $p > 0$, there exists a constant $K_p > 0$ such that

$$
E\{\sup_{t \leq T} |\int_0^t \sigma(s)\, dw_s|^p\} \leq K_p E\{\int_0^T Tr\,[\sigma(s)\sigma^*(s)]\, ds\}^{p/2},
\tag{1.26}
$$

provided that $E\{\int_0^T Tr\,[\sigma(s)\sigma^*(s)]\, ds\}^{p/2} < \infty$. $\qquad \square$

1.5   Stochastic Differential Equations

Let $b : \mathbf{R}^d \times [0, T] \rightarrow \mathbf{R}^d$ and $\sigma : \mathbf{R}^d \times [0, T] \rightarrow \mathbf{R}^{d \times m}$ be vector and matrix-valued functions, respectively. Consider Itô's differential equation in $\mathbf{R}^d$:

$$dx(t) = b(x(t), t)\, dt + \sigma(x(t), t)\, dw_t,$$
$$x(0) = \xi, \tag{1.27}$$

where w_t is a Brownian motion in $\mathbf{R}^m$, and ξ is a $\mathcal{F}_0$- measurable $\mathbf{R}^d$-valued random variable. By convention, this differential equation is interpreted as the following stochastic integral equation

$$x(t) = \xi + \int_0^t b(x(s), s)\, ds + \int_0^t \sigma(x(s), s) d\, w_s, \quad 0 \le t \le T. \tag{1.28}$$

Under the usual Lipschitz continuity and linear growth conditions, the following existence theorem holds true.

Theorem 4.1 (Existence and Uniqueness)   Let the coefficients $b(x, t)$ and $\sigma(x, t)$ of the equation (1.28) be measurable functions satisfying the following conditions:

(1) For any $x \in \mathbf{R}^d$ and $t \in [0, T]$, there exists a constant $C > 0$, such that

$$|b(x, t)|^2 + Tr\,[\sigma(x, t)\sigma^*(x, t)] \le C\,(1 + |x|^2).$$

(2) For any $x, y \in \mathbf{R}^d$ and $t \in [0, T]$,

$$|b(x, t) - b(y, t)|^2 + Tr\,\{[\sigma(x, t) - \sigma(y, t)][\sigma(x, t) - \sigma(y, t)]^*\} \le K|x - y|,$$

for some $K > 0$.

Then, given $\xi \in L^2(\Omega; \mathbf{R}^d)$, the equation has a unique solution $x(t)$ which is a continuous $\mathcal{F}_t$-adapted process in $\mathbf{R}^d$ with $E \sup_{t \le T} |x(t)|^2 < \infty$. $\square$

Remarks:

(1) Suppose that the above conditions are satisfied only locally. That is, for any $N > 0$, there exist constants C and K depending on N such that conditions (1) and (2) are satisfied for $|x| \le N$ and $|y| \le N$. Then it can be shown that, by considering a truncated equation with coefficients b_N and σ_N, the equation (1.28) has a unique local solution $x(t)$. That is, there exists an increasing sequence $\{\sigma_n\}$ of stopping times converging to τ a.s. such that $x(t)$ satisfies the equation (1.28) for $t < \tau$.

(2) Instead of the Itô equation, let us consider the Stratonovich equation

$$x(t) = \xi + \int_0^t b(x(s), s)\, ds + \int_0^t \sigma(x(s), s) \circ d\, w_s, \quad 0 \le t \le T. \quad (1.29)$$

The above can be rewritten as the Itô equation:

$$x(t) = \xi + \int_0^t \hat{b}(x(s), s)\, ds + \int_0^t \sigma(x(s), s)\, d\, w_s, \quad (1.30)$$

where $\hat{b}(x, s) = b(x, s) + d(x, s)$ with

$$d_i(x, s) = \frac{1}{2} \sum_{j=1}^d \sum_{k=1}^m [\frac{\partial}{\partial x_j} \sigma_{ik}(x, s)] \sigma_{jk}(x, s), \quad (1.31)$$

for $i = 1, \cdots, d$. If conditions (1), (2) in Theorem 4.1 are satisfied with b replaced by $\hat{b}$, equation (1.29) has a unique solution.

(3) The Itô equation (1.28) can be generalized to the semimartingale case in a straightforward manner. If w_t is replaced by a a continuous martingale $M_t, 0 \le t \le T$, in $\mathbf{R}^m$, it yields

$$x(t) = \xi + \int_0^t b(x(s), s)\, ds + \int_0^t \sigma(x(s), s) d\, M_s, \quad 0 \le t \le T. \quad (1.32)$$

It is possible to give sufficient conditions on the coefficients of the equation and the quadratic variation matrix $[M]_t = [\langle M^i, M^j \rangle_t]$ such that there exists a unique solution. Suppose that there exists a predictable matrix-valued process q_t satisfying

$$[M]_t = \int_0^t q_s ds,$$

and conditions (1), (2) in Theorem 4.1 are replaced by

(1a) For any $x \in \mathbf{R}^d$ and $t \in [0, T]$, there exists a constant $C > 0$, such that

$$|b(x, t)|^2 + Tr\,[\sigma(x, t) q_t \sigma^*(x, t)] \le C\,(1 + |x|^2),$$

for any $x \in \mathbf{R}^d$, $t \in [0, T]$.

(2a) For any $x, y \in \mathbf{R}^d$ and $t \in [0, T]$,

$$|b(x, t) - b(y, t)|^2 + Tr\,\{[\sigma(x, t) - \sigma(y, t)] q_t [\sigma(x, t) - \sigma(y, t)]^*\} \le K|x - y|,$$

for some $K > 0$.

Then equation (1.32) has a unique solution.

1.6 Comments

The material presented in this chapter is intended to motivate the study of stochastic partial differential equations by examples and to give a brief review of some basic definitions and the subject of stochastic differential equations in finite dimensions. There exists a long list of books and articles on the subject, some of which were cited in the introduction. For Brownian motions and martingales, the reader is referred to the books [64, 72]. For a more in-depth study of the stochastic integrals and stochastic differential equations, one can consult the additional books [16, 39] and [75] among others.

As a passage to the main topics of stochastic partial differential equations, in the next chapter, we shall first consider a relatively simple class of equations: first-order scalar partial differential equations with finite-dimensional white-noise coefficients. A general theory of this type of equations was first introduced by Kunita [50] and was generalized to the case of martingale white noises with spatial parameters [51].

Chapter 2

Scalar Equations of First Order

2.1 Introduction

Partial differential equations of the first order involving one single unknown function are regarded as the simplest ones to study. As is well known, in the deterministic case, integration of such an equation can be reduced to solving a family of ordinary differential equations. This approach is known as the method of characteristics [17]. In applications, these equations arise from continuum mechanics to describe a certain type of conservation laws. For instance, consider the dispersion of smoke particles in an incompressible fluid, the conservation of mass in the Euclidean space $\mathbf{R}^3$ leads to the first-order equation [56]

$$\frac{\partial u}{\partial t}(x,t) + \sum_{j=1}^{3} v_j(x,t)\frac{\partial u}{\partial x_j}(x,t) = 0, \quad u(x,0) = u_0(x), \tag{2.1}$$

where $u(x,t)$ is the particle density at time t and position $x = (x_1, x_2, x_3)$, $v = (v_1, v_2, v_3)$ is the flow velocity, and u_0 is the initial density distribution. Given the velocity field v, the Cauchy problem for the linear equation (2.1) can be easily solved by introducing the characteristic equation

$$\frac{dx(t)}{dt} = v(x,t), \quad x(0) = x. \tag{2.2}$$

Let $\phi(x,t)$ denote the solution of the equation (2.2). Then along the solution curve, equation (2.1) yields $du(\phi(t),t)/dt = 0$. Therefore, for a smooth initial density u_0, the solution of equation (2.1) can be written simply as

$$u(x,t) = u_0[\phi^{-1}(x,t)], \tag{2.3}$$

where $\phi^{-1}(\cdot,t)$ denotes the inverse map of $\phi(\cdot,t)$ on $\mathbf{R}^3$, provided that it exists. Now, in a turbulent flow, the fluid velocity is a random field. In the presence of rapid fluctuations, it is plausible to model the turbulent velocity as the sum of a random field and a spatially dependent white noise

$$v_i(x,t) = b_i(x,t) + \sum_{n=1}^{k} \sigma_{ij}(x,t) \circ \dot{w}_t^j,$$

or in the differential form

$$v_i(x,t)dt = b_i(x,t)dt + \sum_{n=1}^{k} \sigma_{ij}(x,t) \circ dw_t^j, \quad i = 1,2,3, \qquad (2.4)$$

where $w_t = (w_t^1, ..., w_t^k)$ is a standard Wiener process in $\mathbf{R}^k$, and $b_i(x,t)$ and $\sigma_{ij}(x,t)$ are random fields adapted to the σ-field generated by the Wiener process. As a matter of convenience, throughout this chapter, the Stratonovich differential $\circ dw_t^j$ is often used instead of the Itô differential dw_t^j. In view of Theorem 1-3.1[1], under some smoothness condition, they differ only by a correction term [see equation (1.31)]. In view of equation (2.4), the corresponding equation (2.1) can be interpreted as the stochastic integral equation

$$\begin{aligned} u(x,t) = u_0(x) &- \sum_{i=1}^{3} \int_0^t b_i(x,s)\frac{\partial u}{\partial x_i}(x,s)ds \\ &- \int_0^t \sum_{i=1}^{3} \sum_{j=1}^{k} \frac{\partial u}{\partial x_i}(x,s)\sigma_{ij}(x,s) \circ dw_s^j. \end{aligned} \qquad (2.5)$$

Similar to (2.2), introduce the stochastic characteristic equations

$$x(t) = x + \int_0^t b(x(s),s)ds + \int_0^t \sigma(x(s),s) \circ dw_s, \qquad (2.6)$$

where $b = (b_1, ..., b_k)$ and $\sigma \circ dw_s = \sum_{j=1}^{k} \sigma_{.j} \circ dw_s^j$. Then, formally, by applying the Itô formula, we have $du[x(t),t] = 0$ so that the corresponding solution $u(x,t)$ of the Cauchy problem (2.1) is given by

$$u(x,t) = u_0[\varphi^{-1}(x,t)], \qquad (2.7)$$

where the solution $\varphi(x,t)$ of equation (2.6) is assumed to have an inverse $\varphi^{-1}(x,t)$ a.s.. With this simple example in mind, we shall extend this idea to treat a more general class of linear and nonlinear first order equations. In this chapter we shall deal with linear and quasilinear first-order equations with coefficients being finite-dimensional, spatially dependent white noises. As will be seen, the solutions are no longer finite dimensional stochastic processes. They need to be described by a certain type of random fields, known as semimartingales with a spatial parameter. To construct a solution by the method of characteristics, two essential tools in stochastic analysis are required: a generalized Itô's formula and the solution of a stochastic equation as a stochastic flow of diffeomorphism. These technical preliminaries are given in Section 2.2. The linear and quasilinear equations are treated in Section 2.3 and Section 2.4, respectively. The equations are integrated along the stochastic characteristic curves. This approach gives a unique path-wise solution in an explicit form. Some general comments are given in Section 2.5.

---

[1]Here and henceforth Theorem (Lemma) n-j.k denotes Theorem (Lemma) j.k given in Chapter n.

2.2 Generalized Itô's Formula

In this section, we shall collect some technical lemmas concerning the Itô integral and the Stratonovich integral depending on a spatial parameter $x \in \mathbf{R}^d$. To this end, let $w_t = (w_t^1, ..., w_t^n)$, $t \geq 0$, be a standard Wiener process in $\mathbf{R}^n$ defined in a complete probability space $(\Omega, \mathcal{F}, P)$ and let $\mathcal{F}_t^s = \sigma\{w_r, s \leq r \leq t\}$ be the sub-σ field of $\mathcal{F}$ over the time interval $[s, t] \subset [0, \infty)$. Let $\Phi(x, t, \omega)$, or simply $\Phi(x, t)$ for $t \in [0, T]$ and $\omega \in \Omega$, be a family of real-valued random processes with parameter $x \in \mathbf{R}^d$. As a random function of x and t, $\Phi(x, t)$ will be called a *random field*. If, for each $x \in \mathbf{R}^d$, $\Phi(x, t)$ is $\mathcal{F}_t$-adapted (predictable), it will be termed an *adapted (predictable) random field*. In general a random field may be scalar, vector or matrix-valued. In the latter cases the components of $\Phi(x, t)$ are labelled as $\Phi_i(x, t)$ and $\Phi_{ij}(x, t)$ etc. In this chapter all random fields are assumed to be continuous in x and t a.s. and $\mathcal{F}_t$-adapted. We say that $\Phi(x, t)$ is a *spatially dependent martingale (semimartingale)*, if, for each $x \in \mathbf{R}^d$, it is a martingale (semimartingale) with respect to the filtration $\{\mathcal{F}_t\}$. By convention, sometimes we may write $\Phi(x, t), \Phi_i(x, t), \cdots$ as $\Phi_t(x), \Phi_t^i(x), \cdots$. We shall deal with semimartingales $V(x, t)$ with components $V_i(x, t), i = 1..., p$, of the special form

$$V(x, t) = \int_0^t b(x, s)ds + \int_0^t \sigma(x, s) \circ dw_s,$$

or, in components,

$$V_i(x, t) = \int_0^t b_i(x, s)ds + \sum_{j=1}^n \int_0^t \sigma_{ij}(x, s) \circ dw_s^j, \qquad (2.8)$$

where $b_i(x, t)$ and $\sigma_{ij}(x, t)$ are continuous $\mathcal{F}_t$-adapted random fields. The mutual variation function of $V_i(x, t)$ and $V_j(y, t)$ is given by

$$< V_i(x, t), V_j(y, t) > = \int_0^t q_{ij}(x, y, s)ds, \qquad (2.9)$$

where $q_{ij}(x, y, s) = \sum_{l=1}^n \sigma_{il}(x, s)\sigma_{jl}(y, s)$. Let $q(x, y, t) = [q_{ij}(x, y, t)]_{p \times p}$ be the covariation matrix. Then the *local characteristic* of $V(x, t)$ is given by the pair $(q(x, y, t), \hat{b}(x, t))$, where $\hat{b}(x, t) = b(x, t) + d(x, t)$ and $d(x, t)$ is the Stratonovich correction term as defined in equation (1.31).

For a smooth function $f : \mathbf{R}^d \to \mathbf{R}$, let $\partial_x^\alpha f(x)$ be defined by

$$\partial_x^\alpha f(x) = \frac{\partial^{|\alpha|}}{(\partial x_1)^{\alpha_1} \cdots (\partial x_d)^{\alpha_d}} f(x),$$

or simply $\partial^\alpha f = \partial_{x_1}^{\alpha_1} ... \partial_{x_d}^{\alpha_d} f$, where $\alpha = (\alpha_1, ..., \alpha_d)$ is a multi-index with $|\alpha| = \alpha_1 + ... + \alpha_d$. Let $\mathbf{C}^m(\mathbf{R}^d; \mathbf{R})$ or $\mathbf{C}^m$ in short, denote the set of all

m-time continuously differential real-valued functions f on $\mathbf{R}^d$ so that the partial derivatives $\partial^\alpha f$ exist for $|\alpha| \leq m$. Define a norm in $\mathbf{C}^m$ by

$$\|f\|_m = \sup_{x \in \mathbf{R}^d} \frac{|f(x)|}{(1 + |x|)} + \sum_{1 \leq |\alpha| \leq m} \sup_{x \in \mathbf{R}^d} |\partial^\alpha f(x)|.$$

For $\delta \in (0, 1]$, let $\mathbf{C}^{m,\delta}(\mathbf{R}^d, \mathbf{R})$ or simply $\mathbf{C}^{m,\delta}$ denotes the set of $\mathbf{C}^m$-functions f such that all of its partial derivatives $\partial^\alpha f$ with $|\alpha| \leq m$ are Hölder-continuous with exponent δ. The corresponding norm on $\mathbf{C}^{m,\delta}$ are given by

$$\|f\|_{m+\delta} = \|f\|_m + \sum_{|\alpha|=m} \|\partial^\alpha f\|_\delta,$$

where

$$\|f\|_\delta = \sup_{\substack{x,y \in \mathbf{R}^d \\ x \neq y}} \frac{|f(x) - f(y)|}{|x - y|^\delta}.$$

Let $\mathbf{C}_b^m$ denote the set of $\mathbf{C}^m$-bounded functions and $\mathbf{C}_b^{m,\delta}$, the set of $\mathbf{C}^{m,\delta}$-bounded functions.

Similarly, let $g(x, y)$ be real-valued function on $\mathbf{R}^d \times \mathbf{R}^d$. Using the same notation as before, we say that $g \in \mathbf{C}^m$ if the mixed derivatives $\partial_x^\alpha \partial_y^\beta g(x, y)$ exist for $|\alpha|, |\beta| \leq m$. Define the $\mathbf{C}^m$-norm of g as

$$\|g\|_m = \sup_{(x,y) \in \mathbf{R}^{2d}} \frac{|g(x,y)|}{(1+|x|)(1+|y|)} + \sum_{1 \leq |\alpha|,|\beta| \leq m} \sup_{(x,y) \in \mathbf{R}^{2d}} |\partial_x^\alpha \partial_y^\beta g(x,y)|,$$

Analogously, the space $\mathbf{C}^{m,\delta}$ consists of $\mathbf{C}^m$-functions g such that

$$\|g\|_{m+\delta} = \|g\|_m + \sum_{|\alpha|,|\beta|=m} \|\partial_x^\alpha \partial_y^\beta g\|_\delta < \infty,$$

where

$$\|g\|_\delta = \sup_{\substack{(x,y),\,(x',y') \in \mathbf{R}^{2d} \\ x \neq x',\, y \neq y'}} \frac{|g(x,y) - g(x',y) - g(x,y') + g(x',y')|}{|x - x'|^\delta |y - y'|^\beta}.$$

The spaces of bounded functions g in $\mathbf{C}^m$ and $\mathbf{C}^{m,\delta}$ will be denoted by $\mathbf{C}_b^m$ and $\mathbf{C}_b^{m,\delta}$, respectively, as before.

Let $\sigma(x, t)$ be a continuous, $\mathcal{F}_t$-adapted, real-valued random field such that

$$\int_0^t |\sigma(x, s)|^2 ds < \infty \quad a.s. \text{ for each } x,$$

and w_t is a Brownian motion in one dimension. Then the Itô integral $\int_0^t \sigma(x, s) dw_s$ and the Stratonovich integral $\int_0^t \sigma(x, s) \circ dw_s$ are well defined for each fixed *x*. However they may not be defined as a random field, since the union of null sets N_x over all x may not be a null set. For instance, in order to define the Itô integral $I_t(x) = \int_0^t \sigma(x,s)dw_s$ as a continuous random fields, it is necessary to impose certain regularity property, such as Hölder continuity, on the integrand $\sigma(x,t)$. Then, by applying Kolmogorov's continuity theorem for a random field, such an integral may be defined for all (x,t) as a continuous version of the family of random variables $\{I_t(x)\}$. More generally, consider the semimartingale $V(x,t)$ defined by equation (2.8). By means of Theorem 3.1.2 in [51], it can be shown that the following lemma holds.

Lemma 2.1 Let $b_i(\cdot,t)$ and $\sigma_{ij}(\cdot,t)$ be continuous, $\mathcal{F}_t$-adapted $\mathbf{C}^{m,\delta}$-processes for $m \geq 1$ such that

$$\sup_{i,j} \int_0^T \{\|b_i(\cdot,t)\|_{m+\delta} + \|\sigma_{ij}(\cdot,t)\|_{m+1+\delta}^2\}dt < \infty \quad \text{a.s.}$$

Then, for each i, with $i = 1,...,p$, the spatially dependent Stratonovich process

$$V_i(x,t) = \int_0^t b_i(x,s)ds + \sum_{j=1}^n \int_0^t \sigma_{ij}(x,s) \circ dw_s^j,$$

has a regular version which is a continuous $\mathbf{C}^{m,\varepsilon}$-semimartingale for any $\varepsilon < \delta$. Moreover the following differentiation formula holds:

$$\partial_x^\alpha V_i(x,t) = \int_0^t \partial_x^\alpha b_i(x,s)ds + \sum_{j=1}^n \int_0^t \partial_x^\alpha \sigma_{ij}(x,s) \circ dw_s^j \quad \text{a.s.,} \qquad (2.10)$$

for $|\alpha| \leq m$. $\qquad\qquad\square$

Let $\Phi(x,t)$ be a continuous $\mathbf{C}^m$-semimartingale with local characteristic $(\gamma(x,y,t), \beta(x,t))$ and let $g_t \in \mathbf{R}^d$ be a continuous predictable process such that

$$\int_0^T \{|\beta(g_t,t)| + \gamma(g_t,g_t,t)\}dt < \infty \quad \text{a.s.}$$

Then we can define the Itô (line) integral over g_t based on $\Phi(x,t)$ as the stochastic limit:

$$\int_0^t \Phi(g_s,ds) = \lim_{|\Delta_T|\to 0} \sum_{j=0}^{N-1} \{\Phi(g_{t_j \wedge t}, t_{j+1} \wedge t) - \Phi(g_{t_j \wedge t}, t_j \wedge t)\}, \qquad (2.11)$$

where the convergence is uniform over $[0,T]$ in probability, for any partition $\Delta_T = \{0 = t_0 < t_1 < \cdots < t_N = T\}$ with mesh size $|\Delta_T|$. Correspondingly, the Stratonovich integral is defined as

$$\int_0^t \Phi(g_s,\circ ds) = \lim_{|\Delta_T|\to 0} \sum_{j=0}^{N-1} \tfrac{1}{2}\{\Phi(g_{t_{j+1}\wedge t}, t_{j+1} \wedge t) + \Phi(g_{t_j \wedge t}, t_{j+1} \wedge t)$$
$$-\Phi(g_{t_{j+1}\wedge t}, t_j \wedge t) - \Phi(g_{t_j \wedge t}, t_j \wedge t)\}. \qquad (2.12)$$

In particular, we have

$$\int_0^t V(g_s, \circ ds) = \int_0^t b(g_s, s)ds + \int_0^t \sigma(g_s, s) \circ dw_s. \qquad (2.13)$$

Lemma 2.2 Let $\Phi(x,t)$, $x \in \mathbf{R}^d$, $t \in [0,T]$, be a continuous $\mathbf{C}^1$- semi-martingale with local characteristic satisfying

$$\int_0^T \{\|\beta(\cdot,t)\|_1 + \|\gamma(\cdot,\cdot,t)\|_{2+\delta}\}dt < \infty \quad a.s.,$$

and let $g_t = (g_t^1, ..., g_t^d)$ be a continuous semimartingale in $\mathbf{R}^d$. Then both the Itô and the Stratonovich integrals given above are well defined and they are related by the following formula:

$$\int_0^t \Phi(g_s, \circ ds) = \int_0^t \Phi(g_s, ds) + \frac{1}{2}\sum_{i=1}^d \langle \int_0^t \frac{\partial \Phi}{\partial x_i}(g_s, ds), g_t^i \rangle. \quad \square \qquad (2.14)$$

The Itô formula plays an important role in the stochastic analysis. If $\Phi(x,t)$ were a deterministic function, the formula for $\Phi(g_t, t)$ is well known. As it turns out, the formula can be generalized to the semimartingale case (see Theorem 3.3.2, [51]). The following theorem gives a generalized Itô formula.

Theorem 2.3 Let $\Phi(x,t)$ be a continuous $\mathbf{C}^3$-process and a continuous $\mathbf{C}^2$- semimartingale, for $x \in \mathbf{R}^d$, and $t \in [0,T]$. Suppose that its local characteristic satisfies

$$\int_0^T \{\|\beta(\cdot,t)\|_1 + \|\gamma(\cdot,\cdot,t)\|_2\}dt < \infty \quad a.s.$$

If $g_t = (g_t^1, ..., g_t^d)$ is a continuous semimartingale, then $\Phi(g_t, t)$ is a continuous semimartingale and the following formula holds

$$\Phi(g_t, t) = \Phi(g_0, 0) + \int_0^t \Phi(g_s, \circ ds) + \sum_{j=1}^d \int_0^t \frac{\partial \Phi}{\partial x_j}(g_s, s) \circ dg_s^j. \qquad (2.15)$$

which, when written in terms of Itô integrals, gives

$$\Phi(g_t, t) = \Phi(g_0, 0) + \int_0^t \Phi(g_s, ds) + \sum_{i=1}^d \int_0^t \frac{\partial \Phi}{\partial x_i}(g_s, s)dg_s^i$$

$$+ \frac{1}{2}\sum_{i,j=1}^d \int_0^t \frac{\partial^2 \Phi}{\partial x_i \partial x_j}(g_s, s)d\langle g^i, g^j \rangle_s + \sum_{i=1}^d \langle \int_0^t \frac{\partial \Phi}{\partial x_i}(g_s, ds), g_t \rangle. \quad \square$$

$$(2.16)$$

As a matter of convenience, in the differential form, the Stratonovich formula (2.15) will be written symbolically as

$$\circ d\Phi(g_t, t) = \Phi(g_t, \circ dt) + \sum_{j=1}^{d} \frac{\partial \Phi}{\partial x_j}(g_t, t) \circ dg_t^j. \qquad (2.17)$$

Suppose that g_t is replaced by a continuous $\mathbf{C}^r$-semimartingale $V(x, t)$, $x \in \mathbf{R}^d$ as given by equation (2.8). By changing the spatial parameter, let $\Phi(\xi, t)$, $\xi \in \mathbf{R}^p$, be the semimartingale as in the previous lemma. We need to differentiate random fields such as $\Phi[V(x, t), t]$ and $J(x, t) = \int_0^t \Phi[V(x, s), \circ ds]$ with respect to the parameter ξ. The next lemma, a simplified version of Theorem 3.3.4 in [51], provides the desired chain rules of differentiation.

Lemma 2.4 Let $\Phi(\xi, t)$, $\xi \in \mathbf{R}^p$ with local characteristic (γ, β) such that

$$\int_0^T \{\|\beta(\cdot, t)\|_{m+\delta} + \|\gamma(\cdot, \cdot, t)\|_{m+1+\delta}\} dt < \infty \quad a.s.$$

Let $V(x, t)$, $x \in \mathbf{R}^d$ be a continuous $\mathbf{C}^{r,\delta'}$-semimartingale in $\mathbf{R}^p$ as given by (2.8) such that

$$\sup_{i,j} \int_0^T \{\|b_i(\cdot, t)\|_{m+\delta} + \|\sigma_{ij}(\cdot, t)\|_{m+1+\delta}^2\} dt < \infty \quad a.s.$$

Then the random field $\Phi[V(x, t), t]$ and the Stratonovich integral $J(x, t) = \int_0^t \Phi[V(x, s), \circ ds]$ are well defined as continuous $\mathbf{C}^{m_0, \varepsilon}$-semimartingales with $m_0 = m \wedge r$ and $\varepsilon < \delta \wedge \delta'$. Moreover the following differentiation formulas hold

$$\frac{\partial \Phi}{\partial x_i}[V(x, t), t] = \sum_{j=1}^{p} \frac{\partial \Phi}{\partial \xi_j}[V(x, t), t] \frac{\partial V_j}{\partial x_i}(x, t),$$

and

$$\frac{\partial J}{\partial x_i}(x, t) = \sum_{j=1}^{p} \int_0^t \frac{\partial V_j}{\partial x_i}(x, s) \frac{\partial \Phi}{\partial \xi_j}[V(x, s), \circ ds]. \qquad \square$$

Now we consider the following system of stochastic equations:

$$d\phi_t^i = V_i(\phi_t, \circ dt)$$
$$= b_i(\phi_t, t) dt + \sum_{j=1}^{n} \sigma_{ij}(\phi_t, t) \circ dw_t^j, \qquad (2.18)$$
$$\phi_0^i = x_i, \quad i = 1, \cdots, d, \quad 0 < t < T.$$

If b_i and σ_{ij} are smooth, then, under the usual linear growth and the Lipschitz continuity assumptions, the stochastic system has a unique continuous solution $\phi_t(x)$ with $\phi_t = (\phi_t^1, \cdots, \phi_t^d)$. Here we regard $\phi_t(x)$ as a random field and are interested in the mapping $\phi_t(\cdot) : \mathbf{R}^d \to \mathbf{R}^d$. Suppose the mapping is continuous and it has a continuous inverse $\phi_t^{-1}(\cdot)\,a.s.$ Then the composition maps $\phi_{s,t} = \phi_t \circ \phi_s^{-1}$, for $0 \le s \le t \le T$, form a *stochastic flow of homeomorphism*. If $\phi_{s,t}(\cdot) \in \mathbf{C}^m$ and all of its partial derivatives $\partial_x^\alpha \phi_{s,t}(x)$ with $|\alpha| \le m$ are continuous in $(x, s, t)\,a.s.$, then $\{\phi_{s,t}\}$ is said to be a *stochastic flow of $\mathbf{C}^m$-diffeomorphism*. The following is a more precise statement of the key theorem.

Theorem 2.5   For $i = 1, ..., d$, $j = 1, ..., n$, let $b_i(x,t)$ and $\sigma_{ij}(x,t)$ be continuous $\mathcal{F}_t$-adapted $\mathbf{C}_b^m$−random fields such that, for $m \ge 1$,

$$\sup_{i,j} \int_0^T \{\|b_i(\cdot,t)\|_{m+\delta} + \|\sigma_{ij}(\cdot,t)\|_{m+1+\delta}\}dt < \infty \quad a.s.$$

Then the equation (2.18) has a continuous solution $\phi_t(x)$ which is a $\mathbf{C}^{m,\varepsilon}$-semimartingale for any $\varepsilon < \delta$. Moreover it generates a stochastic flow of $\mathbf{C}^m$-diffeomorphism over $[0,T]$. $\square$

This theorem follows from Theorem 4.7.3 in [51]. Since the stochastic flow is generated by Wiener processes, it is known as the *Brownian flow of diffeomorphism*.

2.3   Linear Stochastic Equations

Consider a linear partial differential equation of first order in the variables t and x as follows:

$$\frac{\partial u}{\partial t} = \sum_{i=1}^d v_i(x,t)\frac{\partial u}{\partial x_i} + v_{d+1}(x,t)u + v_{d+2}(x,t),$$

for $t > 0$ and $x = (x_1, ..., x_d)$ in $\mathbf{R}^d$, where v_i, $i = 1, ..., d+2$, are given functions of x and t. Let $V(x,t) = (V_1(x,t), ..., V_{d+2}(x,t))$ be a continuous spatially dependent semimartingale given by (2.8) with $p = d + 2$. In the differential form, it is written as

$$V_i(x, \circ dt) = b_i(x,t)dt + \sum_{j=1}^n \sigma_{ij}(x,t) \circ dw_t^j, \qquad (2.19)$$

or formally,

$$\dot{V}_i(x,t) = b_i(x,t) + \sum_{j=1}^n \sigma_{ij}(x,t) \circ \dot{w}_t^j, \quad j = 1, ..., d+2,$$

which is a spatially dependent white noise with a random drift. Now we replace each coefficient $v_i(x,t)$ of the linear partial differential equation by a white noise $\dot{V}_i(x,t)$ given above, and consider the associated Cauchy problem:

$$\frac{\partial u}{\partial t} = \sum_{i=1}^{d} \dot{V}_i(x,t)\frac{\partial u}{\partial x_i} + \dot{V}_{d+1}(x,t)u + \dot{V}_{d+2}(x,t),$$

$$u(x,0) = u_0(x),$$

(2.20)

where $u_0(x)$ is a given function. By convention, equation (2.19) is interpreted as the following stochastic integral equation

$$u(x,t) = u_0(x) + \sum_{i=1}^{d} \int_0^t \frac{\partial u}{\partial x_i}(x,s)V_i(x,\circ ds)$$

$$+ \int_0^t u(x,s)V_{d+1}(x,\circ ds) + V_{d+2}(x,t).$$

(2.21)

To construct a solution to the Cauchy problem (2.19), we introduce the associated characteristic system

$$\varphi_t^i(x) = x_i - \int_0^t V_i[\varphi_s(x),\circ ds]$$

$$= x_i - \int_0^t b_i[\varphi_s(x),s]ds - \sum_{j=1}^{n} \int_0^t \sigma_{ij}[\varphi_s(x),s]\circ dw_s^j,$$

(2.22)

for $i = 1,\cdots,d$, and

$$\eta_t(x,r) = r + \int_0^t \eta_s(x,r)V_{d+1}[\varphi_s(x),\circ ds]$$

$$+ \int_0^t V_{d+2}[\varphi_s(x),\circ ds],$$

(2.23)

where $\varphi_t(x) = (\varphi_t^1(x),...,\varphi_t^d(x))$ for $t \in [0,T]$ is a characteristic curve starting from x at $t = 0$. Let $\phi_t(\xi) = (\varphi_t(x),\eta_t(\xi))$ with $\xi = (x,r)$. Then the system (2.22) and (2.23) can be written as a single equation of the form

$$\phi_t(\xi) = \xi + \int_0^t K[\phi_s(\xi),s]V[\phi_s(\xi),\circ ds],$$

where $K(\xi,s)$ is a $(d+1) \times (d+2)$ matrix. Under some suitable conditions, the above equation has a unique solution which defines a stochastic flow of diffeomorphism.

Lemma 3.1 For $i = 1,...,d+2$ and $j = 1,...,n$, let $b_i(x,t)$ and $\sigma_{ij}(x,t)$ be continuous $\mathcal{F}_t$-adapted C_b^m-processes with $m \geq 1, \delta \in (0,1]$, such that

$$\int_0^t \|b_i(\cdot, t)\|_{m+\delta}^2 dt \le c_1 \text{ and } \int_0^t \|\sigma_{ij}(\cdot, t)\|_{m+1+\delta}^2 dt \le c_2 \quad \text{a.s.},$$

for some positive constants c_1 and c_2. Then the system (2.22) and (2.23) has a unique solution $\phi_t(\xi) = (\varphi_t(x), \eta_t(x, r))$ over $[0, T]$. Moreover the solution is a continuous $\mathbf{C}^m$-semimartingale which defines a stochastic flow of $\mathbf{C}^m$-diffeomorphism. □

This lemma follows immediately from Theorem 2.5. In particular, $\varphi_t : \mathbf{R}^d \to \mathbf{R}^d$, $0 \le t \le T$, is a stochastic flow of $\mathbf{C}^m$-diffeomorphism. Let $\psi_t = \varphi_t^{-1}$ denote the inverse of φ_t and define $\tilde{\eta}_t(x) = \eta_t[x, u_0(x)]$. With the aid of the solution of the characteristic system, we can construct a path-wise solution to the Cauchy problem (3.4) as follows.

Theorem 3.2 Let the conditions in Lemma 3.1 hold true for $m \ge 3$. For $u_0 \in \mathbf{C}^{m,\delta}$, the Cauchy problem (2.20) has a unique solution $u(\cdot, t)$ for $0 \le t \le T$ such that it is a $\mathbf{C}^m$-semimartingale which can be represented as

$$\begin{aligned}
u(x, t) &= u_0[\psi_t(x)] \exp\{\int_0^t V_{d+1}[\varphi_s(y), \circ ds]\}|_{y=\psi_t(x)} \\
&\quad + \int_0^t \exp\{\int_\tau^t V_{d+1}[\varphi_s(y), \circ ds]\} V_{d+2}[\varphi_\tau(y), \circ d\tau]|_{y=\psi_t(x)}.
\end{aligned} \tag{2.24}$$

Proof. Let $\tilde{\eta}(x, t) = \eta_t[x, u_0(x)]$ and $\mu(x, t) = \tilde{\eta}[\psi_t(x), t]$. We are going to show that $\mu(x, t)$ is a solution of the Cauchy problem (2.20). From (2.21) we have

$$\tilde{\eta}(x, t) = u_0(x) + \int_0^t \tilde{\eta}(x, s) V_{d+1}[\varphi_s(x), \circ ds] + \int_0^t V_{d+2}[\varphi_s(x), \circ ds], \quad (2.25)$$

so that

$$\begin{aligned}
\tilde{\eta}[\psi_t(x), \circ dt] &= \tilde{\eta}[\psi_t(x), t] V_{d+1}[(\varphi_t \circ \psi_t)(x), \circ dt] \\
&\quad + V_{d+2}[(\varphi_t \circ \psi_t)(x), \circ dt] \\
&= \mu(x, t) V_{d+1}(x, \circ dt) + V_{d+2}(x, \circ dt),
\end{aligned} \tag{2.26}$$

due to the fact: $(\varphi_t \circ \psi_t)(x) = x$. By the generalized Itô formula (2.15),

$$\begin{aligned}
\mu(x, t) &= \tilde{\eta}[\psi_t(x), t] \\
&= u_0(x) + \int_0^t \tilde{\eta}[\psi_s(x), \circ ds] + \int_0^t (\partial\tilde{\eta})[\psi_s(x), s] \circ d\psi_s(x),
\end{aligned} \tag{2.27}$$

where

$$(\partial\tilde{\eta})[\psi_s(x), s] \circ d\psi_s(x) = \sum_{i=1}^d \{\frac{\partial\tilde{\eta}(y, s)}{\partial y_i}|_{y=\psi_s(x)}\} \circ d\psi_s^i(x).$$

Since $\circ d(\varphi_t \circ \psi_t) = dx = 0$, by setting $\varphi_t \circ \psi_t = \varphi(\psi_t, t)$ and noticing (2.17), we get

$$\circ d\varphi(\psi_t, t) = \varphi(\psi_t, \circ dt) + \partial\varphi(\psi_t, t) \circ d\psi_t(x) = 0.$$

The above equation gives

$$\circ d\psi_t = -[(\partial\varphi)^{-1}(\psi_t, t)]\varphi(\psi_t, \circ dt),$$

which, in view of (2.22), yields

$$\circ d\psi_t = [(\partial\varphi)^{-1}(\psi_t, t)]\bar{V}(x, \circ dt), \tag{2.28}$$

where we set $\bar{V} = (V^1, ..., V^d)$ and the inverse $(\partial\varphi_t)^{-1}$ exists by Lemma 3.1. On the other hand, it is clear that

$$\partial_x(\varphi_t \circ \psi_t)(x) = [(\partial\varphi_t)(\psi_t)]\partial_x\psi_t(x) = I_d,$$

where I_d is the $(d \times d)$-identity matrix. It follows that

$$(\partial\varphi)^{-1}[\psi_t(x), t] = \partial_x\psi_t(x). \tag{2.29}$$

In view of (2.28) and (2.29),

$$(\partial\tilde{\eta})[\psi_s(x), s] \circ d\psi_s(x) = \partial_y\tilde{\eta}(y, s)|_{y=\psi_s(x)}[\partial_x\psi_s(x)]\bar{V}(x, \circ ds)$$
$$= \partial_x\mu(x, s)\bar{V}(x, \circ ds). \tag{2.30}$$

By taking the equations (2.26) and (2.30) into account, equation (2.27) can be written as

$$\mu(x, t) = u_0(x) + \sum_{i=1}^{d} \int_0^t \partial_{x_i}\mu(x, s)V_i(x, \circ ds)$$
$$+ \int_0^t \mu(x, s)V_{d+1}(x, \circ ds) + V_{d+2}(x, t). \tag{2.31}$$

Hence $\mu(x, t)$ is a solution of the Cauchy problem (2.20) as claimed. Since the equation is linear, the solution $u = \mu(x, t)$ is easily shown to be unique. The fact that $u(\cdot, t)$ is a continuous $\mathbf{C}^m$-semimartingale is a consequence of Theorem 2.5. To verify the representation (2.24), by the Itô formula, one can verify that the solution of equation (2.25) is given by

$$\tilde{\eta}(y, t) = u_0(y) \exp\{\int_0^t V_{d+1}[\varphi_s(y), \circ ds]\}$$
$$+ \int_0^t \exp\{\int_\tau^t V_{d+1}[\varphi_s(y), \circ ds]\}V_{d+2}[\varphi_\tau(y), \circ d\tau],$$

which yields the representation (2.24) by setting $y = \psi_t(x)$. $\qquad\square$

As an example, consider the one-dimensional problem:

$$\frac{\partial u}{\partial t} = \dot{V}(x,t)\frac{\partial u}{\partial x} + \alpha u + f(x,t),$$

$$u(x,0) = u_0(x),$$

(2.32)

where α is a constant and f is a bounded continuous function, and the random coefficient is given by

$$\dot{V}(x,t) = x[b(t) + \sigma(t)] \circ \dot{w}_t,$$

in which $b(t)$ and $\sigma(t)$ are continuous $\mathcal{F}_t$-adapted processes and w_t is a scalar Brownian motion. Then the characteristic equation (2.22) yields

$$\varphi_t(x) = x - \int_0^t b(s)\varphi_s(x)ds - \int_0^t \sigma(s)\varphi_s(x) \circ dw_s,$$

which can be solved to give

$$\varphi_t(x) = x \, \exp\{-\int_0^t b(s)ds - \int_0^t \sigma(s) \circ dw_s\}.$$

Its inverse is simply

$$\psi_t(x) = x \, \exp\{\int_0^t b(s)ds + \int_0^t \sigma(s) \circ dw_s\}.$$

By the representation formula (2.24), we obtain

$$u(x,t) = u_0\{x \exp[\int_0^t b(s)ds + \int_0^t \sigma(s) \circ dw_s]\} e^{\alpha t}$$

$$+ \int_0^t e^{\alpha(t-s)} f\{x \, \exp[\int_s^t b(\tau)d\tau + \int_s^t \sigma(\tau) \circ dw_\tau], s\} \, ds.$$

We wish to point out that the first-order Stratonovich equation (2.21) is actually a parabolic Itô equation. This can be seen by rewriting the Stratonovich integrals in (2.21) in terms of the associated Itô integrals with proper correction terms. For instance, by applying Lemma 2.2 and taking equation (2.19) into account, we have

$$\int_0^t u(x,s)V_{d+1}(x, \circ ds)$$

$$= \int_0^t u(x,s)V_{d+1}(x,ds) + \frac{1}{2}\langle V_{d+1}(x,t), u(x,t)\rangle$$

$$= \int_0^t u(x,s)V_{d+1}(x,ds) + \frac{1}{2}\{\sum_{j=1}^d \int_0^t q_{d+1,j}(x,x,s)\frac{\partial u}{\partial x_j}(x,s)ds$$

$$+ \int_0^t [q_{d+1,d+1}(x,x,s)u(x,s) + q_{d+1,d+2}(x,x,s)]ds\},$$

(2.33)

where

$$q_{i,j}(x,y,t) = \frac{\partial}{\partial t}\langle V_i(x,t), V_j(y,t)\rangle = \sum_{k=1}^{n}\sigma_{ik}(x,t)\sigma_{jk}(y,t),$$

for $i,j = 1,...,d$. Similarly, by differentiating equation (2.21) with respect to x_i, we can obtain

$$\int_0^t \frac{\partial u}{\partial x_i}(x,s)V_i(x,\circ ds)$$

$$= \int_0^t \frac{\partial u}{\partial x_i}(x,s)V_i(x,ds) + \frac{1}{2}\langle V_i(x,t), \frac{\partial u}{\partial x_i}(x,t)\rangle$$

$$= \int_0^t \frac{\partial u}{\partial x_i}(x,s)V_i(x,ds) + \frac{1}{2}\{\sum_{j=1}^{d}\int_0^t q_{i,j}(x,x,s)\frac{\partial^2 u}{\partial x_i \partial x_j}(x,s)ds \qquad (2.34)$$

$$+ \int_0^t [\sum_{j=1}^{d} q'_{i,j}(x,x,s) + q_{i,d+1}(x,x,s)]\frac{\partial u}{\partial x_i}(x,s)]ds$$

$$+ \int_0^t [q'_{i,d+1}(x,x,s)u(x,s) + q'_{i,d+2}(x,x,s)]ds\},$$

where $q'_{i,j}(x,y,t) = \frac{\partial q_{i,j}}{\partial y_i}(x,y,t)$. When the above results (2.33) and (2.34) are used in equation (2.21), we obtain the parabolic Itô equation

$$u(x,t) = u_0(x) + \int_0^t L_s u(x,s)ds + \sum_{i=1}^{d}\int_0^t \frac{\partial u}{\partial x_i}(x,s)V_i(x,ds)$$

$$+ \int_0^t u(x,s)V_{d+1}(x,ds) + \tilde{V}_{d+2}(x,t), \qquad (2.35)$$

where L_t is a second-order elliptic operator defined by

$$L_t u = \frac{1}{2}\sum_{i,j=1}^{d} q_{i,j}(x,x,t)\frac{\partial^2 u}{\partial x_i \partial x_j} + \sum_{i=1}^{d} p_i(x,t)\frac{\partial u}{\partial x_i} + r(x,t)u,$$

$$p_i(x,t) = \frac{1}{2}\sum_{j=1}^{d} q'_{i,j}(x,x,t) + q_{i,d+1}(x,x,t) + q_{d+1,i}(x,x,t),$$

$$r(x,t) = \frac{1}{2}\{q_{d+1,d+1}(x,x,t) + \sum_{j=1}^{d} q'_{j,d+1}(x,x,t)\},$$

and

$$\tilde{V}_{d+2}(x,t) = V_{d+2}(x,t) + \frac{1}{2}\int_0^t \{\sum_{j=1}^{d} q'_{j,d+2}(x,x,s) + q_{d+1,d+2}(x,x,s)\}ds.$$

Corollary 3.3   Let the conditions in Lemma 3.1 hold true for $m \geq 3$. For $u_0 \in \mathbf{C}^{m,\delta}$, the Cauchy problem (2.35) has a unique solution $u(\cdot, t)$ for $0 \leq t \leq T$ such that it is a $\mathbf{C}^m$-semimartingale which can be represented by the equation (3.10). $\qquad\square$

2.4   Quasilinear Equations

Suppose that the random coefficients $\dot{V}_i$ in equation (2.20) may also depend on u. Then one is led to consider a quasilinear equation of the form:

$$\frac{\partial u}{\partial t} = \sum_{i=1}^{d} \dot{V}_i(x, u, t)\frac{\partial u}{\partial x_i} + \dot{V}_{d+1}(x, u, t),$$

$$u(x, 0) = u_0(x),$$

(2.36)

where, formally, $\dot{V}_i(x, u, t) = \frac{\partial V_i}{\partial t}(x, u, t)$. Similar to (2.19), $V_i(x, u, t)$, $(x, u) \in \mathbf{R}^{d+1}$, is a continuous semimartingale given by

$$V_i(x, u, t) = \int_0^t b_i(x, u, s)ds + \sum_{j=1}^{n} \int_0^t \sigma_{ij}(x, u, s) \circ dw_s^j, \quad i = 1, ..., d+1, \quad (2.37)$$

where $b_i(x, u, t)$ and $\sigma_{ij}(x, u, t)$ are continuous adapted random fields. Of course, the equation (2.36) is equivalent to the integral equation

$$u(x, t) = u_0(x) + \sum_{i=1}^{d} \int_0^t \frac{\partial u}{\partial x_i}(x, s)V_i(x, u(x, s), \circ ds)$$

$$+ \int_0^t V_{d+1}(x, u(x, s), \circ ds).$$

(2.38)

Introduce the associated characteristic system

$$\varphi_t^i(x, r) = x_i - \int_0^t V_i[\varphi_s(x, r), \eta_s(x, r), \circ ds], \quad i = 1, ..., d, \quad (2.39)$$

and

$$\eta_t(x, r) = r + \int_0^t V_{d+1}[\varphi_s(x, r), \eta_s(x, r), \circ ds]. \quad (2.40)$$

Notice that, in contrast with the system (2.22) and (2.23) for the linear case, the equation (2.39) is now coupled with (2.40). However, in the semilinear case when $V_i(x, u, t)$, $i = 1, ..., d$, do not depend on u, equation (2.39) can be solved independently as in the linear case. Similar to Lemma 3.1, the following lemma holds for the above system.

Lemma 4.1   For $i = 1, ..., d+1$ and $j = 1, ..., n$, let $b_i(x, r, t)$ and $\sigma_{ij}(x, r, t)$ be continuous $\mathcal{F}_t$-adapted $\mathbf{C}_b^m$-processes with $m \geq 1, \delta \in (0, 1]$, such that

$$\int_0^t \|b_i(\cdot, \cdot, t)\|_{m+\delta}^2 \, dt \leq c_1 \quad \text{and} \quad \int_0^t \|\sigma_{ij}(\cdot, \cdot, t)\|_{m+1+\delta}^2 \, dt \leq c_2 \quad \text{a.s.,}$$

for some positive constants c_1 and c_2. Then the system (2.39) and (2.40) has a unique solution $\phi_t(x, r) = (\varphi_t(x, r), \eta_t(x, r))$ over $[0, T]$. Moreover the solution is a continuous $\mathbf{C}^m$-semimartingale which defines a stochastic flow of $\mathbf{C}^m$-diffeomorphism.   □

Let $\tilde{\varphi}_t(x) = \varphi_t[x, u_0(x)]$ and $\tilde{\eta}(x, t) = \eta_t[x, u_0(x)]$. By Lemma 4.1, the inverse of $\tilde{\varphi}_t$ exists and is denoted by $\tilde{\psi}_t$. Then the following existence theorem holds.

Theorem 4.2   Let the conditions in Lemma 3.1 hold true with $m \geq 3$. For $u_0 \in \mathbf{C}^{m,\delta}$, the Cauchy problem (2.36) has a unique solution given by $u(x, t) = \tilde{\eta}[\tilde{\psi}_t(x), t]$ for $0 \leq t \leq T$. Moreover it is a $\mathbf{C}^m$-semimartingale.

Proof.   Since the proof is quite similar to that of the linear case, it will only be sketched. In view of (2.38) and (2.39), $\tilde{\varphi}_t(x)$ and $\tilde{\eta}(x, t)$ satisfies the following equations:

$$\tilde{\varphi}_t^i(x) = x_i - \int_0^t V_i[\tilde{\varphi}_s(x), \tilde{\eta}(x, s), \circ ds], \quad i = 1, ..., d, \tag{2.41}$$

and

$$\tilde{\eta}(x, t) = u_0(x) + \int_0^t V_{d+1}[\tilde{\varphi}_s(x), \tilde{\eta}(x, s), \circ ds], \tag{2.42}$$

which are both continuous $\mathbf{C}^m$-semimartingales by Theorem 2.5. We wish to show that $\mu(x, t) = \tilde{\eta}[\tilde{\psi}_t(x), t]$ is a solution of the Cauchy problem. To this end, we apply the generalized Itô formula to $\tilde{\eta}(\cdot, t)$ to get

$$\tilde{\eta}[\tilde{\psi}_t(x), t] = u_0(x) + \int_0^t \tilde{\eta}[\tilde{\psi}_s(x), \circ ds] + \int_0^t (\partial \tilde{\eta})[\tilde{\psi}_s(x), s] \circ d\tilde{\psi}_s(x),$$

which, by noticing (2.42), implies that

$$\mu(x, t) = u_0(x) + \int_0^t V_{d+1}[x, \mu(x, s), \circ ds] + \int_0^t (\partial \tilde{\eta})[\tilde{\psi}_s(x), s] \circ d\tilde{\psi}_s(x). \tag{2.43}$$

Similar to (2.28) and (2.30), it can be shown that

$$\circ d\tilde{\psi}_t = [(\partial \tilde{\varphi}_t)^{-1}(\tilde{\psi}_t)]\bar{V}(\tilde{\varphi}_t, \circ dt), \tag{2.44}$$

and

$$(\partial \tilde{\eta})[\tilde{\psi}_t(x), t](\partial \tilde{\varphi}_t)^{-1}[\tilde{\psi}_t(x)] = \partial_x \mu(x, t), \tag{2.45}$$

where we set $\bar{V} = (V^1, ..., V^d)$ and the inverse $(\partial \tilde{\varphi}_t)^{-1}$ exists by Lemma 4.1. With the aid of (2.44) and (2.45), equation (2.43) can be reduced to

$$\mu(x,t) = u_0(x) + \int_0^t \partial_x \mu(x,s) \bar{V}[x, \mu(x,s) \circ ds]$$
$$+ \int_0^t V_{d+1}[x, \mu(x,s), \circ ds], \tag{2.46}$$

which shows that $\mu = \tilde{\eta}[\tilde{\psi}_t(x), t]$ solves the Cauchy's problem (2.36). By Theorem 2.5, we can deduce that $\mu(\cdot, t)$ is a continuous $\mathbf{C}^m$-semimartingale.

To prove the uniqueness, suppose u is another solution. We let $\hat{u} = u - \mu$ and show that $\hat{u} = 0 \, a.s.$ It follows from (2.38) and (2.46) that $\hat{u}$ satisfies

$$\hat{u}(x,t) = \int_0^t \partial_x u(x,s) \bar{V}[x, u(x,s) \circ ds] - \int_0^t \partial_x \mu(x,s) \bar{V}[x, \mu(x,s) \circ ds]$$
$$+ \int_0^t V_{d+1}[x, u(x,s), \circ ds] - \int_0^t V_{d+1}[x, \mu(x,s), \circ ds],$$

which, by letting $u = \hat{u} + \mu$, can be regarded as a stochastic equation in $\hat{u}$ for a given μ. By the assumptions of uniform boundedness conditions on the local characteristics of the semimartigales $V_i's$ as depicted in Lemma 3.1, it can be shown that the coefficients of the resulting equation are Lipschitz-continuous and of a linear growth as required for a standard existence and uniqueness theorem to hold. Since $\hat{u} \equiv 0$ is a solution, by uniqueness, it is the only solution. $\qquad \square$

For the existence of a global solution, we have assumed that the coefficients of the quasilinear equation (2.36) are uniformly bounded in a Hölder norm. Under some milder conditions, such as bounded on compact sets, it is possible to show the existence of a local solution, which may explode in finite time. For instance, consider the following simple example in one space dimension:

$$\frac{\partial u}{\partial t} + \frac{\partial u}{\partial x} = u^2 \circ \dot{w}_t, \quad u(x,0) = u_0(x), \tag{2.47}$$

where w_t is one-dimensional Brownian motion and $u_0(x)$ is a bounded continuous function. It is readily seen that the characteristic equations are

$$\varphi_t(x) = x + t, \quad \tilde{\eta}(x,t) = u_0(x) + \int_0^t \tilde{\eta}^2(x,s) \circ dw_s,$$

which yield $\psi_t(x) = x - t$ and $\tilde{\eta}(x,t) = \dfrac{u_0(x)}{1 - u_0(x)w_t}$. So the solution of (2.47) is given by

$$u(x,t) = \tilde{\eta}(x - t, t) = \frac{u_0(x-t)}{1 - u_0(x-t)w_t}.$$

For instance, suppose that $u_0 \equiv c$ is a nonzero constant. Then the solution will explode a.s. as t approaches $\tau = \inf\{t > 0 : w_t = 1/c\}$.

2.5　General Remarks

The material in this chapter is taken from the well-known book by Kunita [51]. For simplicity we have assumed that the random coefficients $V_i(x,t)'s$ involved are finite dimensional Wiener processes of the form (2.8) as done in his original paper [50]. In fact the results in the previous sections hold true when $V_i(\cdot,t)$ is a general continuous $\mathbf{C}^m$-semimartingale decomposable as a bounded variation part $B_i(x,t)$ plus a spatially dependent martingale $M_i(x,t)$. The Itô and Stratonovich line integrals can be defined in a similar fashion. By confining to finite-dimensional white-noises, we showed that some first-order stochastic partial differential equations can be solved by the method of ordinary Itô equations. Most theorems were stated without proofs, which can be found in [51]. In the subsequent chapters, we shall deal with infinite dimensional Wiener processes appearing as random coefficients in partial differential equations of higher orders.

The method of stochastic characteristics was introduced by Krylov and Rozovsky [48] and by Kunita [50] to solve a certain type of linear parabolic Itô equations, such as the Zakai equation in the nonlinear filtering theory as mentioned before. This approach, when applied to such equations, has led to a probabilistic representation of the solution such as equation (2.24). This representation is known as the stochastic *Feynman-Kac formula*. This will be discussed briefly in the Chapter Four. The aforementioned authors used both the forward and the backward stochastic flows in the construction of solutions. However, to avoid undue technical complication, the backward stochastic integrals were not considered here.

In the deterministic case the method of characteristics is commonly applied to the first-order hyperbolic systems [77]. The generalization of the scalar equation (2.36) to a stochastic hyperbolic system has not been done. In particular it is of interest to determine what type of random coefficients are admitted so that the Cauchy problems of a stochastic hyperbolic system is well posed. Some special type of stochastic hyperbolic systems will be treated in Chapter Five by the so-called energy method.

Chapter 3

Stochastic Parabolic Equations

3.1 Introduction

In this chapter we shall study stochastic partial differential equations of parabolic type, or simply, stochastic parabolic equations in a bounded domain. As a relatively simple problem, consider the heat conduction over a thin wire $(0 \leq x \leq L)$ with a constant thermal diffusivity $\kappa > 0$. Let $u(x,t)$ denote the temperature distribution at point x and time $t > 0$ due to a randomly fluctuating heat source $q(x,t,\omega)$. Suppose both ends of the wire are maintained at the freezing temperature. Then, given an initial temperature $u_0(x)$, the temperature field $u(x,t)$ is governed by the initial-boundary value problem for the stochastic heat equation:

$$
\begin{aligned}
\frac{\partial u(x,t)}{\partial t} &= \kappa \frac{\partial^2 u}{\partial x^2} + q(x,t,\omega), \quad x \in (0,L), \ t > 0, \\
u(0,t) &= u(L,t) = 0, \\
u(x,0) &= u_0(x).
\end{aligned}
\tag{3.1}
$$

Suppose that the noise term q is a regular random field, say,

$$
E \int_0^T \int_0^L q^2(x,t)dtdx < \infty,
$$

where by convention ω in q was omitted, and $u_0 \in L^2(0,L)$. The equation (3.1) can be solved for almost every ω similar to the deterministic case. In fact it can be solved by the well-known method of Fourier series or the eigenfunctions expansion. Now let $q(\cdot,t)$ be a spatially dependent white noise. Formally, it is a generalized Gaussian random field with mean zero and the correlation function:

$$
E\{q(x,t)q(y,s)\} = \delta(t-s)r(x,y),
$$

for $t \geq s$ and $x,y \in (0,L)$, where $\delta(t)$ denotes the Dirac delta function and $r(x,y)$ is the spatial correlation function. In contrast with the previous chapter, the spatially dependent white noise term, written formally as $q(x,t) = \dot{W}(x,t)$ is not finite-dimensional. Even for this simple problem, the mathematical meaning of the equation and the related questions, such as, in what sense the problem has a solution, are no longer clear. Therefore the whole book will be devoted exclusively to the subject of stochastic PDEs involving spatially dependent white noises or white-noise random fields.

In the next section, we shall first present some basic facts concerning the elliptic operator and define the Wiener random field and the related stochastic integrals. Then, in Section 3.3, the heat equation driven by a spatially dependent white noise will be solved by the method of eigenfunctions expansion. The results will then be generalized in Section 3.4 to study the linear parabolic equations with additive white noise. In Section 3.5 the regularity properties of the solutions will be investigated. In the subsequent two sections, the existence, uniqueness and the regularity questions for linear and semilinear stochastic parabolic equations will be treated separately.

3.2 Preliminaries

Let $D \subset \mathbf{R}^d$ be a bounded domain with a smooth boundary ∂D. Denote $L^2(D)$ by H with inner product $(\cdot, \cdot)$ and norm $\|\cdot\|$. For any integer $k \geq 0$, let H^k denote the L^2-Sobolev space of order k, which consists of L^2-functions ϕ with k-time generalized partial derivatives with norm

$$\|\phi\|_k = \{\sum_{j=0}^{k} \sum_{i=1}^{d} \int_D |\frac{\partial^j \phi(x)}{\partial x_i^j}|^2 dx\}^{1/2}. \tag{3.2}$$

Let H^{-k} denote the dual of H^k with norm $\|\cdot\|_{-k}$, and the duality pairing between H^k and H^{-k} be denoted by $\langle \cdot, \cdot \rangle_k$. In particular we set $\langle \cdot, \cdot \rangle_1 = \langle \cdot, \cdot \rangle$, $H^0 = H$ and $\langle \cdot, \cdot \rangle_0 = (\cdot, \cdot)$.

Consider the second-order elliptic operator

$$Lu = \sum_{j,k=1}^{d} a_{jk}(x)\frac{\partial^2 u}{\partial x_j \partial x_k} + \sum_{j=1}^{d} a_j(x)\frac{\partial u}{\partial x_j} + c(x)u \tag{3.3}$$

in domain D. Assume that, $a_{jk} = a_{kj}$ and, for simplicity, all of the coefficients a_{jk}, a_j, c are $\mathbf{C}_b^m$-smooth with $m \geq 2$ in the closure $\bar{D}$ of D. Then, clearly, we can rewrite the operator L in the divergence form:

$$Lu = \sum_{j,k=1}^{d} \frac{\partial}{\partial x_j}[a_{jk}(x)\frac{\partial u}{\partial x_k}] + \sum_{j=1}^{d} b_j(x)\frac{\partial u}{\partial x_j} + c(x)u,$$

where $b_j(x,t) = a_j(x,t) + \sum_{k=1}^{d} \frac{\partial a_{jk}(x)}{\partial x_k}$. The operator L is said to be (uniformly) *strongly elliptic* if for any vector $\xi \in \mathbf{R}^d$ there exist constants $\alpha_1 \geq$

$\alpha > 0$ such that

$$\alpha|\xi|^2 \leq \sum_{j,k=1}^{d} a_{jk}(x)\xi_j\xi_k \leq \alpha_1|\xi|^2. \tag{3.4}$$

In particular, by dropping the first-order terms, it yields the operator $A = A(x, \partial_x)$ given by

$$Au = \sum_{j,k=1}^{d} \frac{\partial}{\partial x_j}[a_{jk}(x)\frac{\partial u}{\partial x_k}] + c(x)u. \tag{3.5}$$

Let $\mathbf{C}_0^\infty(D)$ denote the set of $\mathbf{C}^\infty$-functions on D with compact support. A is a *formally self-adjoint* strongly elliptic operator in the sense that, for $\phi, \psi \in \mathbf{C}_0^\infty$ we have $(A\phi, \psi) = (\phi, A\psi)$. This follows from an integration by parts,

$$\int_D \{\sum_{j,k=1}^{d} \frac{\partial}{\partial x_j}[a_{jk}(x)\frac{\partial\phi(x)}{\partial x_k}]\psi(x) + c(x)\phi(x)\psi(x)\}\,dx$$

$$= \int_D \{\sum_{j,k=1}^{d} \frac{\partial}{\partial x_j}[a_{jk}(x)\frac{\partial\psi(x)}{\partial x_k}]\phi(x) + c(x)\phi(x)\psi(x)\}\,dx.$$

Let H_0^m denote the closure of $\mathbf{C}_0^\infty(D)$ in H^m. By making use of the ellipticity condition, the boundedness of $c(x)$ and an integration by parts, there exists a constant $\beta > 0$ such that, for any $\phi \in H_0^1$,

$$\langle (A - \beta I)\phi, \phi \rangle \leq -\gamma\|\phi\|_1^2, \tag{3.6}$$

for all $\beta > c_0$ with $c_0 = \sup_{x \in D} |c(x)|$, where, by setting $k = 1$ in (3.2), the H^1-norm is given by

$$\|\phi\|_1^2 = \int_D \{\sum_{j=1}^{d} |\frac{\partial\phi(x)}{\partial x_k}|^2 + |\phi(x)|^2\}\,dx.$$

In view of (3.6), by definition, the linear operator $A : H_0^1 \to H^{-1}$ is said to satisfy the *coercivity condition*.

Now consider the elliptic eigenvalue problem:

$$\begin{aligned} A(x, \partial_x)\phi(x) &= -\lambda\phi(x), \quad x \in D, \\ B\phi(x) &= 0, \quad x \in \partial D, \end{aligned} \tag{3.7}$$

where B denotes a boundary operator. It is assumed to be either one of the three types: $B\phi = \phi$ for the Dirichlet boundary condition, $B\phi = \frac{\partial\phi}{\partial n}$ for the Neumann boundary condition and $B\phi = \frac{\partial\phi}{\partial n} + \gamma\phi$ for the mixed boundary condition, where $\frac{\partial}{\partial n}$ denotes the outward normal derivative to ∂D and $\gamma > 0$.

In the variational formulation, regarding A as a linear operator in H with domain $H^2 \cap H_0^1$, the eigenvalue problem (3.7) can be written simply as

$$A\phi = -\lambda\phi, \tag{3.8}$$

where A is self-adjoint and coercive. The following theorem is the basis for the method of eigenfunctions expansion which will be employed later on, (see, e.g., Theorem 7.22, [27]).

Theorem 2.1 For the eigenvalue problem (3.7) or (3.8), there exists a sequence of real eigenvalues $\{\lambda_k\}$ of finite multiplicity such that

$$-\infty < \lambda_1 \leq \lambda_2 \leq \cdots \leq \lambda_k \leq \cdots$$

and $\lambda_k \to \infty$ as $k \to \infty$. Corresponding to each eigenvalue λ_k, there is an eigenfunction e_k such that the set $\{e_k, \ k = 1, 2, \cdots\}$ forms a complete orthonormal basis for H. Moreover, if the coefficients of A are $\mathbf{C}^\infty$-smooth, so is the eigenfunction e_k for each $k \geq 1$. $\qquad\square$

Suppose that A is strictly negative, or there is $\alpha > 0$ such that $\langle A\phi, \phi \rangle \leq -\alpha\|\phi\|^2$ for all $\phi \in H_0^1$. Then clearly we have all eigenvalues $\lambda_k \geq \alpha$. As a special case, let A be given by

$$A\phi = (\kappa\Delta - \alpha)\phi, \tag{3.9}$$

which is clearly a strictly negative, self-adjoint, strongly elliptic operator for $\alpha, \kappa > 0$. Therefore, as far as the method of eigenfunctions expansion is concerned, the elliptic operators defined by (3.5) and (3.9) are similar. For the ease of discussion, the elliptic operator to appear in the parabolic and hyperbolic equations will be taken to be of the special form (3.9) in the subsequent analysis.

Let $R : H \to H$ be a linear integral operator with a symmetric kernel $r(x, y) = r(y, x)$ defined by

$$(R\phi)(x) = \int_D r(x, y)\phi(y)\, dy, \quad x \in D.$$

If r is square integrable so that

$$\int_D \int_D |r(x, y)|^2 dx dy < \infty,$$

then R on H is a self-adjoint *Hilbert-Schmidt* operator which is compact, and the eigenvalues μ_k of R are positive such that [86]

$$\sum_{k=1}^{\infty} \mu_k^2 < \infty,$$

and the normalized eigenfunctions ϕ_k form a complete orthonormal basis for H. The Hilbert-Schmidt norm $\| \cdot \|_{\mathcal{L}_2}$ is defined by

$$\|R\|_{\mathcal{L}_2}^2 = \int_D \int_D |r(x,y)|^2 dx dy = \sum_{k=1}^{\infty} \mu_k^2.$$

In case the eigenvalue μ_k goes to zero faster such that

$$\sum_{k=1}^{\infty} \mu_k < \infty,$$

the operator R is said to be a *nuclear (or trace class) operator*. The norm of such an operator is given by its *trace*:

$$\|R\|_{\mathcal{L}_1} = Tr\, R = \sum_{k=1}^{\infty} \mu_k = \int_D r(x,x) dx.$$

Let $\{w_t^k\}$ be a sequence of independent, identically distributed standard Brownian motions in one dimension. Assume that R is a positive nuclear operator on H defined as above. Define the random field:

$$W_t^n = W^n(\cdot, t) = \sum_{k=1}^{n} \sqrt{\mu_k} w_t^k \phi_k. \tag{3.10}$$

It is easy to check that $W^n(\cdot, t)$ is a Gaussian random field in H with mean $E\, W^n(\cdot, t) = 0$ and the covariance:

$$E\, W^n(x,t) W^n(y,s) = r_n(x,y)\, (t \wedge s), \tag{3.11}$$

where

$$r_n(x,y) = \sum_{k=1}^{n} \mu_k \phi_k(x) \phi_k(y).$$

Note that, for $n > m \geq 1$,

$$\|W_t^n - W_t^m\|^2 = \sum_{k=m+1}^{n} \mu_k (w_t^k)^2$$

is a $\mathcal{F}_t$-submartingale. By the B-D-G inequality (Theorem 2-3.3), we get

$$E \sup_{0 \leq t \leq T} \|W_t^n - W_t^m\|^2 \leq CE\, \|W_T^n - W_T^m\|^2 = CT \sum_{k=m+1}^{n} \mu_k$$

which goes to zero as $n, m \to \infty$. Therefore the sequence $\{W_t^n\}$ of H-valued Wiener processes converges in $L^2(\Omega; \mathbf{C}([0,T]; H))$ to a limit W_t given by

$$W_t = \lim_{n \to \infty} W_t^n = \sum_{k=1}^{\infty} \sqrt{\mu_k} w_t^k \phi_k. \tag{3.12}$$

In the meantime the integral operator R_n with kernel r_n converges to R in the trace norm. Hence W_t is called a Wiener process in H with *covariance operator R*, or a R-Wiener process in H. In fact, by means of an exponential inequality (Lemma 7-6.2) and the Borel-Cantelli lemma, it can be shown that the series in (3.12) converges uniformly with probability one. A H-valued process X_t is called a *Gaussian process in H* if, for each $\phi \in H$, (X_t, ϕ) is a real-valued Gaussian process. In this sense W_t is a Gaussian process in H with covariance operator R.

Theorem 2.2   The R-Wiener process $W_t, t \geq 0$, is a continuous H-valued Gaussian process in H with $W_0 = 0$ such that

(1) For any $s, t \geq 0$ and $g, h \in H$,

$$E(W_t, g) = 0, \quad E(W_t, g)(W_s, h) = (t \wedge s)(Rg, h),$$

(2) W_t has stationary independent increments.

(3) W_t, $t \geq 0$, is a continuous L^2-martingale with values in H and there exists C such that

$$E \sup_{0 \leq t \leq T} \|W_t\|^2 \leq C\,T(Tr\,R),$$

for any $T > 0$.

(4) $W(x, t)$ is a continuous random field for $x \in D, t \geq 0$, if the covariance function $r(x, y)$ is Hölder-continuous in $D \times D$.

Proof.   Inherited from the properties of its finite-sum approximation W_t^n, the properties (1) and (2) can be easily verified. Since W_t^n is a continuous H-valued L^2-martingale, as mentioned before, it converges to W_t in $L^2(\Omega; H)$ uniformly in $t \in [0, T]$. Hence W_t is a continuous L^2-martingale in H and, by a Doob's inequality for submartingale (Theorem 1-2.2),

$$E \sup_{0 \leq t \leq T} \|W_t\|^2 \leq CE\,\|W_T\|^2 = CT\,(Tr\,R),$$

for some positive constant C. To verify property (4), we notice that, for any $x, y \in D$, the Gaussian process $M(x, y, t) = W(x, t) - W(y, t)$ is a martingale with quadratic variation $[M(x, y, \cdot)]_t = t\{r(x, x) - 2r(x, y) + r(y, y)\}$. Hence, by the B-D-G inequality,

$$E \sup_{0 \leq t \leq T} |W(x, t) - W(y, t)|^{2p} = E \sup_{0 \leq t \leq T} |M(x, y, t)|^{2p} \leq C_1 E\,[M(x, , y, \cdot)]_T^p$$

$$\leq C_2 T |r(x, x) - 2r(x, y) + r(y, y)|^p \leq C_3(p, T)|x - y|^{p\alpha},$$

where C_1, C_2, C_3 are some positive constants and α is the Hölder-exponent for r. Therefore, by taking $p > d/\alpha$ and applying Kolmogorov's continuity criterion for random fields (Theorem 1.4.1, [51]), we can conclude that, as a real-valued random field, $W(x,t)$ is continuous in $D \times [0,T]$ as asserted. □

In Chapter Two, some stochastic integrals with respect to a finite-dimensional Wiener process were introduced. Here we shall introduce a stochastic integral in H with respect to the Wiener random field W_t. To this end, let $f(x,t)$ be a continuous, $\mathcal{F}_t$-adapted random field such that

$$E \int_0^T \|f(\cdot,t)\|^2 dt = E \int_0^T \int_D |f(x,t)|^2 dx dt < \infty. \tag{3.13}$$

We first consider the stochastic integral of f in H with respect to a one-dimensional Wiener process $w_t = w_t^1$ given by

$$I(x,t) = \int_0^t f(x,s) dw_s.$$

By convention we write $I_t = I(\cdot,t), f_t = f(\cdot,t)$ etc.; this integral can be written as

$$I_t = \int_0^t f_s dw_s. \tag{3.14}$$

Similar to the usual Itô integral, under the condition (3.13), this integral can be defined as the limit in $L^2(\Omega; H)$ of an approximating sum:

$$I_t = \lim_{|\triangle| \to 0} \sum_{j=0}^{n-1} f_{t_j \wedge t} [w_{t_{j+1} \wedge t} - w_{t_j \wedge t}],$$

where $|\triangle|$ is the mesh of the partition $\triangle = \{0 = t_0 < t_1 < t_2 < \cdots < t_n = T\}$. It is straightforward to show that I_t thus defined, for $t \in [0,T]$, is a continuous L^2-martingale in H with mean zero and

$$E \|I_t\|^2 = E \int_0^t \|f_s\|^2 ds < \infty. \tag{3.15}$$

Before defining stochastic integral of f_t with respect to the Wiener random field W_t, we first consider the case of a bounded integrand. To this end, let $\sigma(x,t)$ be an essentially bounded, continuous adapted random field such that

$$E \int_0^T \sup_{x \in D} |\sigma(x,t)|^2 dt < \infty. \tag{3.16}$$

For any orthonormal eigenfunction ϕ_k of R, we set

$$\sigma^k(x,t) = \sigma(x,t)\phi_k(x).$$

Then, in view of (3.16), σ^k satisfies the condition (3.13), since

$$E \int_0^T \|\sigma^k(\cdot,t)\|^2 dt = E \int_0^T \int_D |\sigma(x,t)\phi_k(x)|^2 dx dt$$

$$\leq E \int_0^T \sup_{x\in D} |\sigma(x,t)|^2 dt < \infty.$$

Therefore the H-valued stochastic integral

$$I_t^k = \int_0^t \sigma_s^k dw_s^k = \left(\int_0^t \sigma_s dw_s^k\right)\phi_k$$

is well defined and it satisfies

$$E \|I_t^k\|^2 \leq E \int_0^T \sup_{x\in D} |\sigma(x,t)|^2 dt, \qquad (3.17)$$

for any k and $t \in [0,T]$.

Let J_t^n denote the stochastic integral defined by

$$J_t^n = \int_0^t \sigma_s dW_s^n = \sum_{k=1}^n \sqrt{\mu_k}\left(\int_0^t \sigma_s dw_s^k\right)\phi_k = \sum_{k=1}^n \sqrt{\mu_k} I_t^k.$$

In view of (3.16) and (3.17), for $n > m$, there is a constant $C > 0$ such that

$$E \|J_t^m - J_t^n\|^2 = \sum_{k=m+1}^n \mu_k E \|I_t^k\|^2 \leq C \sum_{k=m+1}^n \mu_k,$$

which goes to zero as $n > m \to \infty$. Therefore the sequence $\{J_t^n\}$ of continuous L^2-martingales in H converges to a limit J_t for $t \in [0,T]$. We define

$$J_t = \int_0^t \sigma_s dW_s \qquad (3.18)$$

as the stochastic integral of σ with respect to the R-Wiener process in H. In fact it can be shown that the following theorem holds.

Theorem 2.3   Let $\sigma_t = \sigma(\cdot,t)$ be a continuous adapted random field satisfying the property (3.16). The stochastic integral J_t given by (3.18) is a continuous process in H satisfying the following properties:

(1) For any $g, h \in H$, we have $E(J_t, g) = 0$, and

$$E\{(J_t, g)(J_s, h)\} = E \int_0^{t \wedge s} (Q_\tau g, h) d\tau, \qquad (3.19)$$

where Q_t is defined by

$$(Q_t g, h) = \int_D q(x, y, t) g(x) h(y) dx dy \qquad (3.20)$$

with $q(x, y, t) = r(x, y)\sigma(x, t)\sigma(y, t)$.

(2) In particular, the following holds

$$E \, \|J_t\|^2 = E\,(J_t, J_t) = E \int_0^t Tr\, Q_s ds, \qquad (3.21)$$

where $Tr\, Q_t = \int_D q(x, x, t) dx.$

(3) J_t is a continuous L^2-martingale in H with local variation operator Q_t defined by

$$\langle (J, g), (J, h) \rangle_t = \int_0^t (Q_s\, g, h)\, ds, \qquad (3.22)$$

or, simply, $[J]_t = \int_0^t Q_s\, ds.$ □

The local covariation operator Q_t for a H-valued martingale J_t will also be called a *local characteristic operator*, and the local characteristic $q(x, y, t)$, a *local covariation function*. More generally we need to define a stochastic integral J_t with integrand $\sigma_t \in H$ satisfying condition (3.13) instead of (3.16). For simplicity, we will show this is possible under some stronger conditions than necessary. That is, the integrand is a continuous adapted process and the covariance function of the Wiener random field is bounded.

Theorem 2.4 Let $\sigma_t = \sigma(\cdot, t) \in H$ be a continuous adapted random field satisfying the condition

$$E \int_0^T \|\sigma(\cdot, t)\|^2 dt = E \int_0^T \int_D |\sigma(x, t)|^2 dx dt < \infty. \qquad (3.23)$$

If the covariance function $r(x, y)$ is bounded such that

$$r_0 = \sup_{x \in D} r(x, x) < \infty,$$

then the stochastic integral J_t given by (3.18) is well defined and has the same properties as depicted in Theorem 2.3.

Proof. Since the set of $\mathbf{C}_0^\infty$-functions on $\bar{D}$ is dense in H, there exists a sequence $\{\sigma_t^n\}$ of continuous adapted $\mathbf{C}_0^\infty$-random fields satisfying condition (3.16) such that

$$\lim_{n \to \infty} E \int_0^T \|\sigma_t^n - \sigma_t\|^2 dt = 0. \qquad (3.24)$$

By Theorem 2.3, we can define

$$J_t^n = \int_0^t \sigma_s^n dW_s,$$

and, for $n > m$,

$$E \left\| J_t^n - J_t^m \right\|^2 = E \left\| \int_0^t (\sigma_s^n - \sigma_s^m) dW_s \right\|^2 = E \int_0^t Tr \, Q_s^{mn} ds,$$

where

$$
\begin{aligned}
Tr \, Q_s^{mn} &= \int_D q^{mn}(x, x, t) dx \\
&= \int_D r(x, x) [\sigma^n(x, s) - \sigma^m(x, s)]^2 dx \leq r_0 \| \sigma_s^n - \sigma_s^m \|^2.
\end{aligned}
$$

It follows that

$$E \left\| J_t^n - J_t^m \right\|^2 \leq r_0 \, E \int_0^T \| \sigma_t^n - \sigma_t^m \|^2 dt,$$

which goes to zero as $n > m \to \infty$ due to (3.24). Therefore the sequence $\{J_t^n\}$ converges to the limit denoted by J_t as in (3.18). We can check that all its properties given in Theorem 2.3 are valid. This completes the proof. $\quad\square$

Remarks:

(1) Here we restricted the definition of J_t to continuous adapted integrands and a bounded covariance. The definition can be extended to the case with weaker conditions, such as a right continuous integrand and a L^2-bounded covariance function.

(2) Notice that

$$Tr \, Q_t = \int_D q(x, x, t) dx = \int_D r(x, x) \sigma^2(x, t) dx \leq r_0 \| \sigma_t \|^2,$$

which suggests that we may write $Tr \, Q_t = \| \sigma_t \|_R^2$.

(3) Let $\Phi_t : H \to H$ be a continuous linear operator for $t \in [0, T]$ such that $\| \Phi_t h \| \leq C \| h \|$ for some $C > 0$. Then, similar to the proof of Theorem 2.4, the stochastic evolution integral $\int_0^t \Phi_{t-s} \sigma_s dW_s$ can be defined for each $t \in [0, T]$, which will be shown in Section 6.3 later on.

In the subsequent sections, to prove the existence theorems, we shall often use a fixed point argument, in particular the contraction mapping principle as in the case of deterministic partial differential equations. For the convenience of later applications, we will state the contraction mapping principle for a random equation in a Banach space. To this end, let $\mathcal{B}$ be a separable Banach space and let $\xi(\omega)$ be a $\mathcal{B}$-valued random variable. Denote by $\mathcal{X}$ a Banach space of random variables $\xi(\omega)$ with norm $\|\cdot\|_{\mathcal{X}}$. Let $\mathcal{Y}$ be a closed subset of $\mathcal{X}$. Suppose that the map $\Phi : \mathcal{Y} \to \mathcal{Y}$ is well defined. Consider the equation:

$$\xi = \Phi(\xi), \quad \xi \in \mathcal{Y}. \tag{3.25}$$

By the Banach's fixed point theorem (p. 119, [38]), the following contraction mapping principle holds.

Theorem 2.5 (Contraction Mapping Principle) Suppose, for any $\xi, \eta \in \mathcal{Y}$, there is a constant $\rho \in [0, 1)$ such that

$$\|\Phi(\xi) - \Phi(\eta)\|_{\mathcal{X}} \le \rho\|\xi - \eta\|_{\mathcal{X}}.$$

Then there exists a unique solution $\xi^* \in \mathcal{Y}$ of the equation (3.25) which is the fixed point of the map Φ.   □

Another useful result in differential equations is the well-known Gronwall inequality (p. 36, [34]).

Lemma 2.6 (Gronwall Inequality) If $\alpha \ge 0$ is a constant and $\beta(t) \ge 0$, $\theta(t)$ are real continuous functions on $[a, b] \subset \mathbf{R}$ such that

$$\theta(t) \le \alpha + \int_a^t \beta(s)\theta(s)ds, \quad a \le t \le b,$$

then

$$\theta(t) \le \alpha \exp\{\int_a^t \beta(s)ds\}, \quad a \le t \le b. \qquad □$$

3.3   Solution of Random Heat Equation

Let D be a bounded domain in $\mathbf{R}^d$ with a smooth, say $\mathbf{C}^2$, boundary ∂D. We first consider the simple initial-boundary value problem for the randomly perturbed heat equation:

$$\frac{\partial u}{\partial t} = (\kappa\Delta - \alpha)u + \dot{W}(x, t), \quad x \in D, t \in (0, T),$$

$$u|_{\partial D} = 0, \tag{3.26}$$

$$u(x, 0) = h(x),$$

where $\Delta = \sum_{i=1}^d \frac{\partial^2}{\partial x_i^2}$ is the Laplacian operator; κ and α are positive constants and $h(x)$ is a given function in $H = L^2(D)$. The spatially dependent white noise $\dot{W}(x,t)$, by convention, is the formal time derivative $\frac{\partial}{\partial t} W(x,t)$ of the Wiener random field $W(x,t)$. Let $W(\cdot,t)$ be a H-valued Wiener process with mean zero and covariance function $r(x,y)$ so that

$$E\,W(x,t) = 0, \quad E\{W(x,t)W(y,s)\} = (t \wedge s)r(x,y), \quad t \in [0,T], \quad x \in D.$$

The associated covariance operator R in H with kernel $r(x,y)$ is assumed to be of finite trace, that is,

$$Tr\,R = \int_D r(x,x)dx < \infty.$$

To construct a solution, we apply the well-known method of eigenfunctions expansion. Let $\{\lambda_k\}$ be the set of eigenvalues of $(-\kappa\Delta + \alpha)$ with a homogeneous boundary condition and, correspondingly, let $\{e_k\}$ denote the complete orthonormal system of eigenfunctions such that

$$(-\kappa\Delta + \alpha)e_k = \lambda_k e_k, \quad e_k|_{\partial D} = 0, \quad k = 1, 2, ..., \qquad (3.27)$$

with $\alpha < \lambda_1 \leq \lambda_2 \leq ... \leq \lambda_k < ...$, and $\lambda_k \to \infty$ as $k \to \infty$. By Theorem 2.1, they are known to have a finite multiplicity and the corresponding eigenfunctions $e_k \in \mathbf{C}^\infty$. Moreover the eigenvalue has the asymptotic property [37]:

$$\lambda_k = C_d(D)k^{2/d} + o(k^{2/d}) \quad \text{as} \quad k \to \infty, \qquad (3.28)$$

where the constant $C_d(D)$ is independent of k.

Now we seek a formal solution in terms of the eigenfunctions as follows:

$$u(x,t) = \sum_{k=1}^{\infty} u_t^k e_k(x), \qquad (3.29)$$

where u_t^k, $k = 1, 2, ...$, are unknown processes to be determined. To simplify the computations, we assume that the covariance operator R and A have a common set of eigenfunctions with $Re_k = \sigma_k^2 e_k$. Then, in view of (3.12), we can write

$$W(x,t) = \sum_{k=1}^{\infty} \sigma_k e_k(x)w_t^k, \qquad (3.30)$$

where $Tr\,R = \sum_{k=1}^{\infty} \sigma_k^2 < \infty$ and $\{w_t^k\}$ is a sequence of independent standard Wiener processes in one dimension. Upon substituting (3.29) and (3.30) into (3.26), one obtains an infinite system of Itô equations:

$$du_t^k = -\lambda_k u_t^k dt + \sigma_k dw_t^k, \quad u_0^k = h_k, \quad k = 1, 2, ..., \qquad (3.31)$$

where $h_k = (h, e_k)$ with $h \in L^2(D)$. The solution of (3.31) is given by

$$u_t^k = h_k e^{-\lambda_k t} + \sigma_k \int_0^t e^{-\lambda_k(t-s)}dw_s^k, \quad k = 1, 2, ..., \qquad (3.32)$$

which are independent O-U (Ornstein-Uhlenbeck) processes with mean

$$\hat{u}_t^k = E\,u_t^k = h_k e^{-\lambda_k t}$$

and covariance

$$Cov\,\{u_t^k, u_s^l\} = \delta_{kl}\frac{\sigma_k^2}{2\lambda_k}\{e^{-\lambda_k|t-s|} - e^{-\lambda_k(t+s)}\}, \tag{3.33}$$

where $\delta_{kl} = 1$ if $k = l$ and $= 0$ otherwise. To show the convergence of the series solution (3.29), we write

$$u(x,t) = \hat{u}(x,t) + v(x,t).$$

It consists of the mean solution

$$\hat{u}(x,t) = \sum_{k=1}^{\infty} \hat{u}_t^k e_k(x), \tag{3.34}$$

and the random deviation

$$v(x,t) = \sum_{k=1}^{\infty} v_t^k e_k(x), \tag{3.35}$$

where

$$v_t^k = \sigma_k \int_0^t e^{-\lambda_k(t-s)}\,dw_s^k. \tag{3.36}$$

We know that the mean solution series (3.35) converges in the L^2-sense and it satisfies the problem (3.26) without random perturbation. It suffices to show the random series (3.36) converges in some sense to a solution of (3.26) with $h = 0$. To proceed, consider the partial sum

$$v^n(x,t) = \sum_{k=1}^{n} v_t^k e_k(x). \tag{3.37}$$

In view of (3.36), we have

$$E\,\|v^n(\cdot,t)\|^2 = \sum_{k=1}^{n} E\,|v_t^k|^2 = \sum_{k=1}^{n} \frac{\sigma_k^2}{2\lambda_k}\{1 - e^{-2\lambda_k t}\}$$
$$\leq \sum_{k=1}^{n} \frac{\sigma_k^2}{2\lambda_k}. \tag{3.38}$$

Recall that $\lambda_k > \alpha$ and $\sigma_k^2 = (R\,e_k, e_k)$. The above yields

$$\sup_{t\leq T} E\,\|v^n(\cdot,t)\|^2 \leq \frac{1}{2\alpha}\sum_{k=1}^{n}\sigma_k^2 \leq \frac{1}{2\alpha}Tr\,R, \tag{3.39}$$

which is finite by assumption. It follows that

$$\sup_{t \leq T} E \, \|v(\cdot,t) - v^n(\cdot,t)\|^2 = \frac{1}{2\alpha} \sum_{k=n+1}^{\infty} \frac{\sigma_k^2}{2\lambda_k} \to 0, \quad \text{as} \quad n \to \infty,$$

and $v(\cdot,t)$ is a $\mathcal{F}_t$-adapted H-valued process. In fact the process is mean-square continuous. To see this, we compute

$$E \, \|v(\cdot,t) - v(\cdot,s)\|^2 = \sum_{k=1}^{\infty} E \, |v_t^k - v_s^k|^2. \tag{3.40}$$

For $0 \leq s < t \leq T$,

$$\{v_t^k - v_s^k\} = \sigma_k \{ \int_0^t e^{-\lambda_k(t-r)} dw_r^k - \int_0^s e^{-\lambda_k(s-r)} dw_r^k \}$$

$$= \sigma_k \{ \int_0^s [e^{-\lambda_k(t-r)} - e^{-\lambda_k(s-r)}] dw_r^k + \int_s^t e^{-\lambda_k(t-r)} dw_r^k \},$$

so that, noting the independence of the above integrals,

$$E \, |v_t^k - v_s^k|^2 = \sigma_k^2 \{ E \, |\int_0^s [e^{-\lambda_k(t-r)} - e^{-\lambda_k(s-r)}] dw_r^k|^2 + E \, |\int_s^t e^{-\lambda_k(t-r)} dw_r^k|^2 \}$$

$$= \sigma_k^2 \{ \int_0^s [e^{-\lambda_k(t-r)} - e^{-\lambda_k(s-r)}]^2 dr + \int_s^t e^{-2\lambda_k(t-r)} dr \}$$

$$\leq \frac{\sigma_k^2}{\lambda_k} [1 - e^{-2\lambda_k(t-s)}] \quad \leq \quad \sigma_k^2(t-s).$$

Making use of this bound in (3.40), we obtain

$$E \, \|v(\cdot,t) - v(\cdot,s)\|^2 \leq \sum_{k=1}^{\infty} \sigma_k^2(t-s) = (Tr \, R)(t-s),$$

which shows that $v(\cdot,t)$ is H-continuous in mean-square. A question arises naturally: In what sense does the random series (3.29) represent a solution to the problem (3.26)? To answer this question, we first show that, merely under the condition of finite trace, it may not be a *strict solution* in the sense that $u(\cdot,t)$ is in the domain of $A = (\kappa\Delta - \alpha)$, a subset of H^2, and it satisfies, for each $(x,t) \in D \times [0,T]$,

$$u(x,t) = h(x) + \int_0^t (\kappa\Delta - \alpha)u(x,s)ds + W(x,t), \quad \text{a.s.}$$

To justify our claim, it is enough to show that the sequence $\{A v^n(\cdot,t)\}$ in H may diverge in mean-square. In view of (3.37) and (3.38),

$$E \, \|Av^n(\cdot,t)\|^2 = \sum_{k=1}^{n} \lambda_k^2 E \, |v_t^k|^2$$

$$= \sum_{k=1}^{n} \frac{\lambda_k \sigma_k^2}{2} \{1 - e^{-2\lambda_k t}\} \geq \frac{1}{2}(1 - e^{-2\lambda_1 t}) \sum_{k=1}^{n} \lambda_k \sigma_k^2.$$

For instance, for d=2, we have $\lambda_k \sim C_1 k$ by (3.28). If $\sigma_k \sim C_2/k$ for large k, then the above partial sum diverges so that $E\|Av^n(\cdot,t)\|^2 \to \infty$. In general this rules out the case of a strict solution. In fact it turns out to be a *weak solution* in the PDE sense. This means that $u(\cdot,t)$ is an adapted H-valued process such that for any test function ϕ in $\mathbf{C}_0^\infty$ and for each $t \in [0,T]$, the following equation holds:

$$(u(\cdot,t),\phi) = (h,\phi) + \int_0^t (u(\cdot,s),A\phi)ds + (W(\cdot,t),\phi), \quad \text{a.s.} \tag{3.41}$$

This equation can be easily verified for its n-term approximation u^n given by

$$u^n(x,t) = \sum_{k=1}^n u_t^k e_k(x).$$

By integrating this equation against the test function over D and making use of (3.31), we get

$$(u^n(\cdot,t),\phi) = \sum_{k=1}^n u_t^k(e_k,\phi) = \sum_{k=1}^n \{h_k - \lambda_k \int_0^t u_s^k ds + \sigma_k w_t^k\}(e_k,\phi)$$

$$= (h^n,\phi) + \int_0^t (u^n(\cdot,s),A\phi)ds + (W^n(\cdot,t),\phi),$$

where h^n and W^n are the n-term approximations of h and W, respectively. Since they together with u^n converge strongly in mean-square, by passing to the limit in each term of the above equation as $n \to \infty$, this equation yields equation (3.41). The fact that the weak solution is unique is easy to verify. Suppose that u and $\tilde{u}$ are both solutions. Then it follows from (3.30) that $\mu = (u - \tilde{u})$ satisfies the equation

$$(\mu(\cdot,t),\phi) = \int_0^t (\mu(\cdot,s),A\phi)\,ds.$$

By letting $\phi = e_k$ and $\mu_t^k = (\mu(\cdot,t),e_k)$, the above yields

$$\mu_t^k = -\lambda_k \int_0^t \mu_s^k\,ds, \ k = 1,2,\cdots,$$

which implies

$$|\mu_t^k|^2 \leq \lambda_k^2 T \int_0^t |\mu_s^k|^2\,ds, \ k = 1,2,\cdots.$$

By the Gronwall inequality, we can conclude that $\mu_t^k = (u(\cdot,t)-\tilde{u}(\cdot,t),e_k) = 0$ for every k. Since the set of eigenfunctions is complete, we have $u(\cdot,t) = \tilde{u}(\cdot,t)$ a.s. in H.

We notice that since each term in the series (3.29) is an O-U process, the sum $u(\cdot, t)$ is a H-valued Gaussian process with mean $\hat{u}(x, t)$ given by (3.34) and covariance function

$$Cov.\{u(x,t), u(y,s)\} = \sum_{k=1}^{\infty} \frac{\sigma_k^2}{2\lambda_k}\{e^{-\lambda_k|t-s|} - e^{-\lambda_k(t+s)}\}e_k(x)e_k(y). \quad (3.42)$$

Let us summarize the previous results as a theorem.

Theorem 3.1   Let the eigenfunctions expansion of $u(x,t)$ be given by (3.29). Then, for W_t given by (3.30), $u(\cdot, t), 0 \leq t \leq T$ is an adapted H-valued Gaussian process which is continuous in the mean-square sense. Moreover it is a unique weak solution of the problem (3.26) satisfying the equation (3.41). □

It is well known that the Green's function $G(x, y; t)$ for the heat equation in (3.26) can be expressed as

$$G(x, y; t) = \sum_{k=1}^{\infty} e^{-\lambda_k t} e_k(x) e_k(y). \quad (3.43)$$

Then the solution (3.29) can be rewritten as

$$u(x,t) = \hat{u}(x,t) + v(x,t)$$
$$= \sum_{k=1}^{\infty} h_k e^{-\lambda_k t} e_k(x) + \sum_{k=1}^{\infty} \sigma_k \int_0^t e^{-\lambda_k(t-s)} dw_s^k e_k(x),$$

which shows formally that

$$u(x,t) = \int_D G(x,y;t)h(y)dy + \int_0^t \int_D G(x,y;t-s)W(y,ds)dy. \quad (3.44)$$

Introduce the associated Green's operator G_t defined by

$$(G_t h)(x) = \int_D G(x,y;t)h(y)\,dy. \quad (3.45)$$

Then (3.44) can be written simply as

$$u(\cdot, t) = G_t h + \int_0^t G_{t-s} W(\cdot, ds). \quad (3.46)$$

Also it is easy to check the properties: $G_0 h = h$ and $G_t(G_s h) = G_{t+s} h$, for all $t, s > 0$. As a useful analytical technique, the Green's function representation will be used extensively in the subsequent sections.

We observe that, in view of (3.28), the partial sum may converge even if the trace of the covariance operator R is infinite. This is the case, for instance, when $d = 1, r(x, y) = \delta(x - y)$ and $\dot{W}(x, t)$ becomes a space-time white noise. Further properties of solution regularity and their dependence on the noise will be discussed later on. Also it is worth noting that the theorem holds when the Dirichlet boundary condition is replaced by a Neumann boundary condition for which the normal derivative $\frac{\partial}{\partial n} u|_{\partial D} = 0$, or by a proper mixed boundary condition. However, for simplicity, we shall consider only the first two types of boundary conditions.

3.4 Linear Equations with Additive Noise

Let $f(x, t)$ and $\sigma(x, t)$, $x \in D$, $t \in [0, T]$ be random fields such that $f(\cdot, t)$ and $\sigma(\cdot, t)$ are $\mathcal{F}_t$-adapted such that

$$E \int_0^T \{\|f(\cdot, s)\|^2 + \|\sigma(\cdot, s)\|^2\} ds < \infty. \tag{3.47}$$

Regarded as a H-valued process, $f(\cdot, t)$, $\sigma(\cdot, t)$, $\cdots$ will sometimes be written as f_t, σ_t and so on for brevity. Suppose that $W(x, t)$ is a R-Wiener random field with covariance function $r(x, y)$ bounded by r_0, or $\sup_{x \in D} r(x, x) \le r_0$. By Theorem 2.3, the Itô integral:

$$M(\cdot, t) = \int_0^t \sigma(\cdot, s) W(\cdot, ds) \tag{3.48}$$

is well defined and it will also be written as

$$M_t = \int_0^t \sigma_s dW_s.$$

It is a H-valued martingale with local covariation operator Q_t defined by

$$[Q_t h](x) = \int_D q(x, y; t) h(y) dy$$

with the kernel $q(x, y; t) = r(x, y)\sigma(x, t)\sigma(y, t)$. Notice that, by (3.47) and the bound on r,

$$E \int_0^T Tr\, Q_s ds = E \int_0^T \int_D r(x, x)\sigma^2(x, s) dx ds \tag{3.49}$$
$$\le r_0 E \int_0^T \|\sigma(\cdot, s)\|^2 ds < \infty.$$

Let

$$V(x, t) = \int_0^t f(x, s) ds + M(x, t), \quad \text{or} \quad V_t = \int_0^t f_s ds + M_t, \tag{3.50}$$

which is a spatially dependent semimartingale with local characteristic $(q(x, y; t), f(x, t))$.

Replacing $W(x, t)$ by $V(x, t)$ in (3.26), we now consider the following problem:

$$\frac{\partial u}{\partial t} = (\kappa \Delta - \alpha) u + \dot{V}(x, t), \quad x \in D, \, t \in (0, T),$$

$$Bu|_{\partial D} = 0, \quad u(x, 0) = h(x),$$

(3.51)

where $Bu = u$ or $Bu = \frac{\partial}{\partial n} u$, and $\dot{V}(x, t) = f(x, t) + \sigma(x, t) \dot{W}(x, t)$. Suggested by the representation (3.44) in terms of the Green's function (3.43), we rewrite (3.51) as

$$
\begin{aligned}
u(x, t) &= \int_D G(x, y; t) h(y) dy + \int_0^t \int_D G(x, y; t - s) V(y, ds) dy \\
&= \int_D G(x, y; t) h(y) dy + \int_0^t \int_D G(x, y; t - s) f(y, s) ds dy \\
&\quad + \int_0^t \int_D G(x, y; t - s) \sigma(y, s) W(y, ds) dy,
\end{aligned}
$$

(3.52)

which is the mild solution of (3.51). We claim it is also a weak solution satisfying

$$
\begin{aligned}
(u_t, \phi) &= (h, \phi) + \int_0^t (u_s, A\phi) ds + \int_0^t (f_s, \phi) ds \\
&\quad + \int_0^t (\phi, \sigma_s dW_s),
\end{aligned}
$$

(3.53)

for any $\phi \in \mathbf{C}_0^2$, where $A = (\kappa \Delta - \alpha)$ is defined as before. To verify this fact, we let

$$u(x, t) = \tilde{u}(x, t) + v(x, t),$$

(3.54)

where

$$\tilde{u}(x, t) = \int_D G(x, y; t) h(y) dy + \int_0^t \int_D G(x, y; t - s) f(y, s) dy ds.$$

(3.55)

It is easily shown that $\tilde{u}(x, t)$ is a weak solution of the problem (3.51) when $\sigma(x, t) = 0$. This reduces to showing that

$$v(x, t) = \int_0^t \int_D G(x, y; t - s) \sigma(y, s) W(y, ds) dy$$

(3.56)

must satisfy the equation:

$$(v_t, \phi) = \int_0^t (v_s, A\phi) ds + \int_0^t (\phi, \sigma_s dW_s).$$

(3.57)

In view of (3.43), equation (3.56) can be written in terms of the eigenfunctions as follows

$$v_t = \sum_{k=1}^{\infty} v_t^k e_k, \tag{3.58}$$

where

$$v_t^k = (v_t, e_k) = \int_0^t e^{-\lambda_k(t-s)} dz_s^k, \tag{3.59}$$

and

$$z_t^k = (M_t, e_k) = \int_0^t (e_k, \sigma_s dW_s). \tag{3.60}$$

One can check that the process z_t^k is a continuous martingale with the quadratic variation

$$[z^k]_t = \int_0^t q_s^k ds,$$

where

$$q_t^k = \int_D \int_D q(x, y; t) e_k(x) e_k(y) dx dy = (Q_t e_k, e_k).$$

Let $v^n(\cdot, t)$ be the n-term approximation of v_t:

$$v^n(\cdot, t) = \sum_{k=1}^{n} v_t^k e_k. \tag{3.61}$$

From (3.60), we obtain

$$E|z_t^k|^2 = \int_0^t E(Q_s e_k, e_k) ds. \tag{3.62}$$

Therefore, by condition (3.47), we get

$$
\begin{aligned}
E\|v^n(\cdot, t)\|^2 &= \sum_{k=1}^{n} E|v_t^k|^2 = \int_0^t E \sum_{k=1}^{n} e^{-2\lambda_k(t-s)} (Q_s e_k, e_k) ds \\
&= E \int_0^t \sum_{k=1}^{n} (Q_s e_k, G_{2(t-s)} e_k) ds \\
&\leq \int_0^t E \sum_{k=1}^{n} (Q_s e_k, e_k) ds < E \int_0^t Tr\, Q_s ds < \infty.
\end{aligned}
$$

It follows that the sequence $\{v^n\}$ converges in H to v in mean-square uniformly for $t \in [0, T]$, or

$$\sup_{0 \leq t \leq T} E\|v_t - v^n(\cdot, t)\|^2 \to 0, \qquad as \quad n \to \infty. \tag{3.63}$$

Hence we get

$$E\|v_t\|^2 = E \int_0^t Tr\left[G_{2(t-s)}Q_s\right]ds \le E \int_0^t Tr\,Q_s ds. \qquad (3.64)$$

In fact, by writing

$$v_t^k = z_t^k - \lambda_k \int_0^t e^{-\lambda_k(t-s)} z_s^k ds,$$

with the aid of a Doob's inequality and (3.62), we can obtain the estimate

$$E \sup_{0 \le t \le T} |v_t^k|^2 \le 4E \sup_{0 \le t \le T} |z_t^k|^2$$

$$\le 16E |z_T^k|^2 = 16 \int_0^T E\,(Q_s e_k, e_k)ds.$$

This implies that,

$$E \sup_{0 \le t \le T} \|v_t\|^2 \le \sum_{k=1}^{\infty} E \sup_{0 \le t \le T} |v_t^k|^2 \qquad (3.65)$$
$$\le 16 \int_0^T E\,(Tr\,Q_s)ds < \infty.$$

To show that $v(\cdot, t)$ is mean-square continuous, we first estimate

$$E\,|v_t^k - v_s^k|^2 = \{E\,|\int_0^s [e^{-\lambda_k(t-r)} - e^{-\lambda_k(s-r)}]dz_r^k|^2 + E\,|\int_s^t e^{-\lambda_k(t-r)}dz_r^k|^2\}$$

$$= \{\int_0^s [e^{-\lambda_k(t-r)} - e^{-\lambda_k(s-r)}]^2 q_r^k dr + \int_s^t e^{-2\lambda_k(t-r)} q_r^k dr\}.$$

Since, noticing (3.65), $\|v_t - v_s\|^2 \le 4\sup_{0 \le t \le T} \|v_t\|^2 < \infty$, a.s., by invoking the dominated convergence theorem, the above integrals go to zero as $s \to t$, so that $\lim_{s \to t} E \|v_t - v_s\|^2 = 0$. So the process $v(\cdot, t)$ is mean-square continuous.

Clearly, by (3.59), v_t^k satisfies the equation:

$$v_t^k = -\lambda_k \int_0^t v_s^k ds + z_t^k.$$

Hence the equation (3.61) can be written as

$$v^n(x, t) = \int_0^t Av^n(x, s)ds + M^n(x, t), \qquad (3.66)$$

where $M^n(x, t)$ is the n-term approximation of $M(x, t)$. By taking the inner product with respect to ϕ, this equation yields

$$(v^n(\cdot, t), \phi) = \int_0^t (v_s^n, A\phi)ds + (\phi, M_t^n).$$

Since, as $n \to \infty$, each term in the above equation converges in mean-square to the corresponding term in equation (3.57), we conclude that $v(x,t)$ given by (3.56) is indeed a weak solution. Similar to Theorem 3.1, the uniqueness of solution can be shown easily. Thereby we have proved the following theorem.

Theorem 4.1 Let condition (3.47) hold true and let $u(x,t)$ be defined by (3.52). Then, for $h \in H$ and $t \in [0,T]$, $u(\cdot,t)$ is an adapted H-valued process which is continuous in mean-square. Furthermore it is the unique weak solution of the problem (3.51) satisfying the equation (3.53). □

In fact we can show more, namely the second moment of the solution is uniformly bounded in t over $[0,T]$. More precisely, the following theorem holds.

Theorem 4.2 Under the same conditions as given in Theorem 4.1, the (weak) solution $u(\cdot,t)$ given by (3.53) satisfies the following inequality:

$$E \sup_{0 \leq t \leq T} \|u(\cdot,t)\|^2 \leq C(T) \{ \|h\|^2 + E \int_0^T [\|f(\cdot,s)\|^2 + Tr\, Q_s] ds \} \quad (3.67)$$

where $C(T)$ is a positive constant depending on T.

Proof. Referring to (3.45), (3.54) and (3.55), we can write

$$u_t = (G_t h) + \int_0^t G_{t-s} f_s ds + v_t,$$

so that

$$\|u_t\|^2 \leq 3\{\|G_t h\|^2 + \| \int_0^t G_{t-s} f_s ds\|^2 + \|v_t\|^2\}. \quad (3.68)$$

By making use of (3.43), we find that

$$\|G_t h\|^2 = \sum_{k=1}^{\infty} e^{-2\lambda_k t}(h, e_k)^2 \leq \|h\|^2 \quad (3.69)$$

and

$$\| \int_0^t G_{t-s} f_s ds\|^2 \leq t \int_0^t \|G_{t-s} f_s\|^2 ds \leq t \int_0^t \|f_s\|^2 ds. \quad (3.70)$$

By taking (3.68), (3.69) and (3.70) into account, one obtains

$$E \sup_{0 \leq t \leq T} \|u_t\|^2 \leq 3\{\|h\|^2 + TE \int_0^T \|f_s\|^2 ds + E \sup_{0 \leq t \leq T} \|v_t\|^2\}.$$

With the aid of (3.65), the above is bounded by

$$3\{ \|h\|^2 + TE \int_0^T \|f_s\|^2 ds + 16E \int_0^T Tr\, Q_s ds \}.$$

This yields the bound (3.67) with $C(T) = 3 \max \{16, T\}$.   □

Remarks:

(1) Notice that, since the proofs are based on the methods of eigenfunctions expansion and the Green's function, Theorem 4.1 and Theorem 4.2 still hold true if $A = (\kappa\Delta - \alpha)$ is replaced by a self-adjoint, strongly elliptic operator with smooth coefficients. Keep this fact in mind in the subsequent analysis.

(2) If $\dot{W}(x, t)$ is a space-time white noise with $r(x, y) = \delta(x - y)$, the corresponding Wiener random field $W(\cdot, t)$ is known as a cylindrical Brownian motion in H with the identity operator I as its covariance operator [19]. By (3.28), the eigenvalues λ_k of A grow asymptotically like $k^{2/d}$ for a large k. It is easy to show that

$$E \, \|v(\cdot, t)\|^2 = \sum_k \frac{1}{2\lambda_k} (1 - e^{-2\lambda_k t})$$

converges for $d = 1$ but diverges for $d \geq 2$. This means that, in the case of a space-time white noise, the solution $u(\cdot, t)$ cannot exist as a H-valued process except for $d = 1$. For this reason, in order for the solution to have better regularity properties, we shall consider only a smooth Wiener random field with a finite-trace covariance operator.

3.5   Some Regularity Properties

In this section we will study further regularity of the weak solution $u(x, t)$ to (3.51) given by (3.52). In particular we shall consider the Sobolev H^1-regularity in space and Hölder continuity in time. Recall that, for any integer k, the Sobolev space $H^k = H^k(D)$ denotes the k-th order Sobolev space of functions on D with $H^0 = H = L^2$. As a matter of convenience, instead of the usual H^k-norm defined by (3.2), we shall often use an equivalent norm on H^k with $k \geq 0$, which is still denoted by $\|\cdot\|_k$ defined as follows

$$\|\phi\|_k = \{\sum_{j=1}^{\infty} \lambda_j^k (\phi, e_j)^2\}^{1/2} = \{\langle (-\kappa\Delta + \alpha)^k \phi, \phi \rangle\}^{1/2}, \quad \phi \in H^k. \quad (3.71)$$

Referring back to (3.52)–(3.54), let $u = \tilde{u} + v$, where we rewrite $\tilde{u}$ and v as

$$\tilde{u}_t = G_t h + \int_0^t G_{t-s} f_s ds \quad (3.72)$$

and

$$v_t = \int_0^t G_{t-s} dM_s. \tag{3.73}$$

Recall that $M(x,t) = \int_0^t \sigma(x,s)W(x,ds)$ is a martingale with local characteristic $q(x,y;t) = r(x,y)\sigma(x,t)\sigma(y,t)$, where $\sigma(x,t)$ is a predictable random field and $r(x,y)$ is bounded by r_0. Before proceeding to the regularity question, we shall collect several basic inequalities as technical lemmas which will be useful in the subsequent analysis. Most of these results were derived in the previous section.

Lemma 5.1 Suppose that $h \in H$ and $f(\cdot,t) \in H$ is a predictable random field such that

$$E \int_0^T \|f(\cdot,t)\|^2 dt < \infty.$$

Then $\tilde{u}(\cdot,t)$ given by (3.55) is a continuous adapted process in H such that

$$\tilde{u} \in L^2(\Omega; C([0,T]; H)) \cap L^2(\Omega \times (0,T); H^1).$$

Moreover, the following inequalities hold:

$$\sup_{0 \leq t \leq T} \|G_t h\| \leq \|h\|; \quad \int_0^T \|G_t h\|_1^2 dt \leq \frac{1}{2}\|h\|^2, \tag{3.74}$$

$$E \sup_{0 \leq t \leq T} \|\int_0^t G_{t-s} f(\cdot,s) ds\|^2 dt \leq TE \int_0^T \|f(\cdot,t)\|^2 dt; \tag{3.75}$$

$$E \sup_{0 \leq t \leq T} \|\int_0^t G_{t-s} f(\cdot,s) ds\|^2 dt \leq \frac{1}{2} E \int_0^T \|f(\cdot,t)\|_1^2 dt, \tag{3.76}$$

and

$$E \int_0^T \|\int_0^t G_{t-s} f(\cdot,s) ds\|_1^2 dt \leq \frac{T}{2} E \int_0^T \|f(\cdot,t)\|^2 dt. \tag{3.77}$$

Proof. First suppose that $\tilde{u}$ is regular as indicated. The first inequalities in (3.74) and (3.75) were shown in (3.69) and (3.70), respectively. The remaining ones can be easily verified. For instance, consider the last inequality (3.77) as follows:

$$\int_0^T \|\int_0^t G_{t-s} f_s ds\|_1^2 dt \leq \int_0^T \int_0^t t \|G_{t-s} f_s\|_1^2 ds dt$$

$$\leq T \sum_{k=1}^{\infty} \int_0^T \int_0^t \lambda_k e^{-2\lambda_k(t-s)} (e_k, f_s)^2 ds dt$$

$$= \frac{T}{2} \sum_{k=1}^{\infty} \int_0^T (1 - e^{-2\lambda_k(T-t)})(e_k, f_t)^2 dt \leq \frac{T}{2} \int_0^T \|f_t\|^2 dt,$$

which yields (3.77) after taking the expectation.

It is known that, for $h \in H$, $G_t h \in C([0,T];H) \cap L^2((0,T);H^1)$ [77]. To show that $\tilde{u} \in L^2(\Omega; C([0,T];H)) \cap L^2(\Omega \times (0,T);H^1)$, let $\xi_t = \int_0^t G_{t-s} f_s \, ds$, and $\xi_t^n = \int_0^t G_{t-s} f_s^n \, ds$, where we let

$$f_t^n = \sum_{k=1}^n (f_t, e_k) e_k.$$

Then it is easy to check that $\xi_t^n \in L^2(\Omega; C([0,T];H)) \cap L^2(\Omega \times (0,T);H^1)$. Moreover we can obtain

$$E \sup_{0 \le t \le T} \|\xi_t - \xi_t^n\|^2 \le TE \sum_{k>n} \int_0^T (f_t, e_k)^2 \, dt$$

and

$$E \int_0^T \|\xi_t - \xi_t^n\|_1^2 \, dt \le \frac{T}{2} E \sum_{k>n} \int_0^T (f_t, e_k)^2 \, dt,$$

both of which go to zero as $n \to \infty$. Hence the limit ξ_t belongs to $L^2(\Omega; C([0,T];H)) \cap L^2(\Omega \times (0,T);H^1)$ as claimed. $\qquad \square$

Lemma 5.2 Suppose that

$$v(\cdot, t) = \int_0^t G_{t-s} \sigma(\cdot, s) W(\cdot, ds)$$

as given by (3.73) and assume that

$$E \int_0^T (Tr \, Q_t) dt = E \int_0^T \int_D r(x,x) \sigma^2(x,t) dx dt < \infty. \qquad (3.78)$$

Then $v(\cdot, t)$ is a H^1-process, with a continuous trajectory in H over $[0,T]$, and the following inequalities hold:

$$E \|v(\cdot, t)\|^2 \le E \int_0^t (Tr \, Q_s) ds, \qquad (3.79)$$

$$E \sup_{0 \le t \le T} \|v(\cdot, t)\|^2 \le 16E \int_0^T (Tr \, Q_t) dt, \qquad (3.80)$$

and

$$E \int_0^T \|v(\cdot, t)\|_1^2 dt \le \frac{1}{2} E \int_0^T Tr \, Q_t dt. \qquad (3.81)$$

Proof. Under condition (3.78), the inequality (3.80) was shown in (3.65). Let $v^n(\cdot, t)$ be the n-term approximation of $v(\cdot, t)$ given by (3.61), where v_t^k is continuous in H. Hence, in view of (3.65), $\{v^n(\cdot, t)\}$ is a Cauchy sequence in $L^2(\Omega; C([0, T]; H))$ convergent to v and it has a continuous version.

The inequalities (3.79) and (3.80) were verified by (3.64) and (3.65), respectively. To show (3.81), we have

$$
E \int_0^T \|\int_0^t G_{t-s}\sigma_s dW_s\|_1^2 = \int_0^T \sum_{k=0}^\infty E \int_0^t \lambda_k e^{-2\lambda_k(t-s)}(e_k, \sigma_s dW_s)^2 ds
$$

$$
= \int_0^T \sum_{k=0}^\infty E \int_0^t \lambda_k e^{-2\lambda_k(t-s)}(Q_s e_k, e_k) ds dt
$$

$$
\leq \frac{1}{2} E \int_0^T \sum_{k=0}^\infty (Q_s e_k, e_k) ds = \frac{1}{2} E \int_0^T Tr\, Q_t dt. \qquad \square
$$

With the aid of Lemma 5.1 and Lemma 5.2, we can prove the following theorem, which is a more regular version of Theorem 4.1.

Theorem 5.3 Let the condition (3.47) be satisfied and let $h \in H$. The linear problem (3.51) has a unique weak solution $u(\cdot, t)$ given by (3.52), which is a predictable H^1-valued process with a continuous trajectory in H over $[0, T]$ such that

$$
\begin{aligned}
E \sup_{0 \leq t \leq T} \|u(\cdot, t)\|^2 + E \int_0^T \|u(\cdot, t)\|_1^2 dt \\
\leq C(T) \{ \|h\|^2 + E \int_0^T [\|f(\cdot, s)\|^2 + Tr\, Q_s] ds \},
\end{aligned}
\tag{3.82}
$$

for some constant $C(T) > 0$. Moreover the energy equation holds true:

$$
\begin{aligned}
\|u(\cdot, t)\|^2 = \|h\|^2 + 2 \int_0^t \langle Au(\cdot, s), u(\cdot, s) \rangle ds + 2 \int_0^t (u(\cdot, s), f(\cdot, s)) ds \\
+ 2 \int_0^t (u(\cdot, s), M(\cdot, ds)) + \int_0^t Tr\, Q_s ds,
\end{aligned}
\tag{3.83}
$$

where $M(\cdot, ds) = \sigma(\cdot, s)W(\cdot, ds)$.

Proof. In view of Lemma 5.1 and Lemma 5.2, since the problem is linear, $u = \tilde{u} + v$ is a unique weak solution satisfying the regularity property $u \in L^2(\Omega \times [0, T]; H^1) \cap L^2(\Omega; C([0, T]; H))$. Making use of the simple inequalities:

$$
\|u_t\|^2 \leq 3\{\|G_t h\|^2 + \|\int_0^t G_{t-s} f_s ds\|^2 + \|v_t\|^2\},
$$

and

$$
\begin{aligned}
\int_0^T \|u_t\|_1^2 dt \leq 3\{\int_0^T \|G_t h\|_1^2 dt + \int_0^T \|\int_0^t G_{t-s} f_s ds\|_1^2 dt \\
+ \int_0^T \|v_t\|_1^2 dt\},
\end{aligned}
$$

we can apply the inequalities (3.74), (3.76), (3.80) and (3.81) to obtain the bound in (3.82).

Concerning the energy equation (3.83), recall that $u_t = \tilde{u}_t + v_t$ as defined by (3.54) and denote the n-term approximation: $u^n(\cdot, t) = \tilde{u}_t^n + v^n(\cdot, t)$. By definitions, it is easy to show that $u^n(\cdot, t)$ satisfies equation (3.83). That is,

$$
\begin{aligned}
\|u^n(\cdot, t)\|^2 &= \|h^n\|^2 + 2 \int_0^t (Au^n(\cdot, s), u^n(\cdot, s))ds \\
&+ 2 \int_0^t (u^n(\cdot, s), f_s^n)ds + 2 \int_0^t (u^n(\cdot, s), dM_s^n) + \int_0^t Tr\, Q_s^n ds,
\end{aligned}
\tag{3.84}
$$

where f_t^n and M_t^n are the n-term approximations of f_t and M_t, respectively. Clearly we have

$$
\|u_t - u^n(\cdot, t)\|^2 \leq 2\{\|\tilde{u}_t - \tilde{u}^n(\cdot, t)\|^2 + \|v_t - v^n(\cdot, t)\|^2\}.
\tag{3.85}
$$

From (3.55), with the aid of Lemmas 5.1, we can deduce that

$$
\sup_{0 \leq t \leq T} E \|\tilde{u}_t - \tilde{u}^n(\cdot, t)\|^2 \leq 2\{\|h - h^n\|^2 + E \int_0^T \|f_t - f_t^n\|^2 dt\} \to 0, \text{ as } n \to \infty.
$$

Making use of (3.80), we can get

$$
E \sup_{0 \leq t \leq T} \|v_t - v^n(\cdot, t)\|^2 \leq 16E \int_0^T (Tr\, \delta Q_s^n)ds,
$$

with $\delta Q_t^n = \dfrac{d}{dt}[\delta M^n]_t$ where $\delta M_t^n = (M_t - M_t^n)$ and $[\delta M]_t$ denotes the covariation operator for δM_t. Similar to the derivation of (3.56)–(3.63), it can be shown that

$$
E \int_0^T (Tr\, \delta Q_s^n)ds \to 0, \text{ as } \quad n \to \infty.
\tag{3.86}
$$

In view of the above results, the equation (3.85) yields

$$
\sup_{0 \leq t \leq T} E \|u_t - u^n(\cdot, t)\|^2 \to 0, \text{ as } \quad n \to \infty,
\tag{3.87}
$$

which implies

$$
E \|u^n(\cdot, t)\|^2 \to E \|u_t\|^2, \text{ as } \quad n \to \infty.
\tag{3.88}
$$

By means of Lemmas 5.1 and 5.2, and the fact that $|\langle A\phi, \psi \rangle| \leq \|\phi\|_1 \|\psi\|_1$ for

$\phi, \psi \in H^1$, we can deduce that

$$
\begin{aligned}
E| & \int_0^T \langle Au_t, u_t \rangle \, dt - \int_0^T (Au^n(\cdot, t), u^n(\cdot, t)) dt| \\
& \le E \int_0^T \{|\langle A(u_t - u^n(\cdot, t)), u_t \rangle| + |\langle Au^n(\cdot, t), u_t - u^n(\cdot, t) \rangle|\} dt \\
& \le E \int_0^T (\|u_t\|_1 + \|u^n(\cdot, t)\|_1) \|u_t - u^n(\cdot, t)\|_1 \, dt \\
& \le \{E \int_0^T (\|u_t\|_1 + \|u^n(\cdot, t)\|_1)^2 \, dt\}^{1/2} \{E \int_0^T \|u_t - u^n(\cdot, t)\|_1^2 \, dt\}^{1/2} \\
& \le C \{E \int_0^T \|u_t - u^n(\cdot, t)\|_1^2 \, dt\}^{1/2} \\
& \le 2C \{E \int_0^T (\|\tilde{u}_t - \tilde{u}^n(\cdot, t)\|_1^2 + \|v_t - v^n(\cdot, t)\|_1^2) \, dt\}^{1/2},
\end{aligned}
$$

$$(3.89)$$

for some $C > 0$. It follows that

$$
\lim_{n \to \infty} \int_0^T (Au^n(\cdot, t), u^n(\cdot, t)) dt = \int_0^T \langle Au_t, u_t \rangle dt \tag{3.90}
$$

in $L^1(\Omega)$. This is so because

$$
\begin{aligned}
E \int_0^T \|\tilde{u}_t - \tilde{u}^n(\cdot, t)\|_1^2 dt &= E \sum_{k>n} \lambda_k \int_0^T (G_t h + \int_0^t G_{t-s} f_s ds, e_k)^2 dt \\
&\le 2E \sum_{k>n} \lambda_k \{\int_0^T e^{-2\lambda_k t}(h, e_k)^2 dt + E \int_0^T \int_0^t e^{-2\lambda_k(t-s)}(f_s, e_k)^2 ds dt\} \\
&\le \frac{1}{2} \sum_{k>n} (h, e_k)^2 + \frac{1}{2} E \sum_{k>n} \int_0^T (f_t, e_k)^2 dt,
\end{aligned}
$$

which goes to zero as $n \to \infty$, and

$$
\begin{aligned}
E \int_0^T \|v_t - v^n(\cdot, t)\|_1^2 \, dt &= E \int_0^T \sum_{k>n} \lambda_k (e_k, \int_0^t e^{-\lambda_k(t-s)} \sigma_s dW_s)^2 dt \\
&= E \sum_{k>n} \int_0^T \int_0^t \lambda_k e^{-2\lambda_k(t-s)} (Q_s e_k, e_k) ds dt \\
&\le \frac{1}{2} E \sum_{k>n} \int_0^T (Q_s e_k, e_k) ds \to 0, \quad \text{as} \quad n \to \infty.
\end{aligned}
$$

If we can show that $\int_0^t (u_s^n, dM_s^n)$ and $\int_0^t Tr \, Q_s^n ds$ converge, respectively, to $\int_0^t (u_s, dM_s)$ and $\int_0^t Tr \, Q_s ds$, then by passing the limits in (3.84) termwise, the equation (3.83) follows. For instance, consider

$$E|\int_0^t (u_s, dM_s) - \int_0^t (u^n(\cdot, s), dM_s^n)|$$

$$\leq E|\int_0^t (\delta u^n(\cdot, s), dM_s)| + E|\int_0^t (u^n(\cdot, s), d\delta M_s^n)|,$$

where we set $\delta u^n = (u - u^n)$. First it is easy to show that $\int_0^t Tr\, Q_s^n ds \to \int_0^t Tr\, Q_s ds$ in $L^1(\Omega)$. Next, by a submartingale inequality,

$$E|\int_0^t (\delta u^n(\cdot, s), dM_s)| \leq C_1 E\{\int_0^t (Q_s \delta u^n(\cdot, s), \delta u^n(\cdot, s))ds\}^{1/2}$$

$$\leq C_1\{E \sup_{0\leq t\leq T} \|u_t - u^n(\cdot, t)\|^2 \, E\int_0^T (Tr Q_s)ds\}^{1/2},$$

which, by (3.87), goes to zero as $n \to \infty$. Similarly we have

$$E|\int_0^T (u^n(\cdot, s), d\delta M_s^n)| \leq C_1 E\{\int_0^T (\delta Q_s^n u^n(\cdot, s), u^n(\cdot, s))ds\}^{1/2}$$

$$\leq C_1\{E \sup_{0\leq t\leq T} \|u^n(\cdot, t)(\cdot, t)\|^2\} \, E\int_0^T Tr\,(\delta Q_s^n)ds\}^{1/2} \to 0, \text{ as } n \to \infty.$$

Hence the theorem is proved. $\qquad\square$

Next we consider the L^p-regularity of the solution. Before stating the theorem, we shall present the following two lemmas.

Lemma 5.4 Let $\tilde{u}(x,t)$ be given by (3.72). Suppose that $h \in H$ and $f(\cdot, t)$ is a $\mathcal{F}_t$-adapted process in H such that

$$E\{\int_0^T \|f(\cdot, t)\|^2 dt\}^p < \infty, \tag{3.91}$$

for $p \geq 1$. Then $\tilde{u}(\cdot, t)$ is an adapted H^1-valued process being continuous in H such that

$$E \sup_{0\leq t\leq T} \|\tilde{u}(\cdot, t)\|^{2p} + E\left(\int_0^T \|\tilde{u}(\cdot, t)\|_1^2 dt\right)^p$$

$$\leq C_p\{\|h\|^{2p} + E\left(\int_0^T \|f(\cdot, t)\|^2 dt\right)^p\}, \tag{3.92}$$

for some constant $C_p > 0$. If $h \in H^1$ and condition (3.91) holds for $p > 1$, the process $\tilde{u}(\cdot, t), t \in [0, T]$, is Hölder-continuous in H with exponent $\alpha < (p-1)/2p$.

Proof. It follows from Lemma 5.1 that $\tilde{u}(\cdot, t)$, being continuous in H, is a H^1-valued process, and, for $h \in H$, we have

$$\sup_{0 \leq t \leq T} \|G_t h\|^{2p} + (\int_0^T \|G_t h\|_1^2 dt)^p \leq C_1 \|h\|^{2p}. \tag{3.93}$$

Similarly,

$$E \sup_{0 \leq t \leq T} \| \int_0^t G_{t-s} f_s ds\|^{2p} + E \{ \int_0^T \| \int_0^t G_{t-s} f_s ds\|_1^2 dt\}^p$$
$$\leq C_2 E \{ \int_0^T \|f_s\|^2 dt\}^p, \tag{3.94}$$

for some positive constants C_1 and C_2. Therefore the inequality (3.92) is a consequence of (3.93) and (3.94). To show the Hölder-continuity, for $t, (t+\tau) \in [0, T]$, we have

$$\|\tilde{u}_{t+\tau} - \tilde{u}_t\|^2 \leq 2\{\|(G_{t+\tau} - G_t)h\|^2$$
$$+ \| \int_0^{t+\tau} G_{t+\tau-s} f_s ds - \int_0^t G_{t-s} f_s ds\|^2\}. \tag{3.95}$$

By the series representation (3.43) for the Green function G_t, for $h \in H^1$, it can be shown that

$$\|G_{t+\tau} h - G_t h\|^2 \leq \frac{|\tau|}{2} \|h\|_1^2. \tag{3.96}$$

Clearly,

$$\| \int_0^{t+\tau} G_{t+\tau-s} f_s ds - \int_0^t G_{t-s} f_s ds\|^2 \leq 2\{ \| \int_t^{t+\tau} G_{t+\tau-s} f_s ds\|^2$$
$$+ \| \int_0^t (G_{t+\tau-s} - G_{t-s}) f_s ds\|^2\}, \tag{3.97}$$

and

$$\| \int_t^{t+\tau} G_{t+\tau-s} f_s ds\|^2 \leq (\int_t^{t+\tau} \|f_s\| ds)^2$$
$$\leq |\tau| \int_0^T \|f_s\|^2 ds. \tag{3.98}$$

We claim that the last integral in (3.97) satisfies (see the following Remark (1))

$$\| \int_0^t (G_{t+\tau-s} - G_{t-s}) f_s ds\|^2 \leq C|\tau| \int_0^T \|f_s\|^2 ds, \tag{3.99}$$

for some constant $C > 0$. By taking (3.94)–(3.99) into account, we can deduce that, for any $p > 1$, there exists a constant $C(p, T) > 0$ such that

$$E \| \int_0^{t+\tau} G_{t+\tau-s} f_s ds - \int_0^t G_{t-s} f_s ds\|^{2p}$$
$$\leq C(p, T)|\tau|^p E \{ \int_0^T |f_s|^2 ds)\}^p.$$

By the Kolmogorov continuity criterion, the above shows $\tilde{u}(\cdot,t)$ is Hölder-continuous with exponent $\alpha < (p-1)/2p$. □

Remarks:

(1) To verify (3.99), take $\tau > 0$ and $t, t + \tau \in [0,T]$. By using the series representation of the Green's function and the bound $\lambda e^{-\lambda t} \le e^{-1}(1/t)$ for any $\lambda > 0, t > 0$, it can be shown that

$$\|(G_{t+\tau} - G_t)h\| \le C_1(\tau/t)\|h\|$$

for any $h \in H, t > 0$, and for some $C_1 > 0$. Hence there is a constant $C_2 > 0$ such that

$$\|(G_{t+\tau} - G_t)h\| \le C_2\theta(\tau/t)\|h\|, \quad t > 0,$$

where $\theta(\tau/t) = \{1 \wedge (\tau/t)\}$. It follows that

$$\|\int_0^t (G_{t+\tau-s} - G_{t-s})f_s ds\|^2 \le C_2^2 \{\int_0^t \theta(\frac{\tau}{t-s})\|f_s\| ds\}^2$$

$$\le C_2^2 \int_0^t \theta^2(\tau/s)ds \int_0^t \|f_s\|^2 ds \le C\tau \int_0^T \|f_s\|^2 ds$$

as to be shown.

(2) If the initial state $h \in H$ instead of H^1, one can only show that $\tilde{u}(\cdot,t)$ is Hölder-continuous in $[\epsilon, T]$ for any $\epsilon > 0$.

Lemma 5.5 Suppose that $\sigma(\cdot,t)$ is a predictable H-valued process such that, for $p \ge 1$,

$$E\{\int_0^t Tr\, Q_t dt\}^p = E\{\int_0^T \int_D q(x,x,t)dxdt\}^p < \infty. \tag{3.100}$$

Then $v(\cdot,t)$ is a continuous process in H such that the following inequality holds

$$E \sup_{0 \le t \le T} \|v(\cdot,t)\|^{2p} + E\{\int_0^T \|v(\cdot,t)\|_1^2 dt\}^p < C_p E\{\int_0^T (Tr\, Q_t)dt\}^p, \tag{3.101}$$

for some constant $C_p > 0$. Moreover, if $\sigma(\cdot,t)$ is a predictable H^1-valued process such that

$$E \sup_{0 \le t \le T} (Tr\, Q_t)^p < \infty, \tag{3.102}$$

then, for $p > 1$, the process $v(\cdot,t)$ has a Hölder-continuous version in H with exponent $\alpha < (p-1)/2p$.

Proof. From the energy equation (3.83) with $h = 0$ and $f. = 0$, and by invoking the simple inequality : $(a + b)^p \leq 2^p(|a|^p + |b|^p)$, we obtain

$$\|v_t\|^{2p} < 4^p \{| \int_0^t (v_s, dM_s)|^p + | \int_0^t (Tr\,Q_s)ds|^p\}. \tag{3.103}$$

By the B-D-G inequality,

$$E \sup_{0 \leq t \leq T} | \int_0^t (v_s, dM_s)|^p \leq C_1 E\,\{ \int_0^T (Q_s v_s, v_s)ds\}^{p/2}$$
$$\leq C_1 E \sup_{0 \leq t \leq T} \|v_t\|^p \{ \int_0^T (Tr\,Q_s)ds\}^{p/2} \tag{3.104}$$
$$\leq \epsilon E \sup_{0 \leq t \leq T} \|v_t\|^{2p} + C_\epsilon E\,\{ \int_0^T (Tr\,Q_s)ds\}^p,$$

where use was made of the fact: $C_1(ab) \leq (\epsilon a^2 + C_\epsilon b^2)$ with $C_\epsilon = (C_1^2/4\epsilon)$ for any $\epsilon > 0$. In view of (3.103) and (3.104), we get

$$E \sup_{0 \leq t \leq T} \|v_t\|^{2p} \leq 4^p \{\epsilon E \sup_{0 \leq t \leq T} \|v_t\|^{2p} + (C_\epsilon + 1)E\,[\int_0^T (Tr\,Q_s)ds]^p\}.$$

By choosing $\epsilon = 1/(2 \cdot 4^p)$, the above yields the inequality

$$E \sup_{0 \leq t \leq T} \|v_t\|^{2p} \leq C_2 E\,\{ \int_0^T (Tr\,Q_s)ds\}^p \tag{3.105}$$

with $C_2 = 2 \cdot 4^p(C_\epsilon + 1)$. From the energy equation (3.83), we can deduce that

$$\int_0^t \|v_s\|_1^2 ds \leq C_3 \{| \int_0^t (v_s, dM_s)| + \int_0^t (Tr\,Q_s)ds\}.$$

Following similar estimates from (3.103) to (3.105), we can get

$$E\,\{ \int_0^T \|v_s\|_1^2 ds\}^p \leq C_4 E\,\{ \int_0^T (Tr\,Q_s)ds\}^p. \tag{3.106}$$

Now the inequality (3.101) follows from (3.105) and (3.106) with $C_p = C_2 + C_3$.

To show the Hölder continuity, similar to (3.103), we can obtain, for $s < t$,

$$E \|v_t - v_s\|^{2p} = 4^p \{E\,| \int_s^t (v_s, dM_s)|^p + E\,| \int_s^t (Tr\,Q_s)ds|^p\}. \tag{3.107}$$

By using the B-D-G submartingale inequality and (3.105), we have

$$E \left| \int_s^t (v_r, dM_r) \right|^p \leq E \sup_{s \leq \tau \leq t} \left| \int_s^\tau (v_r, dM_r) \right|^p$$

$$\leq C_5 E \left\{ \int_s^t (Q_r v_r, v_r) dr \right\}^{p/2}$$

$$\leq C_5 E \left\{ \sup_{0 \leq t \leq T} \|v_t\|^2 \int_s^t Tr\, Q_r\, dr \right\}^{p/2}$$

$$\leq C_5 \{ E \sup_{0 \leq t \leq T} \|v_t\|^{2p} \}^{1/2} \{ E \left(\int_s^t Tr\, Q_r\, dr \right)^p \}^{1/2}$$

$$\leq C_6 E \left\{ \int_s^t Tr\, Q_r\, dr \right\}^p,$$

where C_5, C_6 are some positive constants. In view of condition (3.102), the above inequality yields

$$E \left| \int_s^t (v_r, dM_r) \right|^p \leq C_7 |t - s|^p, \tag{3.108}$$

and

$$E \left\{ \int_s^t Tr\, Q_r\, dr \right\}^p \leq C_8 |t - s|^p, \tag{3.109}$$

for positive constants C_7, C_8 depending on p and T. By means of (3.108) and (3.109), one obtains from (3.107) that

$$E \|v_t - v_s\|^{2p} \leq C_9 (t - s)^p, \tag{3.110}$$

for some $C_9 > 0$ and $p > 1$. By applying the Kolmogorov's continuity criterion, we conclude that $v(\cdot, t)$ has a regular version as a Hölder-continuous process in H with exponent $\alpha < (p - 1)/2p$. □

With the aid of Lemma 5.4 and Lemma 5.5, it is not hard to show the following theorem holds true.

Theorem 5.6   Let the conditions (3.91) and (3.100) be satisfied with $p \geq 1$. For any $h \in H$, the solution $u(\cdot, t)$ of the problem (3.51) given by (3.52) is a predictable H^1-valued process which is continuous in H such that the energy equation (3.83) holds and

$$E \sup_{0 \leq t \leq T} \|u(\cdot, t)\|^{2p} + E \left\{ \int_0^T \|u(\cdot, t)\|_1^2 dt \right\}^p$$

$$\leq C \{ \|h\|^{2p} + E \left[\int_0^T \|f(\cdot, t)\|^2 dt \right]^p + E \left[\int_0^T (Tr\, Q_t) dt \right]^p \}, \tag{3.111}$$

where C is a positive constant depending on T and p. If $h \in H^1$ and the conditions (3.91) and (3.102) are satisfied for $p > 1$, then it is Hölder-continuous in H with exponent $\alpha < (p-1)/2p$. □

Remarks:

(1) Here, for simplicity, we assumed that initially $u(\cdot, 0) = h \in H$ is non-random. It is clear from the proof that the theorem holds true if $h(x)$ is a $\mathcal{F}_0$-measurable random field such that $h \in L^{2p}(\Omega, H)$. In the estimate (3.111), we simply change $\|h\|^{2p}$ to $E\|h\|^{2p}$.

(2) Owing to the H^1-regularity of the solution, it is not hard to show that u satisfies the so-called variational equation:

$$
\begin{aligned}
(u_t, \phi) = (h, \phi) + \int_0^t <Au_s, \phi> ds \\
+ \int_0^t (f_s, \phi) ds + \int_0^t (\phi, dM_s),
\end{aligned}
\tag{3.112}
$$

for any $\phi \in H^1$. In contrast with the notion of mild solution, a solution to the variational equation (3.112) is known as a strong solution.

(3) The continuity properties of the solution u in both space and time can also be studied. However, to avoid more technical complication, they will not be discussed here.

3.6 Random Reaction-Diffusion Equations

Consider the nonlinear initial-boundary value problem:

$$
\frac{\partial u}{\partial t} = (\kappa \Delta - \alpha)u + f(u, x, t) + \dot{V}(u, x, t), \quad x \in D, \, t \in (0, T),
\tag{3.113}
$$

$$
Bu|_{\partial D} = 0, \quad u(x, 0) = h(x).
$$

In the above equation, the nonlinear term depends only on u but not on its gradient $\partial_x u$. This is known as a reaction-diffusion equation [78] perturbed by a state-dependent white noise

$$
\dot{V}(u, x, t) = g(x, t) + \sigma(u, x, t) \frac{\partial}{\partial t} W(x, t),
\tag{3.114}
$$

where $g(x, t)$, $x \in D$, and the nonlinear terms $f(u, x, t)$ and $\sigma(u, x, t)$ for $(u, x) \in \mathbf{R} \times D$ are given predictable random fields to be specified later.

Let $G(x, y; t)$ be the Green's function as before with the associated Green's operator G_t. Then the system (3.113) can be converted into the integral equation:

$$
\begin{aligned}
u(x, t) = & \int_D G(x, y; t) h(y) dy + \int_0^t \int_D G(x, y; t - s) g(y, s) ds dy \\
& + \int_0^t \int_D G(x, y; t - s) f(u(y, s), y, s) ds dy \\
& + \int_0^t \int_D G(x, y; t - s) \sigma(u(y, s), y, s) W(y, ds) dy.
\end{aligned} \tag{3.115}
$$

Similar to the linear case, a continuous solution $u(\cdot, t)$ of the above equation in H is called a mild solution for the problem (3.113). Regarding (3.115) as an integral equation in H, we will also write $u_t = u(\cdot, t)$, $F_t(u) = f(u(\cdot, s), \cdot, s)$, $g_t = g(\cdot, t)$, $\Sigma_t(u) = \sigma(u(\cdot, s), \cdot, s)$ and $dW_t = W(\cdot, dt)$, so that it takes the form

$$
\begin{aligned}
u_t = & G_t h + \int_0^t G_{t-s} g_s ds + \int_0^t G_{t-s} F_s(u) ds \\
& + \int_0^t G_{t-s} \Sigma_s(u) dW_s.
\end{aligned} \tag{3.116}
$$

Here, more precisely, we say that $u \in L^2(\Omega \times [0, T]; H)$ is a mild solution of the problem (3.113) if $u(\cdot, t)$ is a predictable process in H which satisfies the integral equation (3.116) for $a.e.$ $(\omega, t) \in \Omega \times [0, T]$ such that

$$
\begin{aligned}
& E \int_0^T \{ \|F_t(u)\|^2 + (R\Sigma_t(u), \Sigma_t(u)) \} dt \\
& = E \int_0^T \int_D \{ |f(u(x, t), x, t)|^2 + r(x, x) \sigma^2(u(x, t), x, t) \} dt dx < \infty.
\end{aligned}
$$

To prove the existence theorem, we impose the following conditions:

(A.1) $f(r, x, t)$ and $\sigma(r, x, t)$ are predictable random fields. There exists a constant $K_1 > 0$ such that

$$
\|f(u, \cdot, t)\|^2 + \|\sigma(u, \cdot, t)\|^2 \leq K_1(1 + \|u\|^2), \quad \text{a.s.}
$$

for any $u \in H$, $t \in [0, T]$.

(A.2) There exists a constant $K_2 > 0$ such that

$$
\|f(u, \cdot, t) - f(v, \cdot, t)\|^2 + \|\sigma(u, \cdot, t) - \sigma(v, \cdot, t)\|^2 \leq K_2 \|u - v\|^2, \quad \text{a.s.}
$$

for any $u, v \in H$, $t \in [0, T]$.

(A.3) $g(\cdot, t)$ is a predictable H-valued process such that

$$
E \left(\int_0^T \|g_t\|^2 dt \right)^p < \infty,
$$

for $p \geq 1$, and $W(x,t)$ is a R-Wiener random field of finite trace and the covariance function $r(x,y)$ is bounded by r_0.

Let $X_{p,T}$ denote the set of all continuous $\mathcal{F}_t$-adapted processes in H for $0 \leq t \leq T$ such that $E \sup_{0 \leq t \leq T} \|u(\cdot, t)\|^{2p} < \infty$, for a given $p \geq 1$. Then $X_{p,T}$ is a Banach space under the norm:

$$\|u\|_{p,T} = \{E \sup_{0 \leq t \leq T} \|u_t\|^{2p}\}^{1/2p}. \tag{3.117}$$

Define an operator Γ in $X_{p,T}$ as follows:

$$\begin{aligned}
\Gamma_t u = G_t h &+ \int_0^t G_{t-s} F_s(u) ds \\
&+ \int_0^t G_{t-s} g_s ds + \int_0^t G_{t-s} \Sigma_s(u) dW_s,
\end{aligned} \tag{3.118}$$

for $u \in X_{p,T}$. In the following two lemmas, we will show that the operator Γ is well defined and Lipschitz continuous in $X_{p,T}$.

Lemma 6.1 Under the conditions (A.1)–(A.3) with $p \geq 1$, the mapping Γ given by (3.118) is a well-defined bounded operator which maps $X_{p,T}$ into itself such that

$$\|\Gamma u\|_{p,T}^{2p} \leq b_1 \{1 + \|h\|^{2p} + E\{(\int_0^T \|g_t\|^2 dt)^p + \|u\|_{p,T}^{2p}\}, \tag{3.119}$$

for some constant $b_1 > 0$, depending only on p, r_0 and T.

Proof. By condition (A.1), we have

$$\begin{aligned}
E(\int_0^T \|F_t(u)\|^2 dt)^p &\leq K_1^p E\{\int_0^T (1 + \|u_t\|^2) dt\}^p \\
&\leq (2K_1)^p \{(T^p + E(\int_0^T \|u_t\|^2 dt)^p\} \\
&\leq (2TK_1)^p \{1 + E \sup_{0 \leq t \leq T} \|u_t\|^{2p}\}.
\end{aligned}$$

Therefore, there exists a positive constant C_1, depending on p and T, such that

$$E(\int_0^T \|F_t(u)\|^2 dt)^p \leq C_1 (1 + \|u\|_{p,T}^{2p}). \tag{3.120}$$

Recall that, in Remark (2) following Theorem 2.4, we introduced the notation

$$\|\Sigma_t(u)\|_R^2 = Tr\, Q_t(u) = \int_D r(x,x) \sigma^2(u(x,t), x, t) dx. \tag{3.121}$$

Then, noting (A.1), we can get

$$\|\Sigma_t(u)\|_R^2 \leq r_0 \|\Sigma_t(u)\|^2 \leq K_1 r_0 (1 + \|u_t\|^2).$$

As before, we can find a constant $C_2 > 0$ such that

$$E\,[\int_0^T \|\Sigma_t(u)\|_R^2 dt\,]^p \leq C_2 (1 + \|u\|_{p,T}^{2p}). \tag{3.122}$$

Now we let $v = \Gamma u$ in (3.118). Due to the inequalities (3.120) and (3.122), we can apply Theorem 5.4 with u replaced by v to assert that $v(\cdot, t)$ is a continuous and adapted process in H and to obtain from (3.111) the estimate:

$$E \sup_{0 \leq t \leq T} \|\Gamma u_t\|^{2p} \leq C\{\|h\|^{2p} + E\,[\int_0^T \|F_t(u)\|^2 dt\,]^p$$
$$+ E\,(\int_0^T \|g_t\|^2 dt)^p + E\,[\int_0^T \|\Sigma_t(u)\|_R^2 dt\,]^p\}. \tag{3.123}$$

By making use of (3.120) and (3.122) in (3.123), the desired bound (3.119) follows with some constant $b_1 > 0$. Therefore the map $\Gamma : X_{p,T} \to X_{p,T}$ is well defined and bounded as asserted. $\qquad \square$

Lemma 6.2 Suppose the conditions (A.1) to (A.3) hold true. Then the map $\Gamma : X_{p,T} \to X_{p,T}$ is Lipschitz continuous. Moreover, for any $u, u' \in X_{p,T}$ with $0 < T \leq 1$, there exists a positive constant b_2 independent of $T \in (0, 1]$ such that

$$\|\Gamma u - \Gamma u'\|_{p,T} \leq b_2 \sqrt{T}\,\|u - u'\|_{p,T}. \tag{3.124}$$

Proof. By condition (A.2), we have

$$E\,[\int_0^T \|F_t(u) - F_t(u')\|^2 dt\,]^p \leq K_2^p E\,(\int_0^T \|u_t - u_t'\|^2 dt)^p \tag{3.125}$$
$$\leq (K_2 T)^p \|u - u'\|_{p,T}^{2p}.$$

Again, by conditions (A.2) and (A.3), we get

$$\|\Sigma_t(u) - \Sigma_t(u')\|_R^2 \leq r_0 \|\Sigma_t(u) - \Sigma_t(u')\|^2$$
$$\leq K_2 r_0 \|u_t - u_t'\|^2.$$

With the aid of this inequality, it is clear that

$$E\,[\int_0^T \|\Sigma_t(u) - \Sigma_t(u')\|_R^2 dt\,]^p \leq (K_2 r_0)^p E\,(\int_0^T \|u_t - u_t'\|^2 dt)^p \tag{3.126}$$
$$\leq [K_2 r_0 T]^p \|u - u'\|_{p,T}^{2p}.$$

Let $v = \Gamma u$, $v' = \Gamma u'$ and $\delta v = v - v'$. Then, in view of (3.118), δv satisfies

$$\delta v_t = \int_0^t G_{t-s}[F_s(u) - F_s(u')]ds$$

$$+ \int_0^t G_{t-s}[\Sigma_t(u) - \Sigma_t(u')]dW_s.$$

By applying Theorem 5.4 to the above equation, the estimate (3.111) gives rise to

$$E \sup_{0 \le t \le T} \|\delta v_t\|^{2p} = E \sup_{0 \le t \le T} \|\Gamma_t(u) - \Gamma_t(u')\|^{2p}$$

$$\le CE \{ [\int_0^T \|F_t(u) - F_t(u')\|^2 dt]^p \qquad (3.127)$$

$$+ [\int_0^T \|\Gamma_t(u) - \Gamma_t(u')\|_R^2 dt]^p \}.$$

By taking (3.125) and (3.126) into account, we can deduce from (3.127) that

$$\|\Gamma u_t - \Gamma u_t'\|_{p,T}^{2p} \le C(1 + r_0^p)(C_2 T)^p \|u - u'\|_{p,T}^{2p}, \qquad (3.128)$$

which implies the inequality (3.124) with $b_2 = \sqrt{C_2}\{C(1 + r_0^p)\}^{1/2p}$. □

With the aid of the above lemmas, it is rather easy to prove the existence theorem for the integral equation (3.115) for a mild solution of (3.113). The proof follows from the contraction mapping principle (Theorem 2.5).

Theorem 6.3 Let the conditions (A.1) to (A.3) be satisfied and let h be a $\mathcal{F}_0$-measurable random field such that $E\|h\|^{2p} < \infty$ for $p \ge 1$. Then the initial-boundary value problem for the reaction-diffusion equation (3.113) has a unique (mild) solution $u(\cdot, t)$ which is a continuous adapted process in H such that $u \in L^{2p}(\Omega; \mathbf{C}([0,T]; H))$ satisfying

$$E \sup_{0 \le t \le T} \|u(\cdot, t)\|^{2p} \le C\{1 + E\|h\|^{2p} + E(\int_0^T \|g_t\|^2 dt)^p\}, \qquad (3.129)$$

for some constant $C > 0$, depending on p, r_0 and T. Moreover, the energy inequality holds

$$E\|u(\cdot, t)\|^2 \le E\{\|h\|^2 + 2\int_0^t (u_s, F_s(u))ds$$

$$+ 2\int_0^t (u_s, g_s)ds + \int_0^t Tr\, Q_s(u)ds\}, \qquad (3.130)$$

for $t \in [0, T]$, where $Tr\, Q_s(u) = \|\Sigma_s(u)\|_R^2$.

Proof. For the first part of the theorem, we need to prove that the integral equation (3.116) has a unique solution in $X_{p,T}$. Now Lemma 6.1 and Lemma

6.2 show that the map $\Gamma : X_{p,T} \to X_{p,T}$ is bounded and Lipschitz-continuous. More precisely it satisfies (3.124) so that

$$\|\Gamma u - \Gamma u'\|_{p,T} \leq \frac{1}{2}\|u - u'\|_{p,T}, \tag{3.131}$$

if $T \leq T_1$ with $T_1 = 1/(4b_2^2)$. Therefore Γ is a contraction mapping in the Banach space X_{p,T_1} and, by Theorem 2.5, it has a unique fixed point u satisfying $u_t = \Gamma_t u$ for $0 \leq t \leq T_1$. This means that u is a unique local solution of the integral equation (3.116) over $[0, T_1]$. The solution can be extended over any finite interval $[0, T]$ by continuing the solution to $[T_1, T_2], [T_2, T_3], \cdots$, and so on.

Now, from (3.123) one can obtain

$$E \sup_{0 \leq t \leq T} \|u_t\|^{2p} \leq CE\{\|h\|^{2p} + [\int_0^T \|g_t\|^2 dt]^p$$
$$+[\int_0^T \|F_t(u)\|^2 dt]^p + [\int_0^T \|\Sigma_t(u)\|_R^2 dt]^p\}. \tag{3.132}$$

By making use of some inequalities leading to (3.120) and (3.122), we can deduce from (3.132) that there is a constant $C_1 > 0$ such that

$$E \sup_{0 \leq t \leq T} \|u_t\|^{2p} \leq C_1 E\{1 + \|h\|^{2p} + [\int_0^T \|g_t\|^2 dt]^p + [\int_0^T \|u_t\|^2 dt]^p\}$$
$$\leq C_1\{1 + \|h\|^{2p} + E[\int_0^T \|g_t\|^2 dt]^p + T^{(p-1)/p}\int_0^T E \sup_{0 \leq s \leq t} \|u_s\|^{2p} dt]\},$$

which, by the Gronwall lemma, implies the inequality (3.129). For $u \in L^{2p}(\Omega; \mathbf{C}([0, T]; H))$, by condition (A.1), it is easy to check that $\tilde{f}_t = F_t(u)$ and $\tilde{\sigma}_t = \Sigma_t(u)$ satisfy the conditions (3.91) and (3.100), respectively. By invoking Lemmas 5.3 and 5.4, the inequality (3.129) follows easily.

To verify the energy inequality, let P_n be a projection operator from H into the linear space $V^n \subset \mathcal{D}(A)$ spanned by the first n eigenfunctions $\{e_1, \cdots, e_n\}$ of A so that, for any $h \in H$,

$$P_n h = h^n = \sum_{k=1}^n (h, e_k)e_k.$$

Let $u^n = P_n u$, $F_s^n = P_n F_s$ and so on. Then, after applying P_n to equation (3.116) and noticing $P_n G_t = G_t P_n$, we get

$$u^n(\cdot, t) = G_t h^n + \int_0^t G_{t-s}g_s^n ds + \int_0^t G_{t-s}F_s^n(u)ds$$
$$+ \int_0^t G_{t-s}\Sigma_s^n(u)\,dW_s.$$

Since $u^n(\cdot, t) \in \mathcal{D}(A)$, the above equation can be rewritten as

$$u^n(\cdot, t) = h^n + \int_0^t A u_s^n ds + \int_0^t F_s^n(u) ds + \int_0^t g_s^n ds$$
$$+ \int_0^t \Sigma_s^n(u)\, dW_s. \tag{3.133}$$

By applying the Itô formula to $\|u^n(\cdot, t)\|^2$ in finite dimensions, taking the expectation, and recalling the fact $(A u^n, u^n) \le 0$, we obtain

$$E \|u^n(\cdot, t)\|^2 \le \|h^n\|^2 + 2E \int_0^t (F_s^n(u), u_s^n) ds + 2 \int_0^t (g_s^n, u_s^n) ds$$
$$+ E \int_0^t Tr\, Q_s^n(u) ds,$$

which yields the energy inequality (3.130) as $n \to \infty$. $\qquad \square$

For example, consider the initial-boundary value problem:

$$\frac{\partial u}{\partial t}(x, t) = (\kappa \Delta - \alpha)u + a \sin u + g(x, t) + \sigma_0(x, t)\, u\, \dot{W}(x, t),$$
$$u(\cdot, t)|_{\partial D} = 0, \quad u(x, 0) = h(x), \tag{3.134}$$

for $x \in D$, $t \in (0, T)$, where a is a constant and $\sigma_0(x, t)$ is a predictable bounded random field such that

$$|\sigma_0(x, t)| \le C, \quad a.s. \quad \text{for each } (x, t) \in D \times [0, T], \tag{3.135}$$

for some constant $C > 0$. In (3.113), $f(u, x, t) = a \sin u$ and $\sigma(u, x, t) = \sigma_0(x, t)\, u$. Clearly the conditions (A.1) and (A.2) are satisfied. So the following is a corollary of Theorem 6.3.

Corollary 6.4 Assume that conditions (3.135) and (A.3) hold. Given a $\mathcal{F}_0$-measurable random field h such that $E \|h\|^{2p} < \infty$ for $p \ge 1$, the initial-boundary value problem (3.134) has a unique mild solution $u(\cdot, t)$ which is a continuous adapted process in H. Furthermore, $u \in L^{2p}(\Omega; C([0, T]; H))$ such that the inequality (3.129) holds. $\qquad \square$

In Theorem 6.3, the global conditions $A.1$ and $A.2$ on linear growth and Lipschitz-continuity can be relaxed to hold locally. However, without additional constraint, they may lead to an explosive solution in finite time. With these in mind, we impose the following conditions:

$(A_n.1)$ $f(r, x, t)$ and $\sigma(r, x, t)$ are predictable random fields. There exists a constant $C_n > 0$ such that

$$\|f(u, \cdot, t)\|^2 + \|\sigma(u, \cdot, t)\|^2 \le C_n, \quad a.s.$$

for any $n > 0$, $u \in H$ with $\|u\| \leq n$, $t \in [0, T]$.

$(A_n.2)$ There exists a constant $K_n > 0$ such that

$$\|f(u, \cdot, t) - f(v, \cdot, t)\|^2 + \|\sigma(u, \cdot, t) - \sigma(v, \cdot, t)\|^2 \leq K_n \|u - v\|^2, \quad \text{a.s.}$$

for any $u, v \in H$, with $\|u\| \vee \|v\| \leq n$, $t \in [0, T]$.

(A.4) There exist constant $C_1 > 0$ such that

$$(u, f(u, \cdot, t)) + \frac{1}{2} Tr\, Q_t(u) \leq C_1(1 + \|u\|^2), \quad \text{a.s.,}$$

for any $u \in H$, $t \in [0, T]$.

It can be shown that, under conditions $(A_n.1)$, $(A_n.2)$ and (A.3), the problem has a local solution. If, in addition, condition (A.4) holds, then the solution exists in any finite time interval. The proof is based on a truncation technique by making use of a mollifer η_n on $[0, \infty)$ to be defined as follows. For $n > 0$, $\eta_n : [0, \infty) \to [0, 1]$ is a $\mathbf{C}^\infty$-function such that

$$\eta_n(r) = \begin{cases} 1, & \text{for } 0 \leq r \leq n \\ 0, & \text{for } r > 2n. \end{cases} \tag{3.136}$$

Theorem 6.5   Let the conditions $(A_n.1)$, $(A_n.2)$ and (A.3) be satisfied and let h be a $\mathcal{F}_0$-measurable random field such that $E \|h\|^2 < \infty$. Then the initial-boundary value problem for the reaction-diffusion equation (3.113) has a unique local solution $u(\cdot, t)$ which is an adapted, continuous process in H. If, in addition, condition (A.4) holds, then the solution exists for $t \in [0, T]$ with any $T > 0$ and $u \in L^2(\Omega; \mathbf{C}([0, T]; H))$ satisfies

$$E \sup_{0 \leq t \leq T} \|u(\cdot, t)\|^2 \leq C\{1 + E \|h\|^2\}, \tag{3.137}$$

for some constant $C > 0$, depending on T.

Proof.   Instead of the equation (3.113), consider the truncated system:

$$\frac{\partial u}{\partial t}(x, t) = (\kappa\Delta - \alpha)u + f_n(u, x, t) + g(x, t) + \sigma_n(u, x, t)\dot{W}(x, t),$$

$$Bu|_{\partial D} = 0, \quad u(x, 0) = h(x), \tag{3.138}$$

where $f_n(u, x, t) = f(J_n u, x, t)$ and $\sigma_n(u, x, t) = \sigma(J_n u, x, t)$ with $J_n u = \eta_n(\|u\|)u$. Then the conditions $(A_n.1)$ and $(A_n.2)$ imply that f_n and σ_n satisfy the global conditions (A.1) and (A.2). For instance, we will show $(A_n.2)$

implies $A.2$. Without loss of generality, let $\|u\| > \|v\|$. Then

$$\|f_n(u,\cdot,t) - f_n(v,\cdot,t)\|^2 + \|\sigma_n(u,\cdot,t) - \sigma_n(v,\cdot,t)\|^2$$
$$= \|f(J_n u,\cdot,t) - f(J_n v,\cdot,t)\|^2 + \|\sigma(J_n u,\cdot,t) - \sigma(J_n v,\cdot,t)\|^2$$
$$\leq K_n \| \eta_n(\|u\|)u - \eta_n(\|v\|)v \|^2)$$
$$\leq K_n \| \eta_n(\|u\|)(u - v) + v [\eta_n(\|u\|) - \eta_n(\|v\|)] \|^2$$
$$\leq K_n \{ \|u - v\| + \|v\| \, | \eta_n(\|u\|) - \eta_n(\|v\|) | \}^2.$$

For $r < s$, by the mean-value theorem, there is $\rho \in (r, s)$ such that

$$|\eta_n(r) - \eta_n(s)| \leq |\eta_n'(\rho)| \, |r - s| \leq 2n\gamma_n |r - s|,$$

noticing the derivative $\eta_n'(\rho) = 0$ for $\rho < n$ or $\rho > 2n$, where we set $\gamma_n = \max_{n \leq \rho \leq 2n} |\eta_n'(\rho)|$. Now condition (A.3) follows from the above two inequalities. Hence, by Theorem 6.3, the system has a unique continuous solution $u^n(\cdot, t)$ in H over $[0, T]$. Introduce a stopping time τ_n defined by

$$\tau_n = \inf\{t > 0 : \|u^n(\cdot, t)\| > n\}$$

if it exists, and set $\tau_n = T$ otherwise. Then, for $t < \tau_n$, $u_t = u^n(\cdot, t)$ is the solution of the problem (3.113). Since τ_n is increasing in n, let $\tau_\infty = \lim_{n \to \infty} \tau_n$ a.s.. For $t < \tau_\infty$, we have $t < \tau_n$ for some $n > 0$, and define $u_t = u^n(\cdot, t)$. Then $\lim_{t \to \tau_\infty} \|u_t\| = \infty$ if $\tau_\infty < T$ and hence u_t is a local (maximal) solution. For uniqueness, suppose that there is another solution $\tilde{u}_t, t < \tau$ for a stopping time τ. Then $\tilde{u}_t = u^n(\cdot, t)$ for $t < \tau_n$. It follows that $\tilde{u}_t = u_t$ for $t < \tau_\infty$ and $\tau = \tau_\infty$.

To show the existence of a global solution, by making use of condition (A.4), similar to the energy inequality (3.130), we can obtain

$$E \|u_{t \wedge \tau_n}\|^2 \leq \|h\|^2 + E \int_0^{t \wedge \tau_n} \|g_s\|^2 ds$$
$$+ \int_0^{t \wedge \tau_n} \|u_s\|^2 ds + C_1 E \int_0^{t \wedge \tau_n} (1 + \|u_s\|^2) \, ds$$
$$\leq \|h\|^2 + C_1 T + E \int_0^T \|g_s\|^2 ds + (C_1 + 1)E \int_0^t \|u_{s \wedge \tau_n}\|^2 \} ds,$$

which, by the Gronwall's inequality, yields the following bound:

$$E \|u_{T \wedge \tau_n}\|^2 \leq C_T, \tag{3.139}$$

for some constant $C_T > 0$ independent of n. On the other hand, we have

$$E \|u_{T \wedge \tau_n}\|^2 \geq E \{I(\tau_n \leq T)\|u_{T \wedge \tau_n}\|^2\} \geq n^2 P\{\tau_n \leq T\}, \tag{3.140}$$

where $I(\cdot)$ denotes the indicator function. In view (3.139) and (3.140), we obtain

$$P\{\tau_n \leq T\} \leq \frac{C_T}{n^2}$$

so that, by the Borel-Cantelli lemma,

$$P\{\tau_\infty > T\} = 1,$$

for any $T > 0$. Hence $u(\cdot, t) = \lim_{n \to \infty} u^n(\cdot, t)$ is a global solution as claimed. □

As an example, consider the stochastic reaction-diffusion equation in $D = (0, 1)$:

$$\frac{\partial u}{\partial t} = \kappa \frac{\partial^2 u}{\partial x^2} - \alpha u + \gamma \phi(\|u\|)u + \sigma_0 u \frac{\partial}{\partial t} W(x, t),$$

$$u(0, t) = u(1, t) = 0, \ 0 < t < T, \tag{3.141}$$

$$u(x, 0) = h(x), \ 0 < x < 1,$$

where κ, α, σ_0 are given positive constants, $\gamma \in \mathbf{R}$, $\phi : [0, \infty) \to [0, \infty)$ is a given continuously differentiable function, $h \in H = L^2(0, 1)$ with norm $\| \cdot \|$, and $W(x, t)$ is a Wiener random field with covariant function $r(x, y)$ bounded by r_0. Then, referring to (3.113), $f(u, x, t) = \gamma \phi(\|u\|)u$, $\sigma(u, x, t) = \sigma_0 u$ and $g(x, t) \equiv 0$. For the existence result, clearly condition (A.3) holds by assumption. To check conditions $(A_n.1)$ and $(A_n.2)$, since σ is linear in u, we need only to show that f is locally bounded and Lipschitz continuous in $L^2(0, 1)$. Since ϕ is a $\mathbf{C}^1$ function on $[0, \infty)$, ϕ and ϕ' are bounded on any finite interval $[0, n]$. It follows that conditions $(A_n.1)$, $(A_n.2)$ and $(A.3)$ are satisfied. Therefore the problem (3.113) has a unique local solution by Theorem 6.5. If the parameter $\gamma \leq 0$, we have

$$(u, f(u, \cdot, t)) + \tfrac{1}{2} Tr\, Q_t(u) = \gamma \phi(\|u\|)(u, u) + \frac{1}{2}\sigma_0^2 \|u\|_R^2$$

$$\leq \frac{1}{2} r_0 \sigma_0^2 \|u\|^2,$$

which shows condition (A.4) is also met. Thus, for $\gamma \leq 0$, the solution exists in any time interval [0,T].

3.7   Parabolic Equations with Gradient-Dependent Noise

So far we have treated parabolic equations for which the nonlinear terms do not depend on the gradient ∂u of u. Now we consider a nonlinear problem as follows:

$$\frac{\partial u}{\partial t} = (\kappa \Delta - \alpha)u + f(u, \partial u, x, t) + \dot{V}(u, \partial u, x, t),$$

$$Bu|_{\partial D} = 0, \quad u(x, 0) = h(x), \tag{3.142}$$

for $x \in D$, $t \in (0, T)$, where

$$\dot{V}(u, \xi, x, t) = g(x, t) + \sigma(u, \xi, x, t)\frac{\partial}{\partial t}W(x, t), \qquad (3.143)$$

$f(u, \xi, x, t)$ and $\sigma(u, \xi, x, t)$ are predictable random fields, with parameter $(u, \xi, x) \in \mathbf{R} \times \mathbf{R}^d \times D$, and $g(x, t)$ and $W(x, t)$ are given as before. In contrast with the previous problem (3.113), the dependence of f and σ on the gradient ∂u will cause some technical complication. To see this, it is instructive to go over an elementary example. Consider the following simple equation in one space dimension:

$$\frac{\partial u}{\partial t} = \kappa\frac{\partial^2 u}{\partial x^2} + \sigma_0\frac{\partial u}{\partial x}\dot{w}(t), \qquad (3.144)$$

where κ and σ_0 are positive constants, and $w(t)$ is a standard Brownian motion in one dimension. Suppose that the equation is subject to the periodic boundary conditions: $u(0, t) = u(2\pi, t)$ and $\partial_x u(0, t) = \partial_x u(2\pi, t)$. The associated eigenfunctions are $\{e_n(x) = \frac{1}{\sqrt{2\pi}}e^{nix}\}$ with $i = \sqrt{-1}$, for $n = 0, \pm 1, \pm 2, \cdots$. Given $u(x, 0) = h(x)$ in $L^2(0, 2\pi)$, this problem can be solved easily by the method of eigenfunctions (Fourier series) expansion:

$$u(x, t) = \sum_{n=-\infty}^{\infty} u_t^n e_n(x), \qquad (3.145)$$

where

$$u_t^n = h_n \exp\{-n^2(\kappa - \frac{1}{2}\sigma_0^2)t + ni[x + \sigma_0 w(t)]\}$$

with $h_n = (h, e_n)$. It is clear that if

$$(\kappa - \frac{1}{2}\sigma_0^2) \geq 0, \qquad (3.146)$$

the solution series converges properly. However, in contrast with the case of gradient-independent noise, if $\kappa - \frac{1}{2}\sigma_0^2 < 0$, the problem is ill-posed in the sense of Hadamard [31]. This can be seen by taking $h = \frac{1}{n\sqrt{2\pi}}e_n$. Then $\|h\| = \frac{1}{n\sqrt{2\pi}} \to 0$ as $n \to \infty$, but $\|u(\cdot, t)\| = |u_t^n| = \frac{1}{n\sqrt{2\pi}}\exp\{n^2|\kappa - \frac{1}{2}\sigma_0^2|t\} \to \infty$ as $n \to \infty$, for any $t > 0$. Therefore the solution does not depend continuously on the initial data. This example suggests that, for a general parabolic initial-boundary value problem containing a gradient-dependent noise, the well-posedness requires the imposition of a stochastic coercivity condition similar to (3.146). Physically, this means that the noise intensity should not exceed a certain threshold set by the diffusion constant κ. Heuristically, a violation of this condition leads to a pseudo heat equation with a negative diffusion coefficient, for which the initial(-boundary) value problem is known to be ill-posed.

Before dealing with the nonlinear problem, we first consider the following linear case:

$$\frac{\partial u}{\partial t} = (\kappa \Delta - \alpha)u + a(x,t)u + [(b(x,t) \cdot \partial_x u]\frac{\partial}{\partial t}W(x,t),$$

$$Bu|_{\partial D} = 0, \qquad (3.147)$$

$$u(x,0) = h(x),$$

for $x \in D$, $t \in (0,T)$, where $a(x,t)$ and $b(x,t)$ are given predictable random fields with $b = (b_1, \cdots, b_d)$. For any $y \in \mathbf{R}^d$, $b \cdot y = \sum_{k=1}^{d} b_k y_k$ denotes the inner product in $\mathbf{R}^d$. To find a mild solution, we convert the system (3.147) into an integral equation:

$$u_t = G_t h + \int_0^t G_{t-s} a_s u_s ds + \int_0^t G_{t-s}(b_s \cdot \partial u_s)dW_s. \qquad (3.148)$$

In order to control the gradient noise term, suggested by the estimate (3.111), introduce a Banach space Y_T equipped with the norm:

$$\|u\|_T = E\{\sup_{0 \le t \le T} \|u_t\|^2 + \int_0^T \|u_t\|_1^2 dt\}^{1/2}. \qquad (3.149)$$

Let Λ denote a mapping in Y_T defined by

$$\Lambda_t u = G_t h + \int_0^t G_{t-s} a_s u_s ds + \int_0^t G_{t-s}[b_s \cdot \partial u_s]dW_s. \qquad (3.150)$$

Theorem 7.1 Assume that $a(x,t)$ and $b(x,t)$ are predictable random fields, and there exist some positive constants α, β such that

$$|a(x,t)| \le \alpha, \quad \text{a.s.} \quad \text{for any } (x,t) \in D \times [0,T],$$

and the following coercivity condition holds

$$\langle Av, v \rangle + \frac{1}{2}\|(b_t \cdot \partial v)\|_R^2 \le -\beta\|v\|_1^2, \quad \text{a.s.} \qquad (3.151)$$

for $\beta \in (0,1)$ and for any $v \in H^1, t \in [0,T]$. Then, given $h \in H$, the linear equation (3.147) has a unique solution $u(\cdot,t)$ as an adapted, continuous process in H. Moreover u belongs to $L^2(\Omega; \mathbf{C}([0,T]; H)) \cap L^2(\Omega \times [0,T]; H^1)$ such that

$$E\{\sup_{0 \le t \le T} \|u_t\|^2 + \int_0^T \|u_t\|_1^2 dt\} < \infty. \qquad (3.152)$$

Proof. The theorem will be proved by the contraction mapping argument. However the usual approach does not work. It is necessary to introduce an equivalent norm to overcome the difficulty as we shall see. As before we consider the map Λ given by (3.150) and first show it is well defined. For $u \in Y_T$, let

$$v_t = \Lambda_t u = G_t h + \nu_t + \xi_t, \tag{3.153}$$

where

$$\nu_t = \int_0^t G_{t-s} a_s u_s ds,$$

and

$$\xi_t = \int_0^t G_{t-s} [b_s \cdot \partial u_s] dW_s.$$

By Lemma 5.1, it is easy to obtain

$$\|G.h\|_T^2 \leq \frac{3}{2} \|h\|^2. \tag{3.154}$$

Furthermore, we have

$$\|\nu_t\|^2 = \| \int_0^t G_{t-s} a_s u_s ds \|^2$$
$$\leq t \int_0^t \|a_s u_s\|^2 ds \leq \alpha T \int_0^t \|u_s\|^2 ds, \tag{3.155}$$

and

$$\int_0^T \|\nu_t\|_1^2 dt = \int_0^T \| \int_0^t G_{t-s} a_s u_s ds \|_1^2 dt$$
$$\leq \frac{T}{2} \int_0^T \|a_s u_t\|^2 dt \leq \alpha \frac{T}{2} \int_0^T \|u_t\|^2 dt. \tag{3.156}$$

Recall that we set $\|v\|_1^2 = \langle -Av, v \rangle$. The coercivity condition (3.151) can be rewritten as

$$\|b_t \cdot \partial v\|_R^2 \leq 2\delta \|v\|_1^2, \tag{3.157}$$

with $\delta = (1-\beta) \in (0,1)$. By means of Lemma 5.2 and (3.157), we can deduce that

$$E \sup_{0 \leq t \leq T} \|\xi_t\|^2 = E \sup_{0 \leq t \leq T} \| \int_0^t G_{t-s} [b_s \cdot \partial u_s] dW_s \|^2$$
$$\leq 16 E \int_0^T \|b_s \cdot \partial u_t\|_R^2 dt \leq 32\delta \, E \int_0^T \|u_t\|_1^2 dt, \tag{3.158}$$

and

$$E \int_0^T \|\xi_t\|_1^2 dt = E \int_0^T \| \int_0^t G_{t-s} [b_s \cdot \partial u_s] dW_s \|_1^2 dt$$
$$\leq \frac{1}{2} E \int_0^T \|b_s \cdot \partial u_t\|_R^2 dt \leq \delta E \int_0^T \|u_t\|_1^2 ds. \tag{3.159}$$

From (3.153), we have

$$\|\Lambda u\|_T^2 \leq 3(\|G.h\|_T^2 + \|\nu\|_T^2 + \|\xi\|_T^2).$$

By taking the inequalities (3.154), (3.155), (3.156), (3.158) and (3.159) into account, we can find a constant $C_1(T) > 0$ such that

$$\|\Lambda u\|_T^2 \leq C_1(T)(\|h\|^2 + \|u\|_T^2).$$

Therefore the linear operator $\Lambda : Y_T \to Y_T$ is well defined and bounded.

It remains to show that Λ is a contraction. To this end, for some technical reason to be seen, we need to introduce an equivalent norm in Y_T, depending on a parameter $\mu > 0$, defined as follows:

$$\|u\|_{\mu,T} = E\{\sup_{0 \leq t \leq T} \|u_t\|^2 + \mu \int_0^T \|u_t\|_1^2 dt\}^{1/2}. \tag{3.160}$$

Let $u, u' \in Y_T$. Then, in view of (3.150), $\eta = (u - u')$ satisfies

$$\eta_t = \int_0^t G_{t-s} a_s \eta_s ds + \int_0^t G_{t-s}[b_s \cdot \partial \eta_s] dW_s. \tag{3.161}$$

In the meantime

$$\|\Lambda u - \Lambda u'\|_{\mu,T}^2 = E\{\sup_{0 \leq t \leq T} \|\Lambda_t \eta\|^2 + \mu \int_0^T \|\Lambda_t \eta\|_1^2 dt\}. \tag{3.162}$$

By making use of (3.161) and the simple inequality $(a+b)^2 \leq C_\varepsilon a^2 + (1+\varepsilon)b^2$ with $C_\varepsilon = (1+\varepsilon)/\varepsilon$, for any $\varepsilon > 0$, we get

$$
\begin{aligned}
E \sup_{0 \leq t \leq T} \|\Lambda_t \eta\|^2 &= E\{\sup_{0 \leq t \leq T} \|\int_0^t G_{t-s} a_s \eta_s ds \\
&\quad + \int_0^t G_{t-s}[b_s \cdot \partial \eta_s])dW_s\|^2\} \\
&\leq E \sup_{0 \leq t \leq T} \{C_\varepsilon \|\int_0^t G_{t-s} a_s \eta_s ds\|^2 \\
&\quad + (1+\varepsilon)\|\int_0^t G_{t-s}[b_s \cdot \partial \eta_s]dW_s\|^2\},
\end{aligned}
\tag{3.163}
$$

and, similarly,

$$
\begin{aligned}
E \int_0^T \|\Lambda_t \eta\|_1^2 dt \\
= E\{\int_0^T \|\int_0^t G_{t-s} a_s \eta_s ds + \int_0^t G_{t-s}[b_s \cdot \partial \eta_s]dW_s\|_1^2 dt\} \\
\leq E\{C_\varepsilon \int_0^T \|\int_0^t G_{t-s} a_s \eta_s ds\|_1^2 dt \\
+ (1+\varepsilon) \int_0^T \|\int_0^t G_{t-s}[b_s \cdot \partial \eta_s]dW_s\|_1^2 dt\}.
\end{aligned}
\tag{3.164}
$$

By applying the estimates (3.155), (3.156), (3.158) and (3.159) to (3.163) and (3.164), and making use of the results in (3.162), one can obtain

$$\|\Lambda\eta\|_{\mu,T}^2 \leq \alpha C_\varepsilon T^2 (1 + \mu/2)\, E \sup_{0 \leq t \leq T} \|\eta_t\|^2$$
$$+ \mu\left[(1+\varepsilon)(1 + 32/\mu)\delta\right] E \int_0^T \|\eta_t\|_1^2 dt. \tag{3.165}$$

Recall that $C_\varepsilon = (1+\varepsilon)/\varepsilon$ and $0 \leq \delta < 1$. Choose $\mu = 32/\varepsilon$ and $\varepsilon = (\sqrt{(1+\delta)/2\delta} - 1)$ so that

$$(1+\varepsilon)(1 + 32/\mu)\delta = (1+\delta)/2 < 1. \tag{3.166}$$

Clearly

$$\alpha C_\varepsilon T^2 (1 + \mu/2) < 1, \tag{3.167}$$

for a sufficiently small T. In view of (3.165), (3.166), and (3.167), we have

$$\|\Lambda\eta\|_{\mu,T} < \rho\|\eta\|_{\mu,T}$$

for some $\rho \in (0, 1)$. Therefore Λ is a contraction in Y_T for a small T. By Theorem 2.5, this implies the existence of a unique local solution in Y_T, which can be continued to any finite time interval [0,T] as mentioned before. □

Now let us return to the nonlinear problem (3.142). Assume that the following conditions are satisfied:

(B.1) For $r \in \mathbf{R}, x, y \in \mathbf{R}^d$ and $t \in [0, T]$, $f(r, x, y, t)$ and $\sigma(r, x, y, t)$ are adapted, continuous random fields. There exist a constant $\alpha > 0$ such that

$$\|f(v, \cdot, \partial v, t)\|^2 + \|\sigma(v, \cdot, \partial v, t)\|_R^2 \leq \alpha(1 + \|v\|^2 + \|v\|_1^2), \quad \text{a.s.}$$

for any $v \in H^1, t \in [0, T]$.

(B.2) There exist positive constants β, γ_1 and γ_2 with $(\gamma_1 + \gamma_2) < 1$ such that

$$\|f(v, \cdot, \partial v, t) - f(v', \cdot, \partial v', t)\|^2 \leq \beta\|v - v'\|^2 + \gamma_1\|v - v'\|_1^2, \quad \text{a.s.}$$

and

$$\|\sigma(v, \cdot, \partial v, t) - \sigma(v', \cdot, \partial v', t)\|_R^2 \leq \beta\|v - v'\|^2 + \gamma_2\|v - v'\|_1^2, \quad \text{a.s.}$$

for any $v, v' \in H^1, t \in [0, T]$.

(B.3) Let $g(\cdot, t)$ be a predictable H-valued process such that

$$E \int_0^T \|g_t\|^2 dt < \infty,$$

and let $W(x, t)$ be a R-Wiener random field as depicted in condition (A.3).

Introduce the operator Φ in Y_T associated with the problem (3.142):

$$\Phi_t u = G_t h + \int_0^t G_{t-s} F_s(u_s, \partial u_s) ds + \int_0^t G_{t-s} g_s ds$$
$$+ \int_0^t G_{t-s} \Sigma_s(u_s, \partial u_s) dW_s. \tag{3.168}$$

Under conditions (B.1) to (B.3), the map $\Phi : Y_T \to Y_T$ is well defined, and it is a contraction mapping as to be shown in the following existence theorem. Since the proof is similar to the linear case, it will be sketchy.

Theorem 7.2   Suppose the conditions (B.1) to (B.3) are satisfied. Then, given $h \in H$, the nonlinear problem (3.142) has a unique solution $u(\cdot, t)$ as an adapted, continuous process in H. Moreover u belongs to $L^2(\Omega; \mathbf{C}([0, T]; H)) \cap L^2(\Omega \times [0, T]; H^1)$ such that

$$E\{\sup_{0 \le t \le T} \|u_t\|^2 + \int_0^T \|u_t\|_1^2 dt\}\} < \infty. \tag{3.169}$$

Proof.   By means of Lemma 5.1 and condition (B.1), we have

$$E \sup_{0 \le t \le T} \| \int_0^t G_{t-s} F_s(u_s, \partial u_s) ds \|^2 \le TE \int_0^T \|F_s(u_s, \partial u_s)\|^2 ds$$
$$\le \alpha TE \int_0^T (1 + \|u_t\|^2 + \|u_t\|_1^2) \, dt \le \alpha T\{T + (T+1)\|u\|_T^2\}, \tag{3.170}$$

and

$$E \int_0^T \| \int_0^t G_{t-s} F_s(u_s, \partial u_s) ds \|_1^2 ds$$
$$\le \frac{T}{2} E \int_0^T \|F_t(u_s, \partial u_s)\|^2 dt \le \frac{1}{2} \alpha T\{T + (T+1)\|u\|_T^2\}. \tag{3.171}$$

Similarly, by making use of Lemma 5.2 and condition (B.1), it is easy to check that

$$E \sup_{0 \le t \le T} \| \int_0^t G_{t-s} \Sigma_s(u_s, \partial u_s) dW_s \|^2 \le 16\alpha\{T + (T+1)\|u\|_T^2\}, \tag{3.172}$$

and

$$E \int_0^T \| \int_0^t G_{t-s} \Sigma_s(u_s, \partial u_s) dW_s \|_1^2 dt \le \frac{1}{2} \alpha\{T + (T+1)\|u\|_T^2\}. \tag{3.173}$$

The fact that $G_t h \in Y_T$ together with (B.3) and the inequalities (3.170)–(3.173) imply that the map $\Phi : Y_T \to Y_T$ is bounded.

To show Φ is a contraction operator in Y_T, we use an equivalent norm $\|\cdot\|_{\mu,T}$ as defined by (3.160). For $u, v \in Y_T$,

$$\|\Phi u - \Phi v\|_{\mu,T}^2 = E\{\sup_{0 \le t \le T} \|\Phi_t u - \Phi_t v\|^2 + \mu \int_0^T \|\Phi_t u - \Phi_t v\|_1^2 dt\}. \quad (3.174)$$

By invoking Lemma 5.1 and Lemma 5.2, similar to the linear case, we have

$$E \sup_{0 \le t \le T} \|\Phi_t u - \Phi_t v\|^2$$

$$\le C_\varepsilon T E \int_0^T \|F_t(u_s, \partial u_s) - F_t(v_s, \partial v_s)\|^2 dt \quad (3.175)$$

$$+16(1+\varepsilon)E \int_0^T \|\Sigma_t(u_s, \partial u_s) - \Sigma_t(v_s, \partial v_s)\|_R^2 dt,$$

and

$$E \int_0^T \|\Phi_t u - \Phi_t v\|_1^2 dt$$

$$\le \frac{T}{2} C_\varepsilon E \int_0^T \|F_t(u_s, \partial u_s) - F_t(v_s, \partial v_s)\|^2 dt \quad (3.176)$$

$$+\frac{1}{2}(1+\varepsilon)E \int_0^T \|\Sigma_t(u_s, \partial u_s) - \Sigma_t(v_s, \partial v_s)\|_R^2 dt.$$

By making use of (3.175), (3.176) and conditions (B.1)–(B.3), the equation (3.174) yields

$$\|\Phi u - \Phi v\|_{\mu,T}^2 \le \rho_1(\varepsilon, \mu, T) E \sup_{0 \le t \le T} \|u_t - v_t\|^2$$

$$+\rho_2(\varepsilon, \mu, T) \mu E \int_0^T \|u_t - v_t\|_1^2 dt, \quad (3.177)$$

where

$$\rho_1 = \beta T[C_\varepsilon T(1 + \mu/2) + (1+\varepsilon)(16 + \mu/2)],$$

$$\rho_2 = \frac{1}{\mu}[C_\varepsilon T(1 + \mu/2)\gamma_1 + (1+\varepsilon)(16 + \mu/2)\gamma_2].$$

As in the linear case, it is possible to choose μ sufficiently large and ε, T sufficiently small such that $\rho = (\rho_1 \wedge \rho_2) < 1$ (see the following Remark (1)). Then

$$\|\Phi u - \Phi v\|_{\mu,T}^2 \le \rho < 1,$$

so that the conclusion of the theorem follows. $\qquad \square$

Remarks:

(1) To show the possibility of choosing the parameters μ, ε, and T to make $\rho_2 < 1$, we first let $T < \varepsilon$ and recall $C_\varepsilon = (1+\varepsilon)/\varepsilon$. Then ρ_2 can be bounded by

$$\rho_2 \le \frac{1}{\mu}(1+\varepsilon)(16 + \mu/2)(\gamma_1 + \gamma_2).$$

Since $(\gamma_1 + \gamma_2) < 1$ by condition (B.2), the above yields $\rho_2 < 3/4$ if we choose $\mu = 32(1 + \varepsilon)/\varepsilon$ and $\varepsilon = 1/4$. It is evident that ρ_1 can be made less than $3/4$ by taking T to be sufficiently small.

(2) Similar to Theorem 6.5, the conditions (B.1) and (B.2) in the above theorem can be relaxed to hold only locally to get a unique local solution, and a global solution can be obtained if there exists a certain energy inequality. This will be discussed in Chapter 6 (Theorem 7.5) for a general class of stochastic evolution equations under weaker conditions.

So far the equations under consideration have been restricted to a single noise term. The previous results can be easily extended to similar equations with multiple-noise terms. In lieu of (3.142), consider the generalized problem:

$$\frac{\partial u}{\partial t} = \kappa \Delta u - \alpha u + f(u, \partial_x u, x, t) + g(x, t)$$
$$+ \sum_{k=1}^{m} \sigma_k(u, \partial_x u, x, t) \dot{W}_k(x, t), \qquad (3.178)$$
$$Bu|_{\partial D} = 0, \quad u(x, 0) = h(x),$$

for $x \in D$, $t \in (0, T)$. Here, for $k = 1, ..., m$, $\sigma_k(r, y, x, t)$ is a continuous $\mathcal{F}_t$-adapted random field and $W_k(x, t)$ is a Wiener random field. Moreover assume that

(B.4) There exist positive constants α, β and γ_2 with $\gamma_1 + \gamma_2 < 1$ such that

$$\sum_{k=1}^{m} \|\sigma_k(v, \cdot, \partial v, t)\|_{R_k}^2 \leq \alpha(1 + \|v\|^2 + \|v\|_1^2), \quad \text{a.s.}$$

and

$$\sum_{k=1}^{m} \|\sigma_k(v, \cdot, \partial v, t) - \sigma_k(v', \cdot, \partial v', t)\|_{R_k}^2$$
$$\leq \beta \|v - v'\|^2 + \gamma_2 \|v - v'\|_1^2, \quad \text{a.s.}$$

for any $v, v' \in H^1$, $t \in [0, T]$, where γ_1 is the same constant as in (A.2) and $\|\sigma_k(\cdots)\|_{R_k}^2$ is similarly defined as before.

(B.5) $\{W_k(x, t), k = 1, ..., m\}$ is a set of independent Wiener random fields and the covariance operator R_k for $W_k(\cdot, t)$ with the corresponding kernel $r_k(x, y)$ is bounded by r_0.

Corollary 7.3 Let $f(u, \partial_x u, x, t)$ and $g(x, t)$ be given as in Theorem 7.2. Suppose the conditions (B.4) and (B.5) are satisfied. Then, given $h \in H$, the nonlinear problem (3.178) has a unique solution $u(\cdot, t)$ as a continuous adapted

process in H. Moreover u belongs to $L^2(\Omega; \mathbf{C}([0,T]; H)) \cap L^2(\Omega \times [0,T]; H^1)$.
□

Except for notational complication, the proof of this corollary follows the same arguments as that of Theorem 7.2 and will therefore be omitted here. The independence of the Wiener fields assumed in condition (B.5) was for simplicity but not necessary. As an example, consider a model problem for turbulent diffusion with chemical reaction in $D \subset \mathbf{R}^3$ as follows:

$$\frac{\partial u}{\partial t} + \sum_{k=1}^{3} v_k(x,t)\frac{\partial u}{\partial x_k} = \kappa \Delta u - \alpha u + b(u,x,t)$$
$$+ \sum_{k=1}^{m} \sigma_k(u, \partial_x u, x, t)\dot{W}_k(x,t), \tag{3.179}$$
$$\frac{\partial u}{\partial n}\Big|_{\partial D} = 0, \quad u(x,0) = h(x),$$

for $x \in D$, $t \in (0,T)$, where $v = (v_1, v_2, v_3)$ denotes the random flow velocity. Suppose that the velocity consists of the mean velocity $g = E\,v$ and a white-noise fluctuation so that

$$v_k(x,t) = g_k(x,t) + \nu_k(x,t)\dot{W}_k(x,t),$$

for $k = 1, 2, 3$, where $\nu_k(x,t)$ is a predictable random field. The reaction term $b(u,x,t)$ is assumed to be non-random. In the notation of equation (3.178), we have

$$f(u, \partial_x u, x, t) = b(u, x, t),$$

and

$$\sigma_k(\partial_x u, x, t) = \nu_k(x,t)\frac{\partial u}{\partial x_k}.$$

Here we impose the following conditions:

(C.1) Let $g_k(x,t)$ and $r_k(x,y)$ be continuous, for $k = 1, 2, 3$, such that there exist positive constants α, r_0 such that

$$\max_{1 \le k \le 3} \sup_{(x,t) \in D_T} |g_k(x,t)| \le \alpha, \quad \max_{1 \le k \le 3} \sup_{x \in D} r_k(x,x) \le r_0,$$

where we let $D_T = D \times [0,T]$.

(C.2) $b(s,x,t)$ is a continuous function, for $s \in \mathbf{R}$, $(x,t) \in D_T$, and there exist positive constants c_1 and c_2 such that

$$\|b(u, \cdot, t)\|^2 \le c_1(1 + \|u\|^2),$$

and

$$\|b(u, \cdot, t) - b(u', \cdot, t)\|^2 \le c_2\|u - u'\|^2,$$

for any $u, u' \in H^1, t \in [0,T]$.

(C.3) For $k = 1, 2, 3$, $\nu_k(x, t)$ is a bounded predictable random field and there exists $\nu_0 > 0$ such that

$$\sup_{(x,t) \in D_T} |\nu_k(x, t)| \leq \nu_0 \quad \text{a.s.} \quad \text{with} \quad \nu_0^2 < \kappa / r_0.$$

Under conditions (C.1) to (C.3), it is easy to verify that the conditions for Theorem 7.2 are met. For instance, to check condition (B.4), we have

$$\sum_{k=1}^{m} \|\sigma_k(\partial u, \cdot, t)\|_{R_k}^2 = \sum_{k=1}^{m} \|\nu_k(\cdot, t) \frac{\partial u}{\partial x_k}\|_{R_k}^2$$

$$\leq r_0 \sum_{k=1}^{m} \|\nu_k(\cdot, t) \frac{\partial u}{\partial x_k}\|^2 \leq \frac{r_0 \nu_0^2}{\kappa} \|u\|_1^2,$$

where we used an equivalent H^1-norm defined as $\|v\|_1 = \{\kappa \sum_{k=1}^{m} \|\frac{\partial u}{\partial x_k}\|^2 + \alpha \|u\|^2\}^{1/2}$. Since $\sigma_k(\partial u, \cdot, t)$ is linear in ∂u, we have $\sigma_k(\partial u, \cdot, t) - \sigma_k(\partial u', \cdot, t) = \sigma_k(\partial(u - u'), \cdot, t)$ so that

$$\sum_{k=1}^{m} \|\sigma_k(\partial u, \cdot, t) - \sigma_k(\partial u', \cdot, t)\|_{R_k}^2 \leq \frac{r_0 \nu_0^2}{\kappa} \|u - u'\|_1^2.$$

Hence, noticing $\gamma_1 = 0$, condition (B.4) is satisfied if $\gamma_2 = \frac{r_0 \nu_0^2}{\kappa} < 1$. By means of Theorem 7.2, we obtain the following result.

Corollary 7.4 Assume that the conditions (C.1), (C.2) and (C.3) are satisfied. Then, given $h \in H$, the problem (3.179) has a unique solution $u(\cdot, t)$ as an adapted, continuous process in H. Moreover u belongs to $L^2(\Omega; \mathbf{C}([0, T]; H)) \cap L^2(\Omega \times [0, T]; H^1)$. □

Chapter 4

Stochastic Parabolic Equations in the Whole Space

4.1 Introduction

This chapter is concerned with some parabolic Itô equations in $\mathbf{R}^d$. In Chapter Three we considered such equations in a bounded domain D. The analysis was based mainly on the method of eigenfunctions expansion for the associated elliptic boundary-value problem. In $\mathbf{R}^d$ the spectrum of such an elliptic operator may be continuous, the method of eigenfunctions expansion is no longer suitable. For an elliptic operator with constant coefficients, the method of Fourier transform is a natural substitution. Consider the initial-value or Cauchy problem for the heat equation on the real line:

$$\frac{\partial u}{\partial t} = \frac{\partial^2 u}{\partial x^2} + f(x,t), \quad -\infty < x < \infty, \quad t > 0,$$

$$u(x,0) = h(x), \tag{4.1}$$

where f and g are given smooth functions. Let $\hat{u}(\lambda, t)$ denote a Fourier transform of u in the space variable defined by

$$\hat{u}(\lambda, t) = \sqrt{\frac{1}{2\pi}} \int e^{-i\lambda x} u(x,t) dx$$

with $i = \sqrt{-1}$, where $\lambda \in \mathbf{R}$ is a parameter and the integration is over the real line. By applying a Fourier transform to the heat equation, it yields a simple ordinary differential equation

$$\frac{d\hat{u}}{dt} = -\lambda^2 \hat{u} + \hat{f}(\lambda, t), \quad \hat{u}(\lambda, 0) = \hat{h}(\lambda), \tag{4.2}$$

which can be easily solved to give

$$\hat{u}(\lambda, t) = \hat{h}(\lambda) e^{-\lambda^2 t} + \int_0^t e^{-\lambda^2 (t-s)} \hat{f}(\lambda, s) ds.$$

By the inverse Fourier transform, we obtain the solution of the heat equation (4.1) as follows:

$$u(x,t) = \sqrt{\frac{1}{2\pi}} \int e^{i\lambda x} \hat{u}(\lambda, t) d\lambda$$

$$= \int K(x-y,t)h(y)dy + \int_0^t \int K(x-y,t-s)f(y,s)dyds,$$

where the Green's function K is the well-known Gaussian or heat kernel given by

$$K(x,t) = \frac{1}{\sqrt{4\pi t}} \exp\{-x^2/4t\}. \tag{4.3}$$

It shows that, similar to the eigenfunctions expansion, the Fourier transform reduces the partial differential equation (4.1) to an ordinary differential equation (4.2), which can be analyzed more easily. With proper care, this method can also be applied to stochastic partial differential equations as one will see.

In the following sections, Section 4.2 is concerned with some basic facts about the Fourier transforms of random fields and a special form of Itô formula. Similar to the case in a bounded domain, the existence and uniqueness questions for linear and semilinear parabolic Itô equations are treated in Section 4.3. For a linear parabolic equation with a multiplicative noise, a stochastic Feynman-Kac formula will be given in Section 4.4 as a probabilistic representation of its solution. As a model equation, solutions to a parabolic Itô equation representing, say, the mass density, are required to be nonnegative on physical grounds. This issue is studied in Section 4.5. Finally, in Section 4.6, the derivation of partial differential equations for the correlation functions of the solution to a certain linear parabolic Itô equation will be provided.

4.2 Preliminaries

Let H be the Hilbert space $L^2(\mathbf{R}^d)$ of real or complex valued functions with the inner product

$$(g,h) = \int g(x)\bar{h}(x)dx,$$

where the integration is over $\mathbf{R}^d$ and the over-bar denotes the complex conjugate. As in Chapter Three, we shall use the same notation for the corresponding Sobolev spaces H^m with norm $\|\cdot\|_m$ and so on.

For $g \in H$, let $\hat{g}(\xi)$ denote the Fourier transform of g defined as follows:

$$\hat{g}(\xi) = (\mathcal{F}g)(\xi) = \left(\frac{1}{2\pi}\right)^{d/2} \int g(x)e^{-ix\cdot\xi}dx, \quad \xi \in \mathbf{R}^d, \tag{4.4}$$

where the dot denotes the inner product in $\mathbf{R}^d$. The inverse Fourier transform of $\hat{g}$ is given by

$$g(x) = (\mathcal{F}^{-1}\hat{g})(x) = (\frac{1}{2\pi})^{d/2} \int \hat{g}(\xi)e^{ix\cdot\xi}d\xi, \quad x \in \mathbf{R}^d. \tag{4.5}$$

By Plancheral's theorem and Parseval's identity, we have

$$\|g\|^2 = \|\hat{g}\|^2; \quad (g,h) = (\hat{g},\hat{h}), \tag{4.6}$$

which shows that $\mathcal{F} : H \to H$ is an isometry. For $g \in \mathbf{C}_0^\infty$, define the H^m-norm $\|g\|_m$ of g by

$$\|g\|_m = \|\rho^{m/2}\hat{g}\| = \{\int \rho^m(\xi)|\hat{g}(\xi)|^2 d\xi\}^{1/2}, \tag{4.7}$$

where

$$\rho(\xi) = (1 + |\xi|^2).$$

For $g \in L^1$ and $h \in L^p$, the *convolution* : $g \star h$ of g and h is defined as

$$(g \star h)(x) = \int g(x-y)h(y)dy. \tag{4.8}$$

Then it satisfies the following property:

$$\widehat{(g \star h)} = \widehat{(h \star g)} = (2\pi)^{d/2}\hat{g}\,\hat{h}, \tag{4.9}$$

and Young's inequality holds:

$$|g \star h|_p \le |g|_1\,|h|_p, \tag{4.10}$$

for $1 \le p \le \infty$, where $|\cdot|_p$ denotes the L^p-norm with $|\cdot|_2 = \|\cdot\|$ (see [27], [77] for details).

Let $M(\cdot,t), t \in [0,T]$ be a continuous H-valued martingale with $M(\cdot,0) = 0$ and the covariation operator

$$[M]_t = \int_0^t Q_s ds,$$

where the local characteristic operator Q_t has $q(x,y;t)$ as its kernel. Suppose that

$$E \sup_{0\le t\le T} \|M(\cdot,t)\|^2 \le CE \int_0^T Tr\,Q_s ds < \infty. \tag{4.11}$$

Let $\hat{M}(\xi,t)$ denote the Fourier transform of $M(x,t)$. Then $\hat{M}(\cdot,t)$ is a continuous martingale in H with local characteristic operator $\hat{Q}_t$ with kernel $\hat{q}(\xi,\eta;t)$ so that

$$\langle \hat{M}(\xi,\cdot), \overline{\hat{M}(\eta,\cdot)}\rangle_t = \int_0^t \hat{q}(\xi,\eta;s)ds.$$

It is easy to check that

$$\hat{q}(\xi, \eta; t) = (\frac{1}{2\pi})^d \int \int q(x, y; t) e^{-i(\xi \cdot x - \eta \cdot y)} dx \, dy, \qquad (4.12)$$

and $Tr \, Q_t = Tr \, \hat{Q}_t$.

Let $K(x, t)$ be a function on $\mathbf{R}^d \times [0, T]$ such that

$$\sup_{0 \le t \le T} \int |K(x, t)| dx \le C_1, \qquad (4.13)$$

for some constant $C_1 > 0$. Consider the stochastic convolution

$$J(x, t) = \int_0^t \int K(x - y, t - s) M(y, ds) dy. \qquad (4.14)$$

To show it is well defined, let $J^n(x, t)$ denote its n-finite sum approximation for $J(x, t)$ so that

$$J^n(x, t) = \sum_{j=1}^n \int K(x - y, t - t_{j-1})[M(y, t_j) - M(y, t_{j-1})] dx,$$

with $0 = t_0 < t_1 < \cdots < t_j < \cdots < t_n = t$. Its Fourier transform is given by

$$\hat{J}^n(\xi, t) = (2\pi)^{d/2} \sum_{j=1}^n \hat{K}(\xi, t - t_{j-1})[\hat{M}(\xi, t_j) - \hat{M}(\xi, t_{j-1})]. \qquad (4.15)$$

By condition (4.13), $\hat{K}(\xi, t)$ is bounded in $\mathbf{R}^d \times [0, T]$. From this and condition (4.11) one can show that $\hat{J}^n(\cdot, t)$ converges in $L^2(\Omega; H)$, uniformly in t, to the limit

$$\hat{J}(\xi, t) = (2\pi)^{d/2} \int_0^t \hat{K}(\xi, t - s) \hat{M}(\xi, ds). \qquad (4.16)$$

Since the Fourier transform $\mathcal{F}$ is a L^2-isometry, we can conclude that $J^n(\cdot, t)$ converges in $L^2(\Omega; H)$ to $J(\cdot, t)$ given by (4.14) and $J(\cdot, t) = \{\mathcal{F}^{-1} \hat{J}\}(\cdot, t)$.

Let

$$V(x, t) = V_0(x) + \int_0^t b(x, s) ds + M(x, t),$$

where $V_0 \in H$ and $b(\cdot, t)$ is a continuous adapted process in H^{-1}. For any $\phi \in H^1$, consider the following equation

$$(V(\cdot, t), \phi) = (V_0(\cdot), \phi) + \int_0^t \langle b(\cdot, s), \phi \rangle ds + (M(\cdot, t), \phi).$$

Define $v(t) = (V(\cdot, t), \phi)$. Then $v(t) - v(0)$ is a real-valued semimartingale with local characteristic $(q(t); b(t))$, where $b(t) = \langle b(\cdot, t), \phi \rangle$ and $q(t) = (Q_t \phi, \phi)$.

For $\phi_j \in H^1$, let $v_j(t) = (V(\cdot, t), \phi_j)$, $b_j(t) = \langle b(\cdot, t), \phi_j \rangle$, $z_j(t) = (M(\cdot, t), \phi_j)$ and $q_{jk}(t) = (Q_t \phi_j, \phi_k)$, for $j, k = 1, \cdots, m$. Let $F(x_1, \cdots, x_m)$ be a $\mathbf{C}^2-$ function on $\mathbf{R}^m$. Set $v^m(t) = (v_1(t), \cdots, v_m(t))$. By the usual Itô formula, we obtain

$$
F[v^m(t)] = F[v^m(0)] + \sum_{j=1}^{m} \int_0^t F_j[v^m(s)] b_j(s) ds
$$
$$
+ \sum_{j=1}^{m} \int_0^t F_j[v^m(s)] dz_j(s) + \frac{1}{2} \sum_{j,k=1}^{m} \int_0^t F_{jk}[v^m(s)] q_{jk}(s) ds,
$$

(4.17)

where we denote $F_j = \dfrac{\partial F}{\partial x_j}$ and $F_{jk} = \dfrac{\partial^2 F}{\partial x_j \partial x_k}$.

4.3   Linear and Similinear Equations

First let us consider the parabolic Itô equation (3.51) with $\nu = \alpha = 1$ and $D = \mathbf{R}^d$:

$$
\frac{\partial u}{\partial t} = (\Delta - 1)u + f(x,t) + \sigma(x,t)\dot{W}(x,t), \quad x \in \mathbf{R}^d,\, t \in (0,T),
$$
$$
u(x,0) = h(x),
$$

(4.18)

where $h \in H$ and $f(\cdot, t)$, $\sigma(\cdot, t)$ are continuous adapted processes in H, and $W(x,t)$ is a continuous Wiener random field with covariance function $r(x,y)$ bounded by r_0. Assume that

$$
E \int_0^T \{ \|g(\cdot, s)\|^2 + \|\sigma(\cdot, s)\|^2 \} ds < \infty,
$$

(4.19)

which implies that

$$
E \int_0^T \|\sigma(\cdot, s)\|_R^2 ds = E \int_0^T \int r(x,x)\sigma^2(x,s) dx ds
$$
$$
\leq r_0 E \int_0^T \|\sigma(\cdot, s)\|^2 \} ds < \infty.
$$

Rewrite (4.18) as an integral equation

$$
u(x,t) = h + \int_0^t Au(x,s) ds + \int_0^t f(x,s) ds + M(x,t),
$$

(4.20)

where $A = (\Delta - 1)$ and

$$
M(x,t) = \int_0^t \sigma(x,s) W(x,ds).
$$

Applying a Fourier transform to (4.20), we obtain

$$\hat{u}(\xi, t) = \hat{h}(\xi) - \rho(\xi) \int_0^t \hat{u}(\xi, s)ds + \int_0^t \hat{f}(\xi, s)ds + \hat{M}(\xi, t), \qquad (4.21)$$

for $\xi \in \mathbf{R}^d$, $t \in (0, T)$, where we recall that $\rho(\xi) = (1 + |\xi|^2)$ and $\hat{M}(\xi, t)$ is the Fourier transform of $M(x, t)$. For each ξ, the above equation has the solution

$$
\begin{aligned}
\hat{u}(\xi, t) &= \hat{h}(\xi)e^{-\rho(\xi)t} + \int_0^t e^{-\rho(\xi)(t-s)} \hat{f}(\xi, s)ds \\
&+ \int_0^t e^{-\rho(\xi)(t-s)} \hat{M}(\xi, ds).
\end{aligned}
\qquad (4.22)
$$

By means of the inverse Fourier transform, it yields

$$u(x, t) = \tilde{u}(x, t) + v(x, t) \qquad (4.23)$$

where

$$\tilde{u}(x, t) = (G_t h)(x) + \int_0^t (G_{t-s} f)(x, s)ds \qquad (4.24)$$

and

$$v(x, t) = \int_0^t (G_{t-s} M)(x, ds). \qquad (4.25)$$

In the above equations, G_t denotes the Green operator defined by

$$(G_t h)(x) = \int K(x - y, t)h(y)dy,$$

where $K(x, t)$ is a Gaussian kernel given by

$$K(x, t) = (2\pi)^{-d} \int e^{ix \cdot \xi - \rho(\xi)t} d\xi = (4\pi t)^{-d/2} \exp -\{\frac{|x|^2}{4t} + t\}. \qquad (4.26)$$

Similar to Lemma 3-5.1, by Plancheral's theorem and Parseval's equality in Fourier transform, the following lemma can be easily verified.

Lemma 3.1 Let $h \in H$ and let $f(\cdot, t)$ be a continuous predictable process in H such that the condition (4.19) is satisfied. Then $\tilde{u}(\cdot, t)$ is a continuous, adapted process in H such that $\tilde{u} \in L^2(\Omega; \mathbf{C}([0, T]; H)) \cap L^2(\Omega \times (0, T); H^1)$. Moreover the following estimates hold:

$$\sup_{0 \leq t \leq T} \|G_t h\| \leq \|h\|; \qquad \int_0^T \|G_t h\|_1^2 dt \leq \frac{1}{2}\|h\|^2, \qquad (4.27)$$

$$E \sup_{0 \leq t \leq T} \| \int_0^t G_{t-s} f(\cdot, s)ds\|^2 ds \leq T \int_0^T E \|f(\cdot, t)\|^2 dt, \qquad (4.28)$$

and

$$E \int_0^T \| \int_0^t G_{t-s} f(\cdot, s) ds \|_1^2 dt \leq \frac{T}{2} E \int_0^T \| f(\cdot, t) \|^2 dt. \qquad (4.29)$$

Proof. The first inequality in (4.27) follows immediately from Young's inequality (4.10) for $p = 2$. To verify the second inequality in (4.27), we have

$$\int_0^T \| G_t h \|_1^2 dt \leq \int_0^T \int \rho(\xi) e^{-2\rho(\xi)t} |\hat{h}(\xi)|^2 d\xi dt$$

$$\leq \frac{1}{2} \| \hat{h} \|^2 = \frac{1}{2} \| h \|^2$$

For the remaining ones, we will only verify the last inequality (4.29). Clearly we have

$$\| \int_0^t G_{t-s} f_s ds \|_1^2 \leq [\int_0^t \| G_{t-s} f_s \|_1 ds]^2 \leq T \int_0^t \| G_{t-s} f_s \|_1^2 ds.$$

Since, in view of (4.8) and (4.9),

$$\| G_{t-s} f_s \|_1^2 = \| \rho^{1/2}(\cdot) e^{-\rho(\cdot)(t-s)} \hat{f}_s \|^2,$$

we get

$$E \int_0^T \| \int_0^t G_{t-s} f_s ds \|_1^2 dt \leq TE \int_0^T \int_0^t \int \rho(\xi) e^{-2\rho(\xi)(t-s)} |\hat{f}(\xi, s)|^2 d\xi ds dt$$

$$\leq \frac{T}{2} E \int_0^T \| f_t \|^2 dt,$$

by interchanging the order of integration and making use of the fact: $\| f_s \| = \| \hat{f}_s \|$. $\qquad \square$

As a counterpart of Lemma 3-5.2, we shall verify the next lemma.

Lemma 3.2   Let $v(\cdot, t) = \int_0^t G_{t-s} M(\cdot, ds)$ be the stochastic integral given by (4.25) with $M(\cdot, ds) = \sigma(\cdot, s) W(\cdot, ds)$ such that

$$E \int_0^T \| \sigma^2(\cdot, t) \|^2 dt < \infty. \qquad (4.30)$$

Then $v(\cdot, t)$ is a H^1−process, with a continuous trajectory in H over $[0, T]$, and the following inequalities hold:

$$E \| v(\cdot, t) \|^2 \leq E \int_0^t \| \sigma(\cdot, s) \|_R^2 ds, \qquad (4.31)$$

$$E \sup_{0 \leq t \leq T} \| v(\cdot, t) \|^2 \leq 16 E \int_0^T \| \sigma(\cdot, t) \|_R^2 dt, \qquad (4.32)$$

and

$$E \int_0^T \|v(\cdot,t)\|_1^2 dt \le \frac{1}{2} E \int_0^T \|\sigma(\cdot,t)\|_R^2 dt. \tag{4.33}$$

Proof.   Applying a Fourier transform to v, we get

$$\hat{v}(\xi,t) = \int_0^t e^{-\rho(\xi)(t-s)} \hat{M}(\xi,ds), \tag{4.34}$$

which can be shown to be a continuous, adapted process in H. Hence, so is $v(x,t)$. To show (4.31), it follows from (4.34) that

$$E \|v_t\|^2 = E \|\hat{v}_t\|^2 = E \int_0^t \int e^{-2\rho(\xi)(t-s)} \hat{q}(\xi,\xi;s) \, d\xi ds$$

$$\le E \int_0^t Tr\, \hat{Q}_s ds = E \int_0^t \|\sigma_s\|_R^2 ds.$$

By rewriting (4.34) as

$$\hat{v}(\xi,t) = \hat{M}(\xi,t) - \rho(\xi) \int_0^t e^{-\rho(\xi)(t-s)} \hat{M}(\xi,s) ds,$$

we can obtain (4.32) similarly to the corresponding inequality in Lemma 3-5.2. As for (4.33), we have

$$E \int_0^T \|v_t\|_1^2 dt = E \int_0^T \int_0^t \int \rho(\xi) e^{-2\rho(\xi)(t-s)} \hat{q}(\xi,\xi,s) d\xi ds dt$$

$$= E \int \int_0^T \int_s^T \rho(\xi) e^{-2\rho(\xi)(t-s)} \hat{q}(\xi,\xi,s) dt ds d\xi$$

$$\le \frac{1}{2} E \int_0^T Tr\, \hat{Q}_s ds = \frac{1}{2} E \int_0^T \|\sigma_t\|_R^2 dt,$$

where we made use of the equality $Tr\, \hat{Q}_s = Tr\, Q_s$ and a change in the order of integrations, which can be justified by invoking Fubini's theorem.   □

By means of Lemmas 3.1 and 3.2, the following theorem can be proved easily.

Theorem 3.3   Let the condition (4.19) be satisfied. For $h \in H$, the solution $u(\cdot,t)$ of the linear problem (4.18) given by (4.23) is a $\mathcal{F}_t$-adapted H^1-valued process over $[0,T]$ with a continuous trajectory in H such that

$$E \sup_{0 \le t \le T} \|u(\cdot,t)\|^2 + E \int_0^T \|u(\cdot,t)\|_1^2 dt$$
$$\le C(T) \{ \|h\|^2 + E \int_0^T [\|g(\cdot,s)\|^2 + \|\sigma(\cdot,s)\|_R^2] ds \}, \tag{4.35}$$

for some constant $C(T) > 0$. Moreover the energy equation holds true:

$$\|u(\cdot,t)\|^2 = \|h\|^2 + 2\int_0^t \langle Au(\cdot,s), u(\cdot,s)\rangle ds + 2\int_0^t (u(\cdot,s), g(\cdot,s)) ds$$
$$+2\int_0^t (u(\cdot,s), M(\cdot,ds)) + \int_0^t \|\sigma(\cdot,s)\|_R^2 ds. \tag{4.36}$$

Proof. In view of (4.23), by making use of Lemmas 3.1 and 3.2, similar to the proof of Theorem 3-5.3, we can show that the solution u has the regularity property: $u \in L^2(\Omega\times[0,T]; H^1)\cap L^2(\Omega; C([0,T]; H))$, and the inequality (4.35) holds true. By applying the Itô formula to $|\hat{u}(\xi,t)|^2$ and noticing (4.21), we get

$$|\hat{u}(\xi,t)|^2 = |\hat{h}(\xi)|^2 - 2\rho(\xi)\int_0^t |\hat{u}(\xi,s)|^2 ds + 2\int_0^t \hat{u}(\xi,s), \overline{\hat{g}(\xi,s)} ds$$
$$+\int_0^t \hat{q}(\xi,\xi,s) ds + 2\int_0^t \hat{u}(\xi,s)\overline{\hat{M}(\xi,ds)},$$

where the over-bar denotes the complex conjugate. By integrating the above equation with respect to ξ, making use of Plancherel's theorem and Parseval's equality, it gives rise to the energy equation (4.36). □

Now consider the semilinear parabolic equation in $\mathbf{R}^d$:

$$\frac{\partial u}{\partial t} = (\Delta - 1)u + f(u, \partial u, x, t) + g(x,t) + \sigma(u, \partial u, x, t)\dot{W}(x,t),$$
$$u(x,0) = h(x), \tag{4.37}$$

for $x \in \mathbf{R}^d$, $t \in (0,T)$, where $g(x,t), f(u,\xi,x,t)$ and $\sigma(u,\xi,x,t)$ are given predictable random fields, as in the case of a bounded domain D in Section 3.7. Analogous to conditions (B.1)–(B.3), we assume that

(C.1) $f(r,\xi,x,t)$ and $\sigma(r,\xi,x,t)$, for $r \in \mathbf{R}; \xi, x \in \mathbf{R}^d$ and $t \in [0,T]$, are predictable random fields. There exists a constant $\alpha > 0$ such that

$$\|f(v,\cdot,\partial v,t)\|^2 + \|\sigma(v,\cdot,\partial v,t)\|_R^2 \le \alpha(1 + \|v\|^2 + \|v\|_1^2), \quad \text{a.s.}$$

for any $v \in H^1, t \in [0,T]$.

(C.2) There exist positive constants β, γ_1 and γ_2 with $(\gamma_1 + \gamma_2) < 1$ such that

$$\|f(v,\cdot,\partial v,t) - f(v',\cdot,\partial v',t)\|^2 \le \beta\|v - v'\|^2 + \gamma_1\|v - v'\|_1^2, \quad \text{a.s.},$$

and

$$\|\sigma(v,\cdot,\partial v,t) - \sigma(v',\cdot,\partial v',t)\|_R^2 \le \beta\|v - v'\|^2 + \gamma_2\|v - v'\|_1^2, \quad \text{a.s.}$$

for any $v, v' \in H^1, t \in [0,T]$.

(C.3) Let $g(\cdot, t)$ be a predictable H-valued process such that

$$E \int_0^T \|g(\cdot, t)\|^2 dt < \infty,$$

and $W(\cdot, t)$ be a R-Wiener random field in H with a bounded covariance function $r(x, y)$.

With the aid of Lemmas 3.1 and 3.2 together with Theorem 3.3, under conditions (C.1)–(C.3), the existence theorem for the problem (4.37) can be proved in a similar fashion as that of Theorem 3-7.2. Therefore we shall simply state the theorem and omit the proof.

Theorem 3.4 Suppose the conditions (C.1), (C.2) and (C.3) are satisfied. Then, given $h \in H$, the semilinear equation (4.37) has a unique solution $u(\cdot, t)$ as an adapted, continuous process in H. Moreover it has the regularity property: $u \in L^2(\Omega; \mathbf{C}([0, T]; H)) \cap L^2(\Omega \times [0, T]; H^1)$ and the energy equation holds:

$$\|u(\cdot, t)\|^2 + 2 \int_0^t \|u(\cdot, s)\|_1^2 ds = \|h\|^2 + 2 \int_0^t (u(\cdot, s), g(\cdot, s)) ds$$
$$+2 \int_0^t (u(\cdot, s), f[u(\cdot, s), \partial u(\cdot, s), \cdot, s)]) ds + \int_0^t Tr \, Q_s(u, \partial u) ds \qquad (4.38)$$
$$+2 \int_0^t (u(\cdot, s), M[u(\cdot, s), \partial u(\cdot, s), \cdot, ds]).$$

Moreover we have

$$E \{ \sup_{0 \leq t \leq T} \|u(t)\|^2 + \int_0^T \|u(\cdot, s)\|_1^2 ds \}$$
$$\leq C \{1 + \|h\|^2 + E \int_0^T \|g(\cdot, t)\|^2 dt \}, \qquad (4.39)$$

for some $C > 0$.   □

4.4   Feynman-Kac Formula

Let us consider the linear parabolic equation with a multiplicative noise:

$$\frac{\partial u}{\partial t} = \frac{1}{2} \Delta u + u \dot{V}(x, t) + g(x, t), \quad x \in \mathbf{R}^d, \, t \in (0, T),$$
$$u(x, 0) = h(x), \qquad (4.40)$$

where $\dot{V}(x,t) = \frac{\partial}{\partial t}V(x,t)$ and

$$V(x,t) = \int_0^t b(x,s)ds + N(x,t); \quad N(x,t) = \int_0^t \sigma(x,s) \circ W(x,ds). \quad (4.41)$$

In order to obtain a path-wise smooth solution, the random fields: $W(\cdot,t)$, $g(\cdot,t)$, $b(\cdot,t)$ and $\sigma(\cdot,t)$ are required to have certain additional regularity properties to be specified. Recall that, in Section 2.3 we gave a probabilistic representation of the solution to a linear parabolic equation, where, to avoid introducing the Wiener random field, the noise is assumed to be finite-dimensional. In fact, for a continuous $\mathbf{C}^m$-semimartingale, under suitable conditions, the probabilistic representation given by Theorem 2-3.2 is also valid. Before stating this result, we have to extend the definitions of the stochastic integrals with respect to the semimartingale, such as (4.41). By Theorem 3.2.5 in [51], the following lemma holds.

Lemma 4.1 Let $\Phi(x,t)$ be a $\mathbf{C}^1$-semimartingale with local characteristic $(\gamma(x,y,t); \beta(x,t))$ for $x,y \in \mathbf{R}^d$, and let ξ_t be a continuous, adapted process in $\mathbf{R}^k$ such that

$$\int_0^T \{|\beta(\xi_s,s)| + \gamma(\xi_s,\xi_s,s)\}dt < \infty, \quad a.s.$$

Then the Itô integral $\int_0^t \Phi(\xi_s,ds)$ and the Stratonovich integral $\int_0^t \Phi(\xi_s, \circ ds)$ can be defined by (2.11) and (2.12), respectively. Moreover, for $m \geq 1$, the following formula holds:

$$\int_0^t \Phi(\xi_s, \circ ds) = \int_0^t \Phi(\xi_s, ds) + \frac{1}{2}\sum_{i=1}^k \langle \int_0^t \frac{\partial \Phi}{\partial x_i}(\xi_s, ds), \xi_t^i \rangle, \quad (4.42)$$

where $\langle \cdot, \cdot \rangle$ denotes the mutual variation symbol. $\square$

Theorem 4.2 Suppose that $W(x,t)$ is a $\mathbf{C}^m$-Wiener random field such that the correlation function $r(x,y)$ belongs to $\mathbf{C}_b^{m,\delta}$. Let $b(\cdot,t), \sigma(\cdot,t)$ and $g(\cdot,t)$ be continuous adapted $\mathbf{C}_b^{m,\delta}$-processes such that

$$\int_0^T \|b(\cdot,t)\|_{m+\delta}^2 \leq c_1, \quad \int_0^T \|g(\cdot,t)\|_{m+\delta}^2 \leq c_2,$$

and

$$\sup_{0 \leq t \leq T} \|\sigma(\cdot,t)\|_{m+1+\delta}^2 \leq c_3, \quad a.s.,$$

for some positive constants $c_i, i = 1, 2, 3$. Then, for $h \in \mathbf{C}_b^{m,\delta}$ with $m \geq 3$, the solution $u(\cdot,t), 0 \leq t \leq T$, to the Cauchy problem (4.40) is a $\mathbf{C}^m$-

semimartingale which has a probabilistic representation given by:

$$u(x,t) = E_z \left\{ h[x + z(t)] \exp\{ \int_0^t V[x + z(t) - z(s), ds] \} \right.$$
$$+ \int_0^t g[x + z(t) - z(s), s] \exp\{ \int_s^t V[x + z(t) - z(r), dr] \} ds \},$$

$$(4.43)$$

where E_z denotes the partial expectation with respect to the standard Brownian motion $z(t) = (z_1(t), \cdots, z_d(t))$ in $\mathbf{R}^d$ independent of $W(x,t)$.

Proof. Consider the first-order equation:

$$\frac{\partial v}{\partial t} = \sum_{i=1}^d \frac{\partial v}{\partial x_i} \circ \dot{z}_i(t) + v\dot{V}(x,t) + g(x,t),$$
$$v(x,0) = h(x), \quad x \in \mathbf{R}^d, \ t \in (0,T),$$

$$(4.44)$$

where $\{z_1, \cdots, z_d\}$ are independent standard Brownian motions as mentioned before. As indicated at the end of Chapter Two, by Theorem 6.2.5 in [51], Theorem 2-3.2 holds when the finite-dimensional noise is replaced by a C^m−Wiener random field. Here we let $V_i = z_i(t), i = 1, \cdots, d; \quad V_{d+1} = V$ and $V_{d+2} = g$. The solution of the first-order equation (4.44) has the probabilistic representation:

$$v(x,t) = \left\{ h[\psi_t(x)] \exp\{ \int_0^t V[\psi_s(y), \circ ds] \} \right.$$
$$+ \int_0^t g[\psi_s(y), s] \exp\{ \int_s^t V[\psi_r(y), \circ dr] \} ds \Big\}_{y=\phi_t(x)},$$

$$(4.45)$$

where $\phi_t(x) = x - z(t)$ and $\psi_t(x) = \phi_t^{-1}(x) = x + z(t)$. By Lemma 4.1,

$$\int_0^t V[\psi_s(y), \circ ds] = \int_0^t V[\psi_s(y), ds] + \frac{1}{2} \sum_{j=1}^d \langle \int_0^t \frac{\partial}{\partial x_j} V[\psi_s(y), ds], z_j(t) \rangle$$

$$= \int_0^t V[\psi_s(y), ds],$$

since $z(t)$ is independent of $W(x,t)$. Therefore (4.45) can be written as

$$v(x,t) = \left\{ h[x + z(t)] \exp\{ \int_0^t V[x + z(t) - z(s), ds] \} \right.$$
$$+ \int_0^t g[x + z(t) - z(s), s] \exp\{ \int_s^t V[x + z(t) - z(r), dr] \} ds \Big\}.$$

$$(4.46)$$

By making use of the fact:

$$\frac{\partial v}{\partial x_i} \circ \dot{z}_i(t) = \frac{1}{2} \triangle v + \frac{\partial v}{\partial x_i} \dot{z}_i(t),$$

and taking a partial expectation of the equation (4.46) with respect to the Brownian motion $z(t)$, it yields the equation (4.43). It follows from this observation and (4.46) that

$$u(x,t) = E_z\{v(x,t)\} = E_z\left\{h[x + z(t)]\exp\{\int_0^t V[x + z(t) - z(s), ds]\}\right.$$
$$\left. + \int_0^t g[x + z(t) - z(s), s]\exp\{\int_s^t V[x + z(t) - z(r), dr]\}ds\right\}$$

is the solution of (4.40) as claimed. $\qquad\qquad\Box$

Remarks: The stochastic Feynman-Kac formula for the equation (4.40) has been obtained by other methods. For instance, the formula can be derived by first approximating the Wiener random field by a finite dimensional Wiener process

$$W_t^n = \sum_{k=1}^n \sqrt{\mu_k} w_t^k \phi_k,$$

where ϕ_k is the eigenfunction of the covariance operator R associated with the eigenvalue μ_k and $\{w_t^k\}$ is a sequence of independent copies of standard Brownian motion in one dimension. Then it can be shown that, under a C^m−regularity condition for W, the corresponding solution $u^n(x,t)$ has a Feynman-Kac representation which converges to the formula (4.43) in a So-belev space as $n \to \infty$. However, as seen from results given in Section 2.3, the present approach allows us to generalize the probabilistic representation of the solution to a stochastic parabolic equation when the Laplacian in equation (4.40) is replaced by an elliptic operator A that is the infinitesimal genera-tor for a diffusion process y_t. In this case, the Feynman-Kac formula can be obtained similarly with z_t replaced by y_t.

4.5   Positivity of Solutions

As physical models, the unknown function $u(x,t)$ often represents the den-sity or concentration of some substance that diffuses in a randomly fluctuating media. In such instances, given the positive data, the solution u must remain positive (non-negative). Due to the presence of noisy coefficients, this fact is far from obvious and needs to be confirmed. For example, consider lin-ear equation (4.40). In view of the Feynman-Kac formula (4.43), if the data $h(x) \geq 0$ and $g(x,t) \geq 0$ a.s., then the solution $u(x,t)$ is clearly positive. In fact it can be shown that

Theorem 5.1   Suppose that the conditions for Theorem 4.2 hold true. If

$g(x,t) \geq 0$ and $h(x) > 0$ for all $(x,t) \in \mathbf{R}^d \times [0,T]$, the solution $u(x,t)$ of the equation (4.40) remains strictly positive in $\mathbf{R}^d \times [0,T]$.

Proof. As mentioned above, the solution is known to be nonnegative, or $u(x,t) \geq 0$. In view of the formula (4.43), it suffices to show that

$$\varphi(x,t) = E_z\{h[x+z(t)] \exp \Phi(x,t)\} > 0, \tag{4.47}$$

where

$$\Phi(x,t) = \int_0^t V[x + z(t) - z(s), ds].$$

This can be shown by applying the Cauchy-Schwarz inequality as follows

$$
\begin{aligned}
\{E_z h^{1/2}[x+z(t)]\}^2 &= \{E_z(h[x+z(t)]e^{\Phi(x,t)})^{1/2} e^{-\Phi(x,t)/2}\}^2 \\
&\leq E_z\{h[x+z(t)]e^{\Phi(x,t)}\}E_z e^{-\Phi(x,t)},
\end{aligned}
$$

or, in view of (4.47), the above yields

$$\varphi(x,t) \geq \frac{\{E_z h^{1/2}[x+z(t)]\}^2}{E_z e^{-\Phi(x,t)}} > 0,$$

for any $(x,t) \in \mathbf{R}^d \times [0,T]$. $\qquad \square$

Now we will show that, under suitable conditions, the positivity of solution holds true for a class of semilinear parabolic Itô equations of the form:

$$
\begin{aligned}
\frac{\partial u}{\partial t} &= (\kappa\Delta - \alpha)u + f(u, \partial u, x, t) + g(x,t) + \sigma(u, \partial u, x, t)\dot{W}(x,t), \\
u(x,0) &= h(x),
\end{aligned}
\tag{4.48}
$$

for $x \in \mathbf{R}^d$, $t \in (0,T)$, where the positive parameters κ, α are reintroduced for a physical reason. Before stating the theorem, we need to introduce some auxiliary functions. For $r \in \mathbf{R}$, let $\eta(r)$ denote the negative part of r, or

$$\eta(r) = \begin{cases} -r, & r < 0, \\ 0, & r \geq 0, \end{cases} \tag{4.49}$$

and put $k(r) = \eta^2(r)$ so that

$$k(r) = \begin{cases} r^2, & r < 0, \\ 0, & r \geq 0. \end{cases} \tag{4.50}$$

For $\epsilon > 0$, let $k_\epsilon(r)$ be a $\mathbf{C}^2$–regularization of $k(r)$ defined by

$$k_\epsilon(r) = \begin{cases} r^2 - \epsilon^2/6, & r < -\epsilon, \\ -\dfrac{r^3}{\epsilon}(\dfrac{r}{2\epsilon} + \dfrac{4}{3}), & -\epsilon \leq r < 0, \\ 0, & r \geq 0. \end{cases} \tag{4.51}$$

Let $u(x,t)$ denote the solution of the parabolic Itô equation (4.48). By convention we set $u_t = u(\cdot, t)$, $[f_t(u)](x) = f[u(x,t), \partial u(x,t), x, t]$ and so on. Also let

$$\tilde{M}[(u(t), dt)](x) = M[u(x,t), \partial u(x,t), x, dt] = \sigma[u(x,t), \partial u(x,t), x, t]W(x, dt).$$

Denote by $q(s, \xi, x; s', \xi', x'; t)$ the local characteristic of the martingale

$$M(s, \xi, x, t), \text{ for } s, s' \in \mathbf{R}; \; \xi, \xi', x, x' \in \mathbf{R}^d; \; t \in [0, T],$$

and set $\tilde{q}(s, \xi, x; t) = q(s, \xi, x; s, \xi, x; t)$. Then we have

$$\tilde{q}(s, \xi, x, t) = r(x, x)\sigma^2(s, \xi, x, t),$$

and, for brevity, denote

$$[\tilde{q}_t(u)](x) = \tilde{q}[u(x,t), \partial u(x,t), x; t].$$

Define

$$\Phi_\epsilon(u_t) = (1, k_\epsilon[u(\cdot, t)]) = \int k_\epsilon[u(x,t)]dx. \tag{4.52}$$

Apply the Itô formula to $k_\epsilon[u(x,t)]$ and then integrate over x to obtain

$$
\begin{aligned}
\Phi_\epsilon(u_t) = \Phi_\epsilon(h) &+ \int_0^t < (\kappa\Delta - \alpha)u_s, k_\epsilon'(u_s) > ds \\
&+ \int_0^t (k_\epsilon'(u_s), f_s(u))ds + \int_0^t (k_\epsilon'(u_s), g_s)ds \\
&+ \frac{1}{2}\int_0^t (k_\epsilon''(u_s), \tilde{q}_s(u))ds + \int_0^t (k_\epsilon'(u_s), \tilde{M}(u_s, ds)),
\end{aligned}
\tag{4.53}
$$

where $k_\epsilon'(r), k_\epsilon''(r)$ denote the first two derivatives of k_ϵ. Since, by an integration by parts,

$$\int_0^t < \Delta u_s, k_\epsilon'(u_s) > ds = -\int_0^t (k_\epsilon''(u_s), |\partial u_s|^2)ds,$$

we have the following lemma:

Lemma 5.2   Let $u_t = u(\cdot, t)$ denote the solution of the parabolic Itô equation (4.48). Then the following equation holds

$$
\begin{aligned}
\Phi_\epsilon(u_t) = \Phi_\epsilon(h) &- \kappa \int_0^t (k_\alpha''(u_s), |\partial u_s|^2)ds \\
&+ \int_0^t (k_\epsilon'(u_s), f_s(u) - \alpha u_s + g_s)ds \\
&+ \frac{1}{2}\int_0^t (k_\epsilon''(u_s), \tilde{q}_s(u))ds + \int_0^t (k_\epsilon'(u_s), \tilde{M}(u_s, ds)). \qquad \square
\end{aligned}
\tag{4.54}
$$

Theorem 5.3   Suppose that the conditions for Theorem 3.4 hold. In addition, assume the following conditions:

(1) $h(x) \geq 0$ and $g(x, t) \geq 0$, a.s., for almost every $x \in \mathbf{R}^d$ and $t \in [0, T]$.

(2) There exists a constant $c_1 \geq 0$ such that

$$\frac{1}{2}\tilde{q}(s, \xi, x; t) - \kappa|\xi|^2 \leq c_1 s^2,$$

for all $s \in \mathbf{R}$, $\xi, x, \in \mathbf{R}^d$, $t \in [0, T]$.

(3) For any $v \in H^1$ and $t \in [0, T]$,

$$(\eta(v), f(v, \partial v, \cdot, t)) \geq 0,$$

where we set $[\eta(v)](x) = \eta[v(x)]$ and $\eta(r)$ is given by (4.49).

Then the solution of the semilinear parabolic equation (4.48) remains positive so that $u(x, t) \geq 0$, a.s., for almost every $x \in \mathbf{R}^d$, $\forall t \in [0, T]$.

Proof.   From Lemma 5.2, after taking an expectation of the equation (4.53), it can be written as

$$E\,\Phi_\epsilon(u_t) = \Phi_\epsilon(h) + \int_0^t E\left(k_\epsilon''(u_s), \frac{1}{2}\tilde{q}_s(u) - \kappa|\partial u_s|^2\right)ds$$
$$+ \int_0^t E\left(k_\epsilon'(u_s), f_s(u) - \alpha u_s + g_s\right)ds.$$

By making use of conditions (1) and (2), and, in view of (4.50), the properties: $k_\epsilon'(r) = 0$ for $r \geq 0$; $k_\epsilon'(r) \leq 0$ and $k_\epsilon''(r) \geq 0$ for any $r \in \mathbf{R}$, the above equation yields

$$E\,\Phi_\epsilon(u_t) \leq c_1 \int_0^t E\left(k_\epsilon''(u_s), |u_s|^2\right)$$
$$+ \int_0^t E\left(k_\epsilon'(u_s), f_s(u_s) - \alpha u_s\right)ds. \tag{4.55}$$

Referring to (4.50) and (4.51), it is clear that, for each $r \in \mathbf{R}$, we have $k_\epsilon(r) \to k(r) = \eta^2(r)$, $k_\epsilon'(r) \to -2\eta(r) \leq 0$, as $\epsilon \to 0$. Also notice that $0 \leq k_\epsilon''(r) \leq 2\theta(r)$ for any $\epsilon > 0$, where $\theta(r) = 1$ for $r < 0$, and $\theta(r) = 0$ for $r \geq 0$. By taking the limits in equation (4.55) as $\epsilon \to 0$, we get

$$E\,\Phi(u_t) = E\,\|\eta(u_t)\|^2$$
$$\leq 2c_1 \int_0^t E\left(\theta(u_s), |u_s|^2\right)ds - 2\int_0^t E\left(\eta(u_s), f_s(u_s)\right)ds$$
$$= 2\,c_1 \int_0^t E\,\|\eta(u_s)\|^2 ds - 2\int_0^t E\left(\eta(u_s), f_s(u_s)\right)ds.$$

By making use of condition (3), the above yields

$$E\,\|\eta(u_t)\|^2 \leq 2\,c_1 \int_0^t E\,\|\eta(u_s)\|^2 ds,$$

which, by means of the Gronwall inequality, implies that

$$E \, \|\eta(u_t)\|^2 = E \int \eta^2[u(x,t)]dx = 0, \text{ for every } t \in [0,T].$$

It follows that $\eta[u(x,t)] = u^-(x,t) = 0$, a.s. for a.e. $x \in \mathbf{R}$, $t \in [0,T]$. This proves the positivity of the solution as claimed. □

Remarks: It should be pointed out that this approach based on the Itô formula applied to a regularized comparison function k_ϵ was introduced by Viot in [82] for a special semilinear parabolic Itô equation. Theorem 5.2 is a generalization of that positivity result to a larger class of stochastic parabolic equations. Notice that condition (2) constitutes a stronger form of coercivity condition, and condition (3) holds if the function f is positive. As mentioned before, the theorem holds for multidimensional noise as well and this can be shown by an obvious modification in the proof. For instance, suppose that

$$\sigma(\cdots,t) \cdot \dot{W}(\cdot,t) = \sum_{i=1}^{d} \sigma_i(\cdots,t)\dot{W}_i(\cdot,t)$$

where $W_i(\cdot,t)$ are Wiener random fields with covariance matrix $[r_{ij}(x,y)]_{d \times d}$. If, in condition (2), we define

$$\tilde{q}(s,\xi,x;t) = \sum_{i,j=1}^{d} r_{ij}(x,x)\sigma_i(s,\xi,x,t)\sigma_j(s,\xi,x,t), \qquad (4.56)$$

then the corresponding solution is positive.

As an example, consider the turbulent diffusion model equation with chemical reaction:

$$\frac{\partial u}{\partial t} = \kappa \Delta u - \alpha u - \beta u^3 + g(x,t) - \sum_{i=1}^{d} \frac{\partial u}{\partial x_i}\sigma_i(x,t)\dot{W}_i(x,t), \qquad (4.57)$$

$$u(x,0) = h(x),$$

where α, β and κ are positive constants. Referring to equation (4.57), suppose that there is constant $q_0 > 0$ such that

$$\tilde{q}(\xi,x;t) = \sum_{i,j=1}^{d} r_{ij}(x,x)\sigma_i(x,t)\sigma_j(x,t)\xi_i\xi_j \le q_0|\xi|^2, \qquad (4.58)$$

which is independent of s for all $\xi, x \in \mathbf{R}$, $t \in [0,T]$.

Corollary 5.4 Let $h(x) \ge 0$ and $g(x,t) \ge 0$, a.e. $x \in \mathbf{R}^d$ and $t \in [0,T]$. If the condition (4.58) holds for $q_0 \le 2\kappa$, then the solution of (4.57) remains positive a.s. for any $t \in [0,T]$, a.e. $x \in \mathbf{R}^d$.

Proof.   By the assumptions and (4.58), in Theorem 5.3, condition (1) is met, and

$$\frac{1}{2}\tilde{q}(\xi, x, t) - \kappa|\xi|^2 \le \frac{1}{2}(q_0 - 2\kappa)|\xi|^2 \le 0,$$

so that condition (2) is also satisfied with $c_1 = 0$. In addition, we have $f_t(v) = -\beta v^3$ so that

$$(\eta(v), f_t(v)) = -\beta(\eta(v), v^3) \ge 0.$$

Hence condition (3) is valid, and the conclusion follows from Theorem 5.2. $\square$

Physically this means that the turbulent diffusion model (4.57) is feasible if the coercivity condition holds. Then the concentration $u(x, t)$ will remain positive as it should.

4.6   Correlation Functions of Solutions

Let $u(\cdot, t)$ be a solution of a parabolic Itô equation. For $x^1, \cdots, x^m \in \mathbf{R}^d, t \in [0, T]$, a function $\Gamma_m(x^1, \cdots, x^m, t)$ is said to the m-th correlation function , or the correlation function of order m, for a solution u if $\Gamma_m(\cdots, t) \in H(\mathbf{R}^{d \times m})$ and, for any $\psi_1, \cdots, \psi_m \in H$,

$$E\left\{(\psi_1, u(\cdot, t)) \cdots (\psi_m, u(\cdot, t))\right\} = (\Gamma(\cdots, t), \psi_1 \otimes \cdots \otimes \psi_m)_m$$

$$= \int \cdots \int \Gamma_m(x^1, \cdots, x^m, t)\psi_1(x^1) \cdots \psi_m(x^m) dx^1 \cdots dx^m,$$

where $(\cdot, \cdot)_m$ denotes the inner product in $H(\mathbf{R}^{m \times d})$ and $\otimes$ means a tensor product. Similarly, the duality pairing between $H^1(\mathbf{R}^{m \times d})$ and $H^{-1}(\mathbf{R}^{m \times d})$ is denoted by $\langle \cdot, \cdot \rangle_m$. Clearly $\Gamma_m(x^1, \cdots, x^m, t)$ is symmetric in the space variables under permutations. We are interested in the possibility of deriving the correlation equations, or the partial differential equations governing $\Gamma_m{'}s$, that can be solved recursively.

In general this is impossible as can be seen from equation (4.48). Formally, if we take an expectation of this equation, it does not lead to an equation for Γ_1. From the computational viewpoint, this is a source of difficulty in dealing with nonlinear stochastic equations. However we will show that, for a certain type of linear parabolic Itô equations, if all correlations up to a certain order exist, then they satisfy a recursive system of partial differential equations. To be specific, consider the linear equation with a multiplicative noise in $\mathbf{R}^d$:

$$\frac{\partial u}{\partial t} = \Delta u + a(x, t)u + g(x, t) + u\,\sigma(x, t)\dot{W}(x, t),$$

$$u(x, 0) = h(x),$$

(4.59)

where the functions $a(x,t), \sigma(x,t)$ and $g(x,t)$ are assumed to be non-random.

Lemma 6.1 Suppose that $a, \sigma \in \mathbf{C}_b(\mathbf{R}^d \times [0,T]; \mathbf{R})$, $g \in L^\infty([0,T]; H)$, and $r_0 = \sup_{x \in \mathbf{R}^d} r(x,x) < \infty$. Then, given $h \in H$, the solution u of the equation (4.59) belongs to $L^p(\Omega \times (0,T); H) \cap L^2(\Omega \times (0,T); H^1)$ for any $p \geq 2$ such that

$$\sup_{0 \leq t \leq T} E \, \|u_t\|^p \leq C_p \{\|h\|^p + \int_0^T \|g_t\|^p dt\}, \tag{4.60}$$

for some constant $C_p > 0$. □

By the regularity assumptions on the data, we can apply Theorem 3.4 to conclude that the equation has a unique solution $u \in L^p(\Omega \times (0,T); H) \cap L^2(\Omega \times (0,T); H^1)$ for any $p \geq 2$. Since u is a strong solution, for any $\phi_k \in H^1$, $k = 1, 2, \cdots$,

$$(u(\cdot,t), \phi_1) = (h, \phi_1) + \int_0^t \{\langle [\Delta + a(\cdot,s)]u(\cdot,s) + g(\cdot,s), \phi_1 \rangle\} ds$$
$$+ \int_0^t (\phi_1, u(\cdot,s)\sigma(\cdot,s)W(\cdot,ds)).$$

By taking an expectation, and changing the order of integration and so on, the above equation yields the equation for the first-order correlation or the mean solution:

$$(\Gamma_1(\cdot,t), \phi_1) = (h, \phi_1) + \int_0^t \langle [\Delta + a(\cdot,s)]\Gamma_1(\cdot,s), \phi_1 \rangle ds$$
$$+ \int_0^t (g(\cdot,s), \phi_1) ds. \tag{4.61}$$

These operations can be justified and will be repeated in later computations without mention. Let

$$v_k(t) = v_k(0) + \int_0^t b_k(s) ds + z_k(t), \tag{4.62}$$

where

$$v_k(t) = (u(\cdot,t), \phi_k), \quad b_k(t) = \langle [\Delta + a(\cdot,s)]u(\cdot,s) + g(\cdot,s), \phi_k \rangle$$

and $z_k(t)$ is the martingale term:

$$z_k(t) = \int_0^t (\phi_k, u(\cdot,s)\sigma(\cdot,t)W(\cdot,ds)).$$

For any j, k, the local characteristic $q_{jk}(u,t)$ of $z_j(t)$ and $z_k(t)$ is given by

$$q_{jk}(u,t) = \int \int q(x,y;t)u(x,t)u(y,t)\phi_j(x)\phi_k(y)dxdy$$

with $q(x, y; t) = r(x, y)\sigma(x, t)\sigma(y, t)$.

Let $F(v_1(t), v_2(t)) = v_1(t)v_2(t)$. In view of (4.62), by the Itô formula (4.17), we have

$$[v_1(t)v_2(t)] = [v_1(0)v_2(0)] + \int_0^t \{b_1(s)v_2(s) + b_2(s)v_1(s)\}ds$$
$$+ \int_0^t \{v_2(s)dz_1(s) + v_1(s)dz_2(s)\} + \int_0^t q_{12}(u, s)ds.$$

In the above equation, after taking an expectation and recalling the definitions of $v_j(t), b_k(t)$ and so on, we can obtain the equation for the second-order correlation equation in the form:

$$(\Gamma_t^2, \phi_1 \otimes \phi_2)_2 = (\Gamma_0^2, \phi_1 \otimes \phi_2)_2$$
$$+ \int_0^t \langle [\triangle \otimes I + I \otimes \triangle]\Gamma_s^2, \phi_1 \otimes \phi_2\rangle_2 ds$$
$$+ \int_0^t ([a_s \otimes I + I \otimes a_s + q(s)I]\Gamma_s^2, \phi_1 \otimes \phi_2)_2 ds \qquad (4.63)$$
$$+ \int_0^t (g_s \otimes \Gamma_s^1 + \Gamma_s^1 \otimes g_s, \phi_1 \otimes \phi_2)_2 ds,$$

where we write $\Gamma_m(\cdots, t)$ as Γ_t^m and I is an identity operator. Explicitly the equation (4.63) reads

$$\int\int \Gamma_2(x^1, x^2, t)\phi_1(x^1)\phi_2(x^2)dx^1 dx^2$$
$$= \int\int \Gamma_2(x^1, x^2, 0), \phi_1(x^1)\phi_2(x^2)dx^1 dx^2$$
$$+ \int_0^t \int\int (\triangle_1 + \triangle_2)\Gamma_2(x^1, x^2, s), \phi_1(x^1)\phi_2(x^2)dx^1 dx^2 ds$$
$$+ \int_0^t \int\int [a(x^1, s) + a(x^2, s) + q(x^1, x^2, s)]$$
$$\times \Gamma_2(x^1, x^2, s)\phi_1(x^1)\phi_2(x^2)dx^1 dx^2 ds$$
$$+ \int_0^t \int\int [g(x^1, s)\Gamma_1(x^2, s) + g(x^2, s)\Gamma_1(x^1, s)]\phi_1(x^1)\phi_2(x^2)dx^1 dx^2 ds,$$

which yields a parabolic equation in $\mathbf{R}^{2d}$ in a distributional sense:

$$\frac{\partial \Gamma_2(x^1, x^2, t)}{\partial t} = (\triangle_1 + \triangle_2)\Gamma_2 + [a(x^1, t) + a(x^2, t) + q(x^1, x^2, t)]\Gamma_2$$
$$+ g(x^1, t)\Gamma_1(x^2, t) + g(x^2, t)\Gamma_1(x^1, t), \qquad (4.64)$$
$$\Gamma_2(x^1, x^2, 0) = h(x^1)h(x^2),$$

where $\triangle_j$ means the Laplacian in variable x^j.

In general let us consider the m-th order correlation $\Gamma_m(x^1, \cdots, x^m, t)$, for $m \geq 1$. We let

$$F(v_1(t), \cdots, v_m(t)) = v_1(t) \cdots v_m(t).$$

As in the case of ($m = 2$), by applying the Itô formula and taking an expectation, we can obtain the m-th correlation equation as follows:

$$
\begin{aligned}
\frac{\partial \Gamma_m(x^1, \cdots, x^m, t)}{\partial t} &= (\sum_{j=1}^{m} \triangle_j)\Gamma_m + \sum_{j,k=1, j\neq k}^{m} [a(x^j, t) + a(x^k, t) \\
&+ q(x^j, x^k, t)]\Gamma_m(x^1, \cdots, x^j, \cdots, x^k, \cdots, x^m, t) \\
&+ \sum_{j,k=1, j\neq k}^{m} g(x^j, t)\Gamma_{m-1}(x^1, \cdots, \check{x}^j, \cdots, x^k, \cdots, x^m, t),
\end{aligned}
\tag{4.65}
$$

$$\Gamma_m(x^1, \cdots, x^m, 0) = h(x^1) \cdots h(x^m),$$

where $\check{x}^j$ in the argument of a function means the variable x^j is deleted. Notice that the sum $\sum_{j=1}^{m} \triangle_j$ is just the Laplacian in $\mathbf{R}^{m\times d}$. Therefore, for any $N \geq 2$, the Nth moment Γ_N can be determined by solving the parabolic equation (4.65) recursively with $m = 1, \cdots, N$. Since a linear deterministic parabolic equation is a special case of the stochastic counterpart, we can apply the existence theorem (Theorem 3.4) to show the existence of solutions to the correlation equations.

Theorem 6.2 Suppose that $a, \sigma \in \mathbf{C}_b(\mathbf{R}^d \times [0, T]; \mathbf{R})$, $g \in L^\infty([0, T]; H)$, and $r_0 = \sup_{x \in \mathbf{R}^d} r(x, x) < \infty$. Then, given $h \in H$, there exists a unique solution $\Gamma_m(\cdots, t)$ of the correlation equation (4.65) such that, for any $m \geq 1$,

$$\Gamma_m \in \mathbf{C}([0, T]; H(\mathbf{R}^{m\times d})) \cap L^2((0, T); H_1(\mathbf{R}^{m\times d})).$$

Proof. The theorem can be easily proved by induction. For $m = 1$, since $h \in H(\mathbf{R}^d)$ and $g \in L^2((0, T); H)$, by Theorem 3.4, there exists a unique solution $\Gamma_1 \in \mathbf{C}([0, T]; H(\mathbf{R}^d)) \cap L^2((0, T); H^1(\mathbf{R}^d))$.

For $m = 2$, it is easy to see that $h \otimes h \in H(\mathbf{R}^{2d})$ and $[g(\cdot, t) \otimes \Gamma_1(\cdot, t) + \Gamma_1(\cdot, t) \otimes f(\cdot, t)] \in L^2((0, T); H(\mathbf{R}^{2d}))$, since

$$\int_0^T \|g(\cdot, t) \otimes \Gamma_1(\cdot, t) + \Gamma_1(\cdot, t) \otimes g(\cdot, t)\|^2 dt \leq 2 \int_0^T \|g(\cdot, t) \otimes \Gamma_1(\cdot, t)\|^2 dt$$

$$= 2 \int_0^T \int \int [g(x^1, t)\Gamma_1(x^2, t)]^2 dx^1 dx^2 dt = 2 \int_0^T \|g(\cdot, t)\|^2 \|\Gamma_1(\cdot, t)\|^2 dt$$

$$\leq 2 \sup_{0 \leq t \leq T} \|g(\cdot, t)\|^2 \int_0^T \|\Gamma_1(\cdot, t)\|^2 dt.$$

Now suppose that the theorem holds for $(m-1)$ such that

$$\Gamma_{m-1} \in \mathbf{C}([0, T]; H(\mathbf{R}^{(m-1)\times d})) \cap L^2((0, T); H(\mathbf{R}^{(m-1)\times d})).$$

Then $\prod_{j=1}^{m} h(x^j)$ belongs to $H(\mathbf{R}^{m \times d})$ and, similar to the case $m = 2$,

$$\int_0^T \int \cdots \int | \sum_{j,k=1,j\neq k}^{m} g(x^j, t)\Gamma_{m-1}(x^1, \cdots, \check{x}^j, \cdots, x^k, \cdots, x^m, t)|^2 dt \, dx^1 \cdots dx^r$$

$$\leq C_m \int_0^T \int \cdots \int |\Gamma_{m-1}(x^1, \cdots, x^{m-1}, t)g(x^m, t)|^2 dt \, dx^1 \cdots dx^m$$

$$= C_m \int_0^T \|\Gamma_{m-1}(\cdots, t)\|^2 \|g(\cdot, t)\|^2 dt$$

$$\leq C_m \sup_{0 \leq t \leq T} \|g(\cdot, t)\|^2 \int_0^T \|\Gamma_{m-1}(\cdots, t)\|^2 dt,$$

for some constant $C_m > 0$. It follows that the equation (4.65) has a unique solution $\Gamma_m \in \mathbf{C}([0,T]; H(\mathbf{R}^{m \times d})) \cap L^2((0,T); H(\mathbf{R}^{m \times d}))$. Therefore, by the induction principle, this is true for any $m \geq 1$.   □

Remarks

(1) Clearly, if, instead of $\mathbf{R}^d$, the domain D is bounded, the corresponding moment functions for the solution are also given by (4.65) subject to suitable homogeneous boundary conditions.

(2) As it stands, the moment functions may not be continuous. According to the theory of parabolic equations [29], if the data h, a, g and q are sufficiently smooth, the correlation equations will have classical solutions that satisfy the parabolic equations point-wise. If the m-th moment $E[u(x,t)]^m$ exists at x, then it equals to $\Gamma_m(x, \cdots, x, t)$.

(3) The correlation equations can be derived for other types of linear stochastic parabolic equations. An interesting example will be provided in what follows.

Consider the following turbulent diffusion model equation (compared with (4.57)) in $\mathbf{R}^d$:

$$\frac{\partial u}{\partial t} = \Delta u + \sum_{i=1}^{d} \frac{\partial u}{\partial x_i} \dot{V}_i(x, t) + g(x, t), \qquad (4.66)$$

$$u(x, 0) = h(x),$$

where the random velocity V_i is of the form

$$V_i(x, t) = \int_0^t v_i(x, s)ds + \int_0^t \sigma_i(x, s)W_i(x, ds),$$

in which $v_i(x,t)$ is the ith component of the mean velocity, and $\sigma_i(x,t)$, $W_i(x,t)$ are given as before. Suppose that v_i and σ_i are bounded and continuous on

$[0, T] \times \mathbf{R}^d$, and the covariance functions r_{ij} for the Wiener fields W_i and W_j are bounded. Then the correlation function Γ_m for the solution can be derived similarly for any m. For instance, the first- and the second-order correlation equations are given by

$$\frac{\partial \Gamma_1(x,t)}{\partial t} = \Delta \Gamma_1 + \sum_{i=1}^{d} v_i(x,t) \frac{\partial \Gamma_1}{\partial x_i} + g(x,t),$$

$$\Gamma_1(x,0) = h(x), \tag{4.67}$$

and

$$\frac{\partial \Gamma_2(x^1, x^2, t)}{\partial t} = (\Delta_1 + \Delta_2)\Gamma_2 + \sum_{i,j=1}^{d} q_{ij}(x^1, x^2, t) \frac{\partial^2 \Gamma_2}{\partial x_i^1 \partial x_j^2}$$

$$+ \sum_{i,j=1}^{d} [v_i(x^1, t) \frac{\partial \Gamma_2}{\partial x_j^2} + v_j(x^1, t) \frac{\partial \Gamma_2}{\partial x_i^1}] \tag{4.68}$$

$$+ [g(x^1, t)\Gamma_1(x^2, t) + g(x^2, t)\Gamma_1(x^2, t)],$$

$$\Gamma_2(x^1, x^2, 0) = h(x^1)h(x^2),$$

where $q_{ij}(x^1, x^2, t) = r_{ij}(x^1, x^2)\sigma_i(x^1, t)\sigma_j(x^2, t)$. The equations for higher-order correlations functions can also be derived but will be omitted here. Notice the extra diffusion and drift terms appear in (4.68) to account for the effects of the turbulent motion of the fluid.

Chapter 5

Stochastic Hyperbolic Equations

5.1 Introduction

Wave motion and mechanical vibration are two of the most commonly observed physical phenomena. As mathematical models, they are usually described by partial differential equations of hyperbolic type. The most well-known one is the wave equation. In contrast with the heat equation, the first-order time derivative term is replaced by a second-order one. Excited by a white noise, a stochastic wave equation takes the form

$$\frac{\partial^2 u}{\partial t^2} = c^2 \Delta u + \sigma_0 \dot{W}(x, t)$$

in some domain D, where c is known as the wave speed and σ_0 is the noise intensity parameter. If the domain D is bounded in $\mathbf{R}^d$, the equation can describe the random vibration of an elastic string for $d = 1$, an elastic membrane for $d = 2$ and a rubbery solid for $d = 3$. When $D = \mathbf{R}^d$, it may be used to model the propagation of acoustic or optical waves generated by a random excitation. For a large-amplitude vibration or wave motion, some nonlinear effects must be taken into consideration. In this case we should include appropriate nonlinear terms in the wave equation. For instance, it may look like

$$\frac{\partial^2 u}{\partial t^2} = c^2 \Delta u + f(u) + \sigma(u) \dot{W}(x, t),$$

where $f(u)$ and $\sigma(u)$ are some nonlinear functions of u. More generally, the wave motion may be modelled by a linear or nonlinear first-order hyperbolic system. A special class of stochastic hyperbolic system will be discussed later on.

In this chapter, we briefly review some basic definitions of hyperbolic equations in Section 5.2. In the following two sections, we study the existence of solutions to some linear and semilinear stochastic wave equations in a bounded domain. As in the parabolic case, we shall employ the Green's function approach based on the method of eigenfunctions expansion. Then, in Section 5.5, this approach coupled with the method of Fourier transform is used to analyze stochastic wave equations in an unbounded domain. Finally, in Sec-

tion 5.6, a similar approach will be adopted to study the solutions of a class
of linear and semilinear stochastic hyperbolic systems.

5.2   Preliminaries

Consider a second-order linear partial differential equation with constant
coefficients:

$$\frac{\partial^2 u}{\partial t^2} = \sum_{j,k=1}^{d} a_{jk} \frac{\partial^2 u}{\partial x_j \partial x_k} u + \sum_{j=1}^{d} b_j \frac{\partial u}{\partial x_j} u + \gamma u + f(x,t), \qquad (5.1)$$

where a_{jk}, b_j, c are constants, and $g(x,t)$ is a given function. Let $A = [a_{jk}]$ be
a $d \times d$ coefficient matrix with entries a_{jk}. According to the standard classification, equation (5.1) is said to be *strongly hyperbolic* if all of the eigenvalues
of A are real, distinct and nonzero. In this case, by a linear transformation of
the coordinates, the equation can be reduced, without changing notation for
u, to a perturbed wave equation

$$\frac{\partial^2 u}{\partial t^2} = c^2 \Delta u + \sum_{j=1}^{d} \tilde{b}_j \frac{\partial u}{\partial x_j} - \alpha u + \tilde{f}(x,t),$$

where $c, \tilde{b}_j, \alpha$ are some new constants, and $\tilde{f}$ is a known function.

Similarly we consider a system of n linear first-order equations:

$$\frac{\partial u_i}{\partial t} = \sum_{j=1}^{n} \sum_{k=1}^{d} a_{ij}^{k} \frac{\partial u_j}{\partial x_k} + \sum_{j=1}^{n} b_{ij} u_j + g_i(x,t), \qquad (5.2)$$

for $i = 1, \cdots, n$. For each k, let $A^k = [a_{ij}^k]$ be the $n \times n$ matrix with entries
a_{ij}^k. Given $\xi \in \mathbf{R}^d$, we form the $n \times n$ matrix $A(\xi) = \sum_{k=1}^{d} \xi_k A^k$. For
$|\xi| = 1$, if all of the eigenvalues of $A(\xi)$ are real, then the system is said to
be *hyperbolic*. The system is *strongly hyperbolic* if the eigenvalues are real,
distinct and nonzero. It is well known that the wave equation can be written
as a hyperbolic system. For the stochastic counterpart, the classification is
similar.

5.3   Wave Equation with Additive Noise

As in Chapter Three, let D be a bounded domain in $\mathbf{R}^d$ with a smooth
boundary ∂D. We first consider the initial-boundary value problem for a

randomly perturbed wave equation in D:

$$\frac{\partial^2 u}{\partial t^2} = (\kappa \Delta - \alpha)u + f(x,t) + \dot{M}(x,t), \quad x \in D, \, t \in (0,T),$$

$$Bu|_{\partial D} = 0, \tag{5.3}$$

$$u(x,0) = g(x), \quad \frac{\partial u}{\partial t}(x,0) = h(x),$$

where $\kappa = c^2$ and α are positive constants; $g(x)$, $h(x)$ are given functions, and, as introduced before, $\dot{M}(x,t)$ is the formal time-derivative of the martingale $M(x,t)$ given by the Itô integral:

$$M(x,t) = \int_0^t \sigma(x,s)W(x,ds). \tag{5.4}$$

Similar to the case of a parabolic equation in Sections 3.3 and 3.4, for $g, h \in H$, if $f(\cdot,t)$ and $\sigma(\cdot,t)$ are predictable processes in H such that

$$E \int_0^T \|f(\cdot,s)\|^2 ds < \infty, \tag{5.5}$$

and

$$E \int_0^T \|\sigma(\cdot,s)\|_R^2 ds \le r_0 \, E \int_0^T \|\sigma(\cdot,s)\|^2 ds < \infty, \tag{5.6}$$

where we recall that $\|\sigma(\cdot,s)\|_R^2 = Tr \, Q_s = \int_D q(x,x,s)dx$, with $q(x,y,s) = r(x,y)\sigma(x,s)\sigma(y,s)$. Then the problem (5.3) can be solved by the method of eigenfunctions expansions. To this end, let λ_k and $e_k(x)$ denote the associated eigenvalues and the eigenfunctions of $(-A) = (-\kappa\Delta + \alpha)$, respectively, for $k = 1, 2, \cdots$. Let

$$u(x,t) = \sum_{k=1}^{\infty} u_t^k e_k(x), \tag{5.7}$$

with $u_t^k = (u(\cdot,t), e_k)$. By expanding the given functions in (5.3) in terms of eigenfunctions, we obtain

$$\frac{d^2 u_t^k}{dt^2} = -\lambda_k u_t^k + f_t^k + \dot{z}_t^k,$$

$$u_0^k = g_k, \quad \frac{d}{dt}u_0^k = h_k, \quad k = 1, 2, \cdots, \tag{5.8}$$

where $f_t^k = (f(\cdot,t), e_k), g_k = (g, e_k), h_k = (h, e_k)$ and $\dot{z}_t^k = \frac{d}{dt}z_t^k$. Recall that $z_t^k = (M(\cdot,t), e_k)$ is a martingale with local characteristic $q_t^k = (Q_t e_k, e_k)$. The equation (5.8), sometimes called a second-order Itô equation, is to be interpreted as the system:

$$du_t^k = v_t^k dt,$$

$$dv_t^k = -\lambda_k u_t^k dt + f_t^k dt + dz_t^k, \tag{5.9}$$

$$u_0^k = g_k, \quad v_0^k = h_k,$$

which can be easily solved to give the solution

$$u_t^k = g_k \cos \sqrt{\lambda_k} t + h_k \frac{\sin \sqrt{\lambda_k} t}{\sqrt{\lambda_k}}$$
$$+ \int_0^t \frac{\sin \sqrt{\lambda_k}(t-s)}{\sqrt{\lambda_k}} f_s^k ds + \int_0^t \frac{\sin \sqrt{\lambda_k}(t-s)}{\sqrt{\lambda_k}} dz_s^k,$$

$$v_t^k = -g_k \sqrt{\lambda_k} \sin \sqrt{\lambda_k} t + h_k \cos \sqrt{\lambda_k} t$$
$$+ \int_0^t \cos \sqrt{\lambda_k}(t-s) f_s^k ds + \int_0^t \cos \sqrt{\lambda_k}(t-s) dz_s^k. \tag{5.10}$$

In view of (5.7) and (5.10), the formal solution of equation (5.3) can be written as

$$u(x,t) = \int_D K'(x,y,t)g(y)dy + \int_D K(x,y,t)h(y)dy$$
$$+ \int_0^t \int_D K(x,y,t-s)f(y,s)dsdy + \int_0^t \int_D K(x,y,t-s)M(y,ds)dy, \tag{5.11}$$

where $K'(x,t) = \frac{\partial}{\partial t} K(x,t)$ and $K(x,t)$ is the Green's function for the wave equation given by

$$K(x,y,t) = \sum_{k=1}^{\infty} \frac{\sin \sqrt{\lambda_k} t}{\sqrt{\lambda_k}} e_k(x) e_k(y). \tag{5.12}$$

In contrast with the heat equation, it can be shown that, for $d \geq 2$, the above series does not converge in $H \times H$. In general, $K(\cdot,\cdot,t)$ is a generalized function or a distribution.

Let G_t denote the Green's operator with kernel $K(x,y,t)$ so that, for any smooth function g,

$$(G_t g)(x) = \int_D K(x,y,t)g(y)dy,$$

and set $G_t' = \frac{d}{dt} G_t$, which is the derived Green's operator with kernel

$$K'(x,y,t) = \sum_{k=1}^{\infty} (\cos \sqrt{\lambda_k} t) e_k(x) e_k(y). \tag{5.13}$$

Then the equation (5.11) can be written simply as

$$u_t = G_t' g + G_t h + \int_0^t G_{t-s} f_s ds + \int_0^t G_{t-s} dM_s. \tag{5.14}$$

In the following lemma, for $k = 0,1,2$, we let $G_t^{(k)} = \frac{d^k}{dt^k} G_t$ with $G_t^{(0)} = G_t$, where the kernel of $G_t^{(k)}$ is given by $K^{(k)}(x,y,t) = \frac{\partial^k}{\partial t^k} K(x,y,t)$. As in Chapter Three, for the Sobelev space $H^m = H^m(D)$, we shall use the equivalent norm $\|g\|_m = \{\sum_{j=1}^{\infty} \lambda_j^m (g, e_j)^2\}^{1/2}$ for $g \in H^m$.

Lemma 3.1 For any function $g \in H^1$ and for nonnegative integers k and m, the following estimates hold:

$$\sup_{0 \le t \le T} \|G_t^{(k)} g\|_m^2 \le \|g\|_{k+m-1}^2, \quad \text{for} \quad 0 \le k + m \le 2. \tag{5.15}$$

Proof. The estimates can be verified by using the series representation (5.12) of the Green's function. We will show that (5.15) holds for $k = m = 1$ so that

$$\sup_{0 \le t \le T} \|G_t' g\|_1^2 \le \|g\|_1^2.$$

In view of (5.13), the above estimate can be easily verified as follows:

$$\sup_{0 \le t \le T} \|G_t' g\|_1^2 = \sup_{0 \le t \le T} \sum_{k=1}^{\infty} \lambda_k (\cos^2 \sqrt{\lambda_k} t)(g, e_k)^2$$
$$\le \sum_{k=1}^{\infty} \lambda_k (g, e_k)^2 = \|g\|_1^2.$$

The other cases can be shown similarly. □

Lemma 3.2 Let $f(\cdot, t) \in L^2(\Omega \times (0, T); H)$ satisfy (5.5). Then $\eta(\cdot, t) = \int_0^t G_{t-s} f(\cdot, s) ds$ is a continuous, adapted H^1-valued process and its time-derivative $\eta'(\cdot, t)$ is a continuous H-valued process such that

$$E \sup_{0 \le t \le T} \|\eta(\cdot, t)\|_k^2 \le TE \int_0^T \|f(\cdot, s)\|_{k-1}^2 ds, \quad \text{for} \quad k = 0, 1, \tag{5.16}$$

and

$$E \sup_{0 \le t \le T} \|\eta'(\cdot, t)\|^2 \le TE \int_0^T \|f(\cdot, s)\|^2 ds. \tag{5.17}$$

Proof. We will verify only the first inequality for $k = 1$. Clearly,

$$\|\eta_t\|_1^2 = \|\int_0^t G_{t-s} f_s ds\|_1^2 \le [\int_0^t \|G_{t-s} f_s\|_1 ds]^2$$
$$\le T \int_0^t \|G_{t-s} f_s\|_1^2 ds,$$

so that, by Lemma 3.1,

$$E \sup_{0 \le t \le T} \|\eta_s\|_1^2 \le TE \int_0^T \|f_s\|^2 ds.$$

The continuity of $\eta(\cdot, t)$ and $\eta'(\cdot, t)$ will not be shown here.   □

Lemma 3.3   Let

$$\mu(\cdot, t) = \int_0^t G_{t-s}M(\cdot, ds) = \int_0^t G_{t-s}\sigma(\cdot, s)W(\cdot, ds), \qquad (5.18)$$

which satisfies the condition (5.6). Then $\mu(\cdot, t)$ is a continuous adapted H^1-valued process in [0,T], and its derivative $\mu'(\cdot, t) = \frac{\partial}{\partial t}\mu(\cdot, t)$ is continuous in H. Moreover, for $k, m = 0, 1$, the following inequalities hold:

$$E \left\| \mu^{(k)}(\cdot, t) \right\|_m^2 \leq C_{k,m} E \int_0^T \|\sigma(\cdot, t)\|_R^2 dt \qquad (5.19)$$

and

$$E \sup_{0 \leq t \leq T} \left\| \mu^{(k)}(\cdot, t) \right\|_m^2 \leq D_{k,m} E \int_0^T \|\sigma(\cdot, t)\|_R^2 dt, \qquad (5.20)$$

with $0 \leq k + m \leq 1$, where $C_{0,0} = 1/\lambda_1, C_{0,1} = C_{1,0} = 1$, and $D_{0,0} = 8/\lambda_1, D_{0,1} = D_{1,0} = 8$.

Proof.   We will first check the inequality (5.19) when $k = m = 0$. Similar to the proof of Lemma 3-5.2, let $\mu^n(\cdot, t)$ be the n-term approximation of $\mu(\cdot, t)$ given by

$$\mu^n(x, t) = \sum_{k=1}^n \mu_t^k e_k(x), \qquad (5.21)$$

where

$$\mu_t^k = (\mu(\cdot, t), e_k) = \int_0^t \frac{\sin\sqrt{\lambda_k}(t-s)}{\sqrt{\lambda_k}} dz_s^k. \qquad (5.22)$$

Since

$$E \left\| \mu(x, t) - \mu^n(x, t) \right\|^2 = E \sum_{k>n} \int_0^t \frac{\sin^2\sqrt{\lambda_k}(t-s)}{\lambda_k} q_s^k ds$$

$$\leq \frac{1}{\lambda_1} E \sum_{k>n} \int_0^t q_s^k ds = \frac{1}{\lambda_1} E \sum_{k>n} \int_0^t (Q_s e_j, e_j) ds,$$

it is easily seen that, by condition (5.6), $\lim_{n\to\infty} E \left\| \mu(x, t) - \mu^n(x, t) \right\|^2 = 0$ uniformly in $t \in [0, T]$. Therefore we have

$$E \left\| \mu(\cdot, t) \right\|^2 = \lim_{n\to\infty} E \left\| \mu^n(\cdot, t) \right\|^2 = \lim_{n\to\infty} E \sum_{k=1}^n \int_0^t \frac{\sin^2\sqrt{\lambda_k}(t-s)}{\lambda_k} q_s^k ds$$

$$\leq \frac{1}{\lambda_1} E \sum_{k=1}^\infty \int_0^t q_s^k ds = \frac{1}{\lambda_1} E \int_0^t \|\sigma(\cdot, s)\|_R^2 ds,$$

which verifies (5.19) with $k = m = 0$.

Next, we will verify the inequality (5.20) with $k = m = 0$. It follows from (5.22) that

$$|\mu_t^k|^2 \leq \frac{2}{\lambda_k}\{|\int_0^t \cos\sqrt{\lambda_k}s)dz_s^k|^2 + |\int_0^t \sin\sqrt{\lambda_k}s)dz_s^k|^2\},$$

which, by invoking a Doob's inequality, yields

$$\begin{aligned}
E\sup_{0\leq t\leq T}|\mu_t^k|^2 &\leq \frac{8}{\lambda_k}E\{\int_0^T \cos^2\sqrt{\lambda_k}s)q_s^k ds \\
&+ \int_0^T \sin^2\sqrt{\lambda_k}s)q_s^k ds\} = \frac{8}{\lambda_k}\int_0^T q_s^k ds.
\end{aligned} \tag{5.23}$$

From the inequality (5.23), (5.21) and condition (5.6), we can deduce that

$$E\sup_{0\leq t\leq T}\|\mu(\cdot,t) - \mu^n(\cdot,t)\|^2 \leq \frac{8}{\lambda_1}E\sum_{k>n}\int_0^T (Q_s e_k, e_k)ds \to 0$$

as $n \to \infty$, and

$$\begin{aligned}
E\sup_{0\leq t\leq T}\|\mu(\cdot,t)\|^2 &= \lim_{n\to\infty}E\sup_{0\leq t\leq T}\|\mu^n(\cdot,t)\|^2 \\
&\leq \frac{8}{\lambda_1}E\sum_{k=1}^\infty\int_0^T (Q_s e_k, e_k)ds = \frac{8}{\lambda_1}E\int_0^T \|\sigma(\cdot,s)\|_R^2 ds.
\end{aligned}$$

Therefore we can conclude that the inequality (5.20) holds. Since $\mu^n(\cdot,t)$ is a continuous adapted process in H^1, so is $\mu(\cdot,t)$. The inequalities in (5.20) with $k + m \neq 0$ can be verified similarly. □

In view of the above lemmas, it can be shown that the formal solution (5.14) constructed by the eigenfunctions expansion is a weak solution. Here a test function $\phi(x,t)$ is taken to be a C^∞–function on $D \times [0,T]$ with $\phi(x,T) = \frac{\partial}{\partial t}\phi(x,T) = 0$ and $\phi(\cdot,t)$ has a compact support in D for each $t \in [0,T]$. Then we say that u is a *weak solution* of (5.3) if $u(\cdot,t)$ is a continuous adapted process in H and, for any test function ϕ, the following equation holds:

$$\begin{aligned}
\int_0^T (u(\cdot,t), \phi''(\cdot,t))dt &= (u(\cdot,0), \phi'(\cdot,0)) - (u'(\cdot,0), \phi(\cdot,0)) \\
&+ \int_0^T (u(\cdot,t), A\phi(\cdot,t))dt + \int_0^T (f(\cdot,t), \phi(\cdot,t))dt \\
&+ \int_0^T (\phi(\cdot,t), M(\cdot,dt)),
\end{aligned} \tag{5.24}$$

in which the prime denotes the time derivative.

Theorem 3.4   Suppose that the conditions in Lemma 3.1 and 3.2 hold true. Then, for $g, h \in H$, $u(\cdot, t)$ defined by the equation (5.14) is the unique weak solution of the problem (5.3) with $u \in L^2(\Omega; \mathbf{C}([0, T]; H))$. Moreover it satisfies the following inequality:

$$
\begin{aligned}
E \sup_{0 \le t \le T} \|u(\cdot, t)\|^2 \le C(T) \{ \|g\|^2 + \|h\|^2 \\
+ E \int_0^T [\, \|f(\cdot, s)\|^2 + \|\sigma(\cdot, s)\|_R^2 \,] ds \,\},
\end{aligned}
\tag{5.25}
$$

where $C(T)$ is a positive constant depending on T.

Proof.   For convenience, let us rewrite the equation (5.14) as

$$
u(\cdot, t) = \tilde{u}(\cdot, t) + \mu(\cdot, t),
$$

where

$$
\tilde{u}(\cdot, t) = G'_t g + G_t h + \int_0^t G_{t-s} f_s ds
$$

and μ is given by (5.18). From Lemmas 3.1 and 3.2, it is easy to see that $\tilde{u}(\cdot, t)$ is a continuous adapted process in H. By a n-term approximation and then taking the limit, one can show that, for any test function ϕ, $\tilde{u}(\cdot, t)$ satisfies the equation

$$
\begin{aligned}
\int_0^T (\tilde{u}(\cdot, t), \phi''(\cdot, t)) dt = (g, \phi'(\cdot, 0)) - (h, \phi(\cdot, 0)) \\
+ \int_0^T (\tilde{u}(\cdot, t), A\phi(\cdot, t)) dt + \int_0^T (f(\cdot, t), \phi(\cdot, t)) dt.
\end{aligned}
\tag{5.26}
$$

According to Lemma 3.3, $\mu(\cdot, t)$ is a continuous process in H. By the linearity of the equation and Lemma 3.3, it remains to show that $\mu(\cdot, t)$ satisfies

$$
\int_0^T (\mu(\cdot, t), \phi''(\cdot, t)) dt = \int_0^T (\mu(\cdot, t), A\phi(\cdot, t)) dt + \int_0^T (\phi(\cdot, t), M(\cdot, dt)). \tag{5.27}
$$

To this end, let us consider the n-term approximation $\mu^n(\cdot, t)$ given by (5.21). In view of (5.22), the expansion coefficient μ_t^k satisfies the Itô equations

$$
d\mu_t^k = v_t^k dt, \qquad dv_t^k = -\lambda_k \mu_t^k dt + dz_t^k,
$$

with $\mu_0^k = v_0^k = 0$. Let $\phi_t^k = (\phi(\cdot, t), e_k)$. By applying an integration by parts and the Itô formula, we can obtain

$$
\int_0^T \mu_t^k \phi_t^{k\prime\prime} dt = -\lambda_k \int_0^T \mu_t^k \phi_t^k dt + \int_0^T \phi_t^k dz_t^k.
$$

Therefore

$$\int_0^T (\mu^n(\cdot,t), \phi''(\cdot,t))dt = \sum_{k=1}^n \int_0^T \mu_t^k \phi_t^{k''} \, dt$$

$$= \sum_{k=1}^n \{-\lambda_k \int_0^T \mu_t^k \phi_t^k \, dt + \int_0^T \phi_t^k \, dz_t^k\},$$

which can be rewritten as

$$\int_0^T (\mu^n(\cdot,t), \phi''(\cdot,t))dt = \int_0^T (\mu^n(t), A\phi(\cdot,t))dt$$

$$+ \int_0^T (\phi(\cdot,t), M^n(\cdot,dt)).$$

As $n \to \infty$, each term in the above equation converges properly in $L^2(\Omega)$ to the expected limit. The equation (5.27) is thus verified. The uniqueness of solution follows from the fact that the difference of any two solutions satisfies the homogeneous wave equation with zero initial-boundary conditions. The inequality (5.25) follows easily from equation (5.14) and the bounds given in Lemmas 3.1 to 3.3. □

If $g \in H^1$, $h \in H$, we will show that the solution $u(\cdot,t)$ is a H^1-valued process and the associated energy equation holds.

Theorem 3.5 Suppose that $g \in H^1, h \in H$ and the random fields $f(\cdot,t)$ and $\sigma(\cdot,t)$ are given as in Lemmas 3.2 and 3.3. Then the solution $u(\cdot,t)$ of (5.3) is a H^1-valued process with the regularity properties: $u \in L^2(\Omega; \mathbf{C}([0,T]; H^1))$ and $v(\cdot,t) = \frac{\partial}{\partial t}u(\cdot,t) \in L^2(\Omega; \mathbf{C}([0,T]; H))$. Moreover the energy equation holds

$$\|v(\cdot,t)\|^2 + \|u(\cdot,t)\|_1^2 = \|h\|^2 + \|g\|_1^2 + 2\int_0^t (f(\cdot,s), v(\cdot,s))ds$$

$$+ 2\int_0^t (v(\cdot,s), M(\cdot,ds)) + \int_0^t Tr\, Q_s \, ds. \tag{5.28}$$

Proof. In view of Theorem 3.4, with the aid of the estimates from Lemmas 3.1 to 3.3, the regularity results: $u \in L^2(\Omega; \mathbf{C}([0,T]; H^1))$ and $v \in L^2(\Omega; \mathbf{C}([0,T]; H))$ can be proved similarly as in the parabolic case (Theorem 3-5.6).

Concerning the energy equation, we first show it holds for the n-term approximate equations

$$du^n(\cdot,t) = v^n(\cdot,t)dt,$$
$$dv^n(\cdot,t) = [Au^n(\cdot,t) + f^n(\cdot,t)]dt + M^n(\cdot,dt),$$

where the expansion coefficients u_t^k and v_t^k for u^n and v^n satisfy the system:

$$du_t^k = v_t^k dt,$$
$$dv_t^k = (-\lambda_k u_t^k + f_t^k)dt + dz_t^k.$$

Apply the Itô formula to get

$$d|v_t^k|^2 = 2v_t^k dv_t^k + q_t^k dt = 2(-\lambda_k u_t^k + f_t^k)v_t^k dt + 2v_t^k dz_t^k + q_t^k dt,$$

which, when written in the integral form, leads to

$$|v_t^k|^2 + \lambda_k |u_t^k|^2 = |v_0^k|^2 + \lambda_k |u_0^k|^2 + 2\int_0^t f_s^k v_s^k ds + 2\int_0^t v_s^k dz_s^k + \int_0^t q_s^k ds.$$

Summing the above over k from 1 to n, we obtain

$$\|v^n(\cdot,t)\|^2 + \|u^n(\cdot,t)\|_1^2 = \|v^n(\cdot,0)\|^2 + \|u^n(\cdot,0)\|_1^2$$
$$+2\int_0^t (f^n(\cdot,s),v^n(\cdot,t))ds + 2\int_0^t (v^n(\cdot,s),M^n(\cdot,ds))$$
$$+\int_0^t Tr\, Q_s^n ds,$$

which converges in the mean to the energy equation (5.28) as $n \to \infty$.   □

5.4   Semilinear Wave Equations

Let $f(s,x,t)$ and $\sigma(s,x,t)$ be predictable random fields depending on the parameters $s \in \mathbf{R}$ and $x \in D$. We consider the initial-boundary value problem for a stochastic wave equation as follows

$$\frac{\partial^2 u}{\partial t^2} = (\kappa\Delta - \alpha)u + f(u,x,t) + \dot{M}(u,x,t), \quad x \in D, t \in (0,T),$$

$$Bu|_{\partial D} = 0, \tag{5.29}$$

$$u(x,0) = g(x), \quad \frac{\partial u}{\partial t}(x,0) = h(x),$$

where

$$\dot{M}(s,x,t) = \sigma(s,x,t)\dot{W}(x,t). \tag{5.30}$$

In terms of the Green's function, we rewrite the system (5.29) as the integral equation

$$u(\cdot,t) = G_t' g + G_t h + \int_0^t G_{t-s} f_s(u)ds + \int_0^t G_{t-s} M(u,ds), \tag{5.31}$$

where we set $f_s(u) = f(u(\cdot, s), \cdot, s)$ and $M(u, ds) = M(u(\cdot, s), \cdot, ds)$. In lieu of (5.29), for $g, h \in H$, if $u(\cdot, t)$ is a continuous adapted process in H satisfying the equation (5.31), then it is called a *mild solution*. Suppose that, for $g \in H^1$, $h \in H$, $u(\cdot, t)$ is a mild solution with the regularity property $u \in L^2([0, T] \times \Omega; H^1)$. Then such a mild solution will be termed a *strong solution*.

For the existence of a strong solution, let $f(\cdot, \cdot, t)$ and $\sigma(\cdot, \cdot, t)$ be predictable random fields that satisfy the usual conditions (to be called conditions A) as follows:

(A.1) There exist positive constants b_1, b_2 such that

$$|f(s, x, t)|^2 + |\sigma(s, x, t)|^2 \leq b_1(1 + |s|^2) \quad \text{a.s.,}$$

(A.2) $|f(s, x, t) - f(s', x, t)|^2 + |\sigma(s, x, t) - \sigma(s', x, t)|^2 \leq b_2|s - s'|^2$ a.s., for any $s, s' \in \mathbf{R}; x \in D$ and $t \in [0, T]$.

(A.3) $W(\cdot, t)$ is a R-Wiener process in H with covariance function $r(x, y)$ bounded by r_0, for $x, y \in D$.

Theorem 4.1 Suppose that conditions A hold true. Then, given $g, h \in H$, the wave equation (5.29) has a unique mild solution $u \in L^2(\Omega; \mathbf{C}([0, T]; H))$ such that

$$E \sup_{0 \leq t \leq T} \|u(\cdot, t)\|^2 < \infty.$$

Proof. To prove the theorem by the contraction mapping principle, let X_T be a Banach space of continuous adapted processes $\eta(\cdot, t)$ in H with the norm

$$\|\eta\|_T = \{E \sup_{0 \leq t \leq T} \|\eta(\cdot, t)\|^2\}^{1/2} < \infty. \tag{5.32}$$

Define the operator Γ in X_T by

$$\Gamma_t \eta = G'_t g + G_t h + \int_0^t G_{t-s} f_s(\eta) ds + \int_0^t G_{t-s} \sigma_s(\eta) dW_s. \tag{5.33}$$

By making use of Lemmas 3.1 to 3.3 together with conditions (A.1) and (A.3), equation (5.33) yields

$$\|\Gamma \eta\|_T^2 \leq C_1\{\|g\|^2 + \|h\|^2 + E \int_0^T [\|f(\eta(\cdot, s), \cdot, t)\|^2 + \|\sigma(\eta(\cdot, s), \cdot, t)\|^2] dt\}$$
$$\leq C_1\{\|g\|^2 + \|h\|^2 + b_1 T(1 + \|\eta\|_T^2)\},$$

for some constants $C_1, C_2 > 0$. This shows that $\Gamma : X_T \to X_T$ is well defined. Let η and $\eta' \in X_T$. Then, by taking (5.33), Lemmas 3.2–3.3 and conditions

A into account, we obtain

$$\|\Gamma\eta - \Gamma\eta'\|_T^2 = E \sup_{0 \le t \le T} \| \int_0^t G_{t-s}[\,f_s(\eta) - f_s(\eta')\,]ds$$

$$+ \int_0^t G_{t-s}[\,\sigma_s(\eta) - \sigma_s(\eta')\,]W(\cdot, ds)\|^2$$

$$\le 2E \{ T \int_0^T \|\,f_s(\eta) - f_s(\eta')\,\|^2 ds + \frac{8}{\lambda_1} \int_0^T \|\sigma_s(\eta) - \sigma_s(\eta')\|_R^2 ds \}$$

$$\le 2b_2(T \vee \frac{8r_0}{\lambda_1})E \int_0^T \|\,\eta_s - \eta_s'\,\|^2 dt.$$

It follows that

$$\|\Gamma\eta - \Gamma\eta'\|_T^2 \le 2b_2(T \vee \frac{8r_0}{\lambda_1})T\|\,\eta - \eta'\,\|_T^2,$$

which shows that Γ is a contraction mapping in X_T for a sufficiently small T. Therefore, by Theorem 3-2.5, there exists a unique fixed point u which is the mild solution as asserted. The solution can then be extended to any finite time interval $[0, T]$. □

More generally, the nonlinear terms in the wave equation may also depend on the velocity $v = \frac{\partial}{\partial t}u$ (in the case of damping) and the gradient ∂u. Then the corresponding initial-boundary value problem is given by

$$\frac{\partial^2 u}{\partial t^2} = (\kappa\Delta - \alpha)u + f(Ju, v, x, t) + \dot{M}(Ju, v, x, t),$$

$$Bu|_{\partial D} = 0, \tag{5.34}$$

$$u(x, 0) = g(x), \quad \frac{\partial u}{\partial t}(x, 0) = h(x),$$

for $x \in D$, $t \in (0, T)$, where we let $Ju = (u; \partial u)$; $f(z, s, x, t)$, $\sigma(z, s, x, t)$ are predictable random fields with parameters $z \in \mathbf{R}^{d+1}$, $x \in D$, $s \in \mathbf{R}$, $t \in [0, T]$, and

$$\dot{M}(Ju, v, x, t) = \sigma(Ju, v, x, t)\dot{W}(x, t).$$

In this case, we rewrite the equation (5.34) as a pair of integral equations:

$$u(\cdot, t) = G_t'g + G_t h + \int_0^t G_{t-s}f_s(Ju, v)ds$$

$$+ \int_0^t G_{t-s}\sigma_s(Ju, v)dW_s,$$

$$v(\cdot, t) = G_t''g + G_t'h + \int_0^t G_{t-s}'f_s(Ju, v)ds \tag{5.35}$$

$$+ \int_0^t G_{t-s}'\sigma_s(Ju, v)dW_s,$$

where we put

$$f_s(Ju, v) = f(Ju(\cdot, s), v(\cdot, s), \cdot, s) \text{ and } \sigma_s(Ju, v) = \sigma(Ju(\cdot, s), v(\cdot, s), \cdot, s).$$

Introduce a product Hilbert space $\mathbb{H} = H^1 \times H$, and let $U_t = (u(\cdot, t); v(\cdot, t))$ with $U_0 = (g; h)$, $F_s(U) = (0; f_s(Ju, v))$ and $\Sigma_s(U) = (0; \sigma_s(Ju, v))$. Then the above system can be regarded as an equation in $\mathbb{H}$:

$$U_t = \mathbf{G}_t U_0 + \int_0^t \mathbf{G}_{t-s} F_s(U) ds + \int_0^t \mathbf{G}_{t-s} \Sigma_s(U) dW_s, \qquad (5.36)$$

where the vectors U_t, F_s and Σ_s are written as a column matrix and $\mathbf{G}_t$ is the Green's matrix:

$$\mathbf{G}_t = \begin{pmatrix} G'_t & G_t \\ G_t'' & G'_t \end{pmatrix}. \qquad (5.37)$$

For $\Phi = (\phi; \psi) \in \mathbb{H}$, let $\mathbf{e}(\Phi) = \mathbf{e}(\phi, \psi)$ be an energy function defined by

$$\mathbf{e}(\Phi) = \|\phi\|_1^2 + \|\psi\|^2, \qquad (5.38)$$

which induces an energy norm $\|\cdot\|_e$ in $\mathbb{H}$ as follows

$$\|\Phi\|_e = [\mathbf{e}(\Phi)]^{1/2}. \qquad (5.39)$$

We will show that the equation (5.34) has a strong solution by proving the existence of a unique solution to the equation (5.36) in $\mathbb{H}$. To this end, let $f(\cdots, t)$ and $\sigma(\cdots, t)$ be predictable random fields satisfying the following conditions B:

(B.1) There exist positive constants c_1, c_2 such that

$$|f(s, y, \eta, x, t)|^2 + |\sigma(s, y, \eta, x, t)|^2 \le c_1\{1 + |s|^2 + |y|^2 + |\eta|^2\} \quad \text{a.s.},$$

(B.2) the Lipschitz condition holds

$$|f(s, y, \eta, x, t) - f(s', y', \eta', x, t)|^2 + |\sigma(s, y, \eta, x, t) - \sigma(s', y', \eta', x, t)|^2 \\ \le c_2\{|s - s'|^2 + |y - y'|^2 + |\eta - \eta'|^2\} \quad \text{a.s.},$$

for any $s, s', \eta, \eta' \in \mathbf{R}; y, y' \in \mathbf{R}^d; x, x' \in D$ and $t \in [0, T]$.

(B.3) $W(\cdot, t)$ is a R-Wiener process in H with covariance function r(x,y) bounded by r_0, for $x, y \in D$.

Theorem 4.2 Suppose the conditions B are satisfied. For $g \in H^1$ and $h \in H$, the semilinear wave equation (5.34) has a unique solution u belongs to $L^2(\Omega; \mathbf{C}([0, T]; H^1))$ with $v = \frac{\partial}{\partial t} u \in L^2(\Omega; \mathbf{C}([0, T]; H))$. Moreover there is a constant $C > 0$, depending on T, c_1 and initial data, such that

$$E \sup_{0 \le t \le T} \{\|u(\cdot, t)\|_1^2 + \|v(\cdot, t)\|^2\} \le C. \qquad (5.40)$$

Proof.   To show the existence of a unique solution, let Y_T denote the Banach space of $\mathcal{F}_t$-adapted, continuous $\mathbb{H}$-valued processes $U_t = (u_t; v_t)$ with the norm

$$\|U\|_T = \{E \sup_{0 \le t \le T} \mathbf{e}(U_t)\}^{1/2} = \{E \sup_{0 \le t \le T} [\,\|u_t\|_1^2 + \|v_t\|^2\,]\,\}^{1/2}. \qquad (5.41)$$

For $U = (u; v) \in Y_T$, define the mapping Λ in Y_T by

$$\Lambda_t(U) = \mathbf{G}_t U_0 + \int_0^t \mathbf{G}_{t-s} F_s(U)ds + \int_0^t \mathbf{G}_{t-s} \Sigma_s(U) W(\cdot, ds). \qquad (5.42)$$

We first show that the operator Λ maps Y_T into itself. For $U \in Y_T$, let $\Lambda_t U = (\mu_t; \nu_t)$. Then, in view of (5.41) and (5.42), we have

$$\|\Lambda U\|_T^2 \le E \sup_{0 \le t \le T} \|\mu(\cdot, t)\|_1^2 + E \sup_{0 \le t \le T} \|\nu(\cdot, t)\|^2. \qquad (5.43)$$

By noticing (5.35) and making use of Lemmas 3.1–3.3, similar to the proof of the previous theorem, it can be shown that there exist constants C_1 and C_2 such that

$$E \sup_{0 \le t \le T} \|\mu(\cdot, t)\|_1^2 = E \sup_{0 \le t \le T} \| G'_t g + G_t h + \int_0^t G_{t-s} f_s(Ju, v)ds$$

$$+ \int_0^t G_{t-s}\sigma_s(Ju, v)dW_s \|_1^2$$

$$\le C_1 \{ \|U_0\|_e^2 + E \int_0^T [\,\|f_s(Ju, v)\|^2 + \|\sigma_s(Ju, v)\|^2\,]ds \},$$

and

$$E \sup_{0 \le t \le T} \|\nu(\cdot, t)\|^2 = E \sup_{0 \le t \le T} \|G_t'' g + G'_t h + \int_0^t G'_{t-s} f_s(Ju, v)ds$$

$$+ \int_0^t G'_{t-s}\sigma_s(Ju, v)dW_s \|^2$$

$$\le C_2 \{ \|U_0\|_e^2 + E \int_0^T [\,\| f_s(Ju, v)\|^2 + \|\sigma_s(Ju, v)\|^2\,]ds \}.$$

Therefore the equation (5.43) yields

$$\|\Lambda_t U\|_T^2 \le (C_1 + C_2)\{ \|U_0\|_e^2 + E \int_0^T [\,\| f_s(Ju, v)\|^2 \qquad (5.44)$$

$$+ \|\sigma_s(Ju, v)\|^2\,]ds\}.$$

By making use of condition (B.1) and the fact that $(\|g\| + \|\partial g\|) \le c_0 \|g\|_1$ for some $c_0 > 0$, we can deduce from (5.44) that there is a constant C, depending on T, $\mathbf{e}(U_0)$ and so on, such that

$$\|\Lambda U\|_T^2 \le C(1 + \|U\|_T^2).$$

Therefore the map $\Lambda : Y_T \to Y_T$ is well defined.

To show that Λ is a contraction map, for $U, U' \in Y_T$, we consider

$$\||\Lambda U - \Lambda U'\||_T^2 = E \sup_{0 \le t \le T} \{\|\mu(\cdot, t) - \mu'(\cdot, t)\|_1^2 + \|\nu(\cdot, t) - \nu'(\cdot, t)\|^2\}.$$

in which both terms on the right-hand side of the equation can be estimated by applying Lemmas 3.1–3.3 to give the upper bound

$$\le C_3 E \int_0^t \{\|f_s(Ju, v) - f_s(Ju', v')\|^2 + \|\sigma_s(Ju, v) - \sigma_s(Ju', v')\|^2\}ds,$$

for some $C_3 > 0$. Now, making use the Lipschitz condition (B.2), the above inequality gives rise to

$$\||\Lambda U - \Lambda U'\||_T^2 \le c_2 C_3 \int_0^T E\{\|u_s - u_s'\|^2 + \|\partial u_s - \partial u_s'\|^2 + \|v_s - v_s'\|^2\}ds$$
$$\le C_4 T \||\,U - U'\||_T^2,$$

for some $C_4 > 0$. Hence, for a small enough T, the map Λ is a contraction which has a fixed point $U = (u; v) \in Y_T$. Hence u is the unique solution with the asserted regularity. $\qquad\square$

5.5 Wave Equations in Unbounded Domain

As in the case of parabolic equations, we will show that the results obtained in the last section hold true when $D = \mathbf{R}^d$ by means of a Fourier transform. First consider the Cauchy problem for the linear wave equation as in (5.3):

$$\frac{\partial^2 u}{\partial t^2} = (\Delta - 1)u + f(x,t) + \dot{M}(x,t), \quad x \in \mathbf{R}^d, \, t \in (0,T),$$
$$u(x,0) = g(x), \quad \frac{\partial u}{\partial t}(x,0) = h(x), \tag{5.45}$$

where we set $\kappa = \alpha = 1$ for convenience, and $\dot{M}(x,t) = \sigma(x,t)\dot{W}(x,t)$.

Assume the random fields f and σ are regular as usual. Following the same notation as in Chapter Four, apply a Fourier transform to the problem (5.45) to get the second-order Itô equation:

$$\frac{d^2}{dt^2}\hat{u}(\xi,t) = -\rho(\xi)\hat{u}(\xi,t) + \hat{f}(\xi,t) + \dot{\widehat{M}}(\xi,t),$$
$$\hat{u}(\xi,0) = \hat{g}(\xi), \quad \frac{d\hat{u}}{dt}(\xi,0) = \hat{h}(\xi), \tag{5.46}$$

for $\xi \in \mathbf{R}^d, t \in (0, T)$, where $\rho(\xi) = (1 + |\xi|^2)$ and $\widehat{M}(\xi, t)$ is the Fourier transform of $M(x, t)$. This initial-value problem can be solved to give

$$\hat{u}(\xi, t) = \hat{g}(\xi)\hat{K}'(\xi, t) + \hat{h}(\xi)\hat{K}(\xi, t) + \int_0^t \hat{K}(\xi, t - s)\hat{f}(\xi, s)ds \tag{5.47}$$
$$+ \int_0^t \hat{K}(\xi, t - s)\widehat{M}(\xi, ds),$$

where the prime denotes the time-derivative and

$$\hat{K}(\xi, t) = \frac{\sin \gamma(\xi)t}{\gamma(\xi)}, \tag{5.48}$$

with $\gamma(\xi) = \sqrt{\rho(\xi)} = (1 + |\xi|^2)^{1/2}$. By an inverse Fourier transform, the equation (5.47) yields

$$u(x, t) = \int K'(x - y, t)g(y)dy + \int K(x - y, t)h(y)dy$$
$$+ \int_0^t \int K(x - y, t - s)f(y, s)dyds \tag{5.49}$$
$$+ \int_0^t \int K(x - y, t - s)M(y, ds)dy,$$

or

$$u(\cdot, t) = G_t'g + G_t h + \int_0^t G_{t-s}f_s ds + \int_0^t G_{t-s}dM_s, \tag{5.50}$$

where G_t is the Green's operator with kernel $G(x, t) = (1/2\pi)^{d/2}K(x, t)$ and K is the inverse Fourier transform of $\hat{K}$ so that:

$$G(x, t) = (\frac{1}{2\pi})^d \int \frac{\sin \gamma(\xi)t}{\gamma(\xi)}e^{ix \cdot \xi}d\xi, \tag{5.51}$$

which exists in the sense of distribution. Recall that, in $\mathbf{R}^d$, we defined the norm $\|\cdot\|_k$ in the Sobolev space H^k in terms of the Fourier transform:

$$\|\phi\|_k^2 = \int (1 + |\xi|^2)^k |\hat{\phi}(\xi)|^2 d\xi = \int \rho^k(\xi)|\hat{\phi}(\xi)|^2 d\xi.$$

Parallel to Lemmas 3.1 to 3.3, we can obtain the same estimates for the terms on the right-hand side of the solution formula (5.50). They will be summarized in the following as a single lemma. Formally, due to the corresponding Fourier analysis, the assertion of the following lemma is not surprising. Since the techniques for the proof are similar, we will only check on a few of them as an illustration of the general idea involved.

Lemma 5.1   Let $f(\cdot, t)$ and $\sigma(\cdot, t)$ be predictable random fields such that

$$E \int_0^T \|f(\cdot, s)\|^2 ds < \infty, \quad \text{and}$$

$$E \int_0^T \|\sigma(\cdot, s)\|_R^2 ds \leq r_0 E \int_0^T \|\sigma(\cdot, s)\|^2 ds < \infty.$$

Let $\eta(\cdot,t) = \int_0^t G_{t-s}f(\cdot,s)ds$ and $\mu(\cdot,t) = \int_0^t G_{t-s}M(\cdot,ds)$. Then all the inequalities in Lemmas 3.1, 3.2 and 3.3 hold true with $\lambda_1 = 1$ in Lemma 3.3.

Proof. In Lemma 3.1, we will only verify the inequality (5.15) holds for $k = m = 1$ so that

$$\sup_{0 \le t \le T} \|G'_t g\|_1^2 \le \|g\|_1^2.$$

In view of (5.48), we have $\hat{K}'(\xi, t) = \cos \gamma(\xi)t$. Hence

$$\|G'_t g\|_1^2 = \|\gamma \widehat{(G'_t g)}\|^2 = \int \rho(\xi)[\cos^2 \gamma(\xi)t]\,|\hat{g}(\xi)|^2 d\xi$$

$$\le \int \rho(\xi)|\hat{g}(\xi)|^2 d\xi = \|g\|_1^2,$$

which verifies the inequality (5.15) for $k = m = 1$.

Next let us verify the inequality (5.17) in Lemma 3.2:

$$E \sup_{0 \le t \le T} \|\eta'(\cdot, t)\|^2 \le T E \int_0^T \|f_s\|^2 ds.$$

Consider

$$\|\eta'(\cdot, t)\|^2 = \|\int_0^t G'_{t-s}f_s ds\|^2$$

$$\le T \int_0^t \|G'_{t-s}f_s\|^2 ds = \int_0^t \|\hat{K}'_{t-s}\hat{f}_s\|^2 ds$$

$$\le T \int_0^t \int \cos^2 \gamma(\xi)(t-s)|\hat{f}(\xi, s)|^2 ds\,d\xi \le T \int_0^T \|f_s\|^2 ds,$$

which implies the desired inequality.

To show the inequality (5.20) with $k = 0, m = 1$, in Lemma 3.3:

$$E \sup_{0 \le t \le T} \|\mu(\cdot, t)\|_1^2 \le 8 E \int_0^T \|\sigma(\cdot, t)\|_R^2 dt,$$

we recall that the Fourier transform of the stochastic integral $\mu(\cdot, t)$ can be written as

$$\hat{\mu}(\xi, t) = (2\pi)^{d/2} \int_0^t \hat{K}(\xi, t-s)\widehat{M}(\xi, ds) = \int_0^t \frac{\sin[\gamma(\xi)(t-s)]}{\gamma(\xi)} \widehat{M}(\xi, s)ds.$$

Therefore, we have

$$\|\mu(\cdot, t)\|_1^2 = \|\gamma\hat{\mu}(\cdot, t)\|^2 = \int |\int_0^t \sin[\gamma(\xi)(t-s)]\widehat{M}(\xi, s)ds|^2 d\xi$$

$$\le 2 \int \{|\int_0^t \cos[\gamma(\xi)s]\widehat{M}(\xi, ds)|^2 + |\int_0^t \sin[\gamma(\xi)s]\widehat{M}(\xi, ds)|^2\}d\xi.$$

It follows that

$$
E \sup_{0 \le t \le T} \|\mu(\cdot, t)\|_1^2 \le 2 \int E \sup_{0 \le t \le T} \{ | \int_0^t \cos[\gamma(\xi)s]\widehat{M}(\xi, ds)|^2
$$

$$
+ | \int_0^t \sin[\gamma(\xi)s]\widehat{M}(\xi, ds)|^2 \} d\xi \le 8 E \int_0^T \int \hat{q}(\xi, \xi, s) d\xi ds
$$

$$
= 8 E \int_0^T Tr Q_s ds = 8 E \int_0^T \|\sigma(\cdot, t)\|_R^2 dt,
$$

as to be shown.   □

With the aid of the estimates given above in Lemma 5.1, it is clear that, by following the proofs of Theorems 4.1 and 4.2 with D replaced by $\mathbf{R}^d$, the following two theorems for the Cauchy problem (5.45) can be proved verbatim. Therefore we shall only state the existence theorems without proof.

Theorem 5.2   Consider the Cauchy problem:

$$
\frac{\partial^2 u}{\partial t^2} = (\kappa \Delta - \alpha)u + f(u, x, t) + \dot{M}(u, x, t), \quad x \in \mathbf{R}^d, \, t \in (0, T),
$$

$$
u(x, 0) = g(x), \quad \frac{\partial u}{\partial t}(x, 0) = h(x),
$$

(5.52)

Suppose that conditions A hold true for Theorem 4.1 with $D = \mathbf{R}^d$. Then, given $g, h \in H$, the Cauchy problem (5.51) has a unique mild solution $u \in L^2(\Omega; \mathbf{C}([0, T]; H))$ such that

$$
E \sup_{0 \le t \le T} \|u(\cdot, t)\|^2 < \infty.
$$
                                                                                   □

Theorem 5.3   Consider the Cauchy problem:

$$
\frac{\partial^2 u}{\partial t^2} = (\kappa \Delta - \alpha)u + f(Ju, v, x, t) + \dot{M}(Ju, v, x, t),
$$

$$
u(x, 0) = g(x), \quad \frac{\partial u}{\partial t}(x, 0) = h(x), \quad x \in D, \, t \in (0, T).
$$

(5.53)

Suppose the conditions B for Theorem 4.2 are satisfied with $D = \mathbf{R}^d$. For $g \in H^1$ and $h \in H$, the Cauchy problem (5.53) has a unique solution $u \in L^2(\Omega; \mathbf{C}([0, T]; H^1))$ with a continuous trajectory in H^1 with $v = \frac{\partial}{\partial t} u \in L^2(\Omega; \mathbf{C}([0, T]; H))$. Moreover there is a constant $C > 0$, depending on T, c_1 and initial data, such that

$$
E \sup_{0 \le t \le T} \{ \|u(\cdot, t)\|_1^2 + \|v(\cdot, t)\|^2 \} \le C.
$$
                                                                                   □

Remarks: A well-known example is given by the Sine-Gordon equation perturbed by a state-dependent noise:

$$\frac{\partial^2 u}{\partial t^2} = (\kappa\Delta - \alpha)u + \beta \sin u + u\dot{M}(x,t)$$

for $x \in \mathbf{R}^d, t \in [0,T]$, where β is a constant. Then, with the initial conditions given in (5.53), Theorem 5.2 is applicable to show the existence of a unique solution as depicted therein. So far we have proved the existence results based on the usual assumption that the nonlinear terms satisfy the global Lipschitz condition. Under weaker assumptions, such as a local Lipschitz condition, a solution may explode in a finite time. Also the analytical techniques used here can be adapted to treating other wave-like equations, such as the elastic plate and the beam equations as well as the Schrödinger equation with noise terms. These matters will be discussed in Chapter Eight.

5.6 Randomly Perturbed Hyperbolic Systems

In the mathematical physics the wave motion in a continuous medium is often modelled by a first-order hyperbolic system [17]. In particular we first consider the following random linear system:

$$\frac{\partial u_i}{\partial t} = \sum_{j=1}^{n}\sum_{k=1}^{d} a_{ij}^{k}\frac{\partial u_j}{\partial x_k} - \alpha u_i + f_i(x,t) + \dot{M}_i(x,t), \quad t \in (0,T),$$

$$u_i(x,0) = h_i(x), \quad i = 1,2,\cdots,n, \; x \in \mathbf{R}^d,$$

(5.54)

where $\alpha > 0$ and the coefficients a_{ij}^{k} are assumed to be some constants,

$$\dot{M}_i(x,t) = \sum_{j=1}^{m} \sigma_{ij}(x,t)\dot{W}_j(x,t),$$

(5.55)

$f_i(x,t)$ and $\sigma_{ij}(x,t)$ are predictable random fields. Also, for simplicity, let $W_1(x,t),\cdots,W_m(x,t)$, be independent, identically distributed Wiener fields in H with a common covariance function $r(x,y)$ bounded by r_0, and $h_1,\cdots,h_n$ are the initial data. Let q_{ij} denote the local covariation function of M_i and M_j so that

$$q_{ij}(x,y,t) = r(x,y)\sum_{k=1}^{m} \sigma_{ik}(x,t)\sigma_{jk}(y,t),$$

(5.56)

with the associated local covariation operator $Q_{ij}(t)$. Let Q_t denote the matrix operator $[Q_{ij}(t)]$. Then

$$Tr\,Q_t = \sum_{i=1}^{n} Tr\,Q_{ii}(t) = \sum_{i=1}^{n}\int q_{ii}(x,x,t)dx.$$

(5.57)

Let the vectors $u = (u_1, \cdots, u_n), \cdots, h = (h_1, \cdots, h_n)$ be regarded as a column matrix. Define the $(n \times n)$ coefficient matrix $A^k = [a_{ij}^k]$, $k = 1, \cdots, d$, and the $n \times m$ diffusion matrix $\Sigma = [\sigma_{ij}]$. Then the above system can be written in the matrix form:

$$\frac{\partial u}{\partial t} = \sum_{k=1}^{d} A^k \frac{\partial u}{\partial x_k} - \alpha u + f(x,t) + \dot{M}(x,t), \quad t \in (0,T),$$

$$u(x,0) = h(x), \quad x \in \mathbf{R}^d,$$

(5.58)

where $\dot{M}(x,t) = \Sigma(x,t)\dot{W}(x,t)$. By applying a Fourier transform to equation (5.58), we obtain

$$\frac{d\hat{u}}{dt}(\xi,t) = [-iA(\xi) - \alpha I]\hat{u}(\xi,t) + \hat{f}(\xi,t) + \widehat{\dot{M}}(\xi,t), \quad t \in (0,T),$$

$$\hat{u}(\xi,0) = \hat{h}(\xi), \quad \xi \in \mathbf{R}^d,$$

(5.59)

where we set $A(\xi) = (\sum_{k=1}^{d} \xi_k A^k)$, I is the identity matrix of order n, and $\widehat{M}(\xi,t)$ is the Fourier transform of $M(x,t)$. The local covariation operator $\hat{Q}_t$ for $\widehat{M}(\cdot,t)$ has the matrix kernel $\hat{q}(\xi,\eta;t) = [\hat{q}_{ij}(\xi,\eta;t)]$ in which the entries are given by

$$\hat{q}_{ij}(\xi,\eta;t) = (\frac{1}{2\pi})^d \int \int e^{-i(x\cdot\xi - y\cdot\eta)} q_{ij}(x,y;t) dx dy.$$

Then it can be shown that $Tr\,\hat{Q}_t = Tr\,Q_t$ as in the scalar case.

The system (5.54) is said to be *strongly hyperbolic* if the eigenvalues $\lambda_1(\xi), \cdots, \lambda_n(\xi)$, of the matrix $A(\xi)$ are real distinct and non-zero for $|\xi| = 1$ [66]. Let $\{\phi_1(\xi), \cdots, \phi_n(\xi)\}$ denote the corresponding set of orthonormal eigenvectors. The solution of the linear Itô equation (5.59) is given by

$$\hat{u}(\xi,t) = \hat{K}(\xi,t)\hat{h}(\xi) + \int_0^t \hat{K}(\xi,t-s)\hat{f}(\xi,s)ds$$

$$+ \int_0^t \hat{K}(\xi,t-s)\widehat{M}(\xi,ds),$$

(5.60)

where $\hat{K}$ is the fundamental matrix

$$\hat{K}(\xi,t) = e^{-tB(\xi)}, \quad \text{with} \quad B(\xi) = \{iA(\xi) + \alpha I\}.$$

(5.61)

In terms of the eigenvectors $\phi_k(\xi)$, we can express the solution (5.54) as

$$\hat{u}(\xi,t) = \sum_{j=1}^{n} \hat{v}_j(\xi,t)\phi_j(\xi),$$

(5.62)

where $\hat{v}_j(\xi,t) = (\hat{u}(\xi,t), \phi_j(\xi))$ and, here, $(\cdot,\cdot)$ denotes the inner product of two complex n-vectors. By taking the inner product of the equation (5.59) with respect to the eigenvector $\phi_j(\xi)$ and making use of the fact in linear algebra:

$$e^{-tB(\xi)}\phi_j(\xi) = e^{-[i\lambda_j(\xi)+\alpha]t}\phi_j(\xi),$$

we obtain

$$
\begin{aligned}
\hat{v}_j(\xi,t) = {}& \hat{h}_j(\xi)e^{-\beta_j(\xi)t} + \int_0^t e^{-\beta_j(\xi)(t-s)}\hat{f}_j(\xi,s)ds \\
& + \int_0^t e^{-\beta_j(\xi)(t-s)}\hat{N}_j(\xi,ds),
\end{aligned}
\tag{5.63}
$$

where $\hat{h}_j(\xi) = (\hat{h}(\xi),\phi_j(\xi))$, $\hat{f}_j(\xi,s) = (\hat{f}(\xi,s),\phi_j(\xi))$, $\beta_j(\xi) = [i\lambda_j(\xi)+\alpha]$ and $\hat{N}_j(\xi,t) = (\hat{M}(\xi,t),\phi_j(\xi))$.

By an inverse Fourier transform, the above result (5.60) leads to the solution of (5.54):

$$
\begin{aligned}
u(x,t) = {}& (G_t g)(x) + \int_0^t (G_{t-s}f)(x,s)ds \\
& + \int_0^t (G_{t-s}M)(x,ds).
\end{aligned}
\tag{5.64}
$$

Here G_t denotes the Green's operator defined by

$$(G_t g)(x) = (\frac{1}{2\pi})^{d/2} \int K(x-y,t)g(y)dy,$$

where

$$K(x,t) = (\frac{1}{2\pi})^{d/2} \int e^{ix\cdot\xi - B(\xi)t}d\xi.$$

The above integral exists in the sense of distribution, that is, for any test function $\varphi \in \mathbf{C}_0^\infty(\mathbf{R}^d;\mathbf{R}^n)$,

$$\int K(x,t)\varphi(x)\,dx = (\frac{1}{2\pi})^{d/2} \int \int e^{ix\cdot\xi - B(\xi)t}\varphi(x)\,d\xi\,dx.$$

To study the regularity of the solution, let $\mathbf{H} = (H)^n$ denote the Cartesian product of n L^2-spaces. We shall use the same notation for the inner product $(\cdot,\cdot)$ and the norm $\|\cdot\|$ in $\mathbf{H}$ as in H, provided that there is no confusion. Therefore, for $g,h \in \mathbf{H}$, we write $(g,h) = \sum_{j=1}^n (g_j,h_j)$ and $\|g\|^2 = \sum_{j=1}^n \|g_j\|^2$. As in the case of the Fourier transform of real-valued functions, the Plancheral's theorem and the Parseval's identity also hold for the vector-valued functions as well. For linear partial differential equations with constant coefficients, the Fourier transform is a useful tool in analyzing the solutions as shown in the previous chapter.

Lemma 6.1 For $h \in \mathbf{H}$, let $f_i(\cdot,t)$ and $\sigma_{ij}(\cdot,t)$ be predictable H-valued processes and let $W_1,\cdots,W_m$, be independent, identically distributed Wiener random fields with a bounded common covariance function $r(x, y)$ as assumed before. Suppose that

$$E\{\int_0^T \|f_i(\cdot, t)\|^2 dt + \int_0^T \|\sigma_{ij}(\cdot, t)\|^2 dt\} < \infty, \qquad (5.65)$$

for $i, j = 1, \cdots, n$. Then the following inequalities hold:

$$\|G_t h\|^2 \leq \|h\|^2, \qquad (5.66)$$

$$E \sup_{0 \leq t \leq T} \|\int_0^t G_{t-s} f(\cdot, s) ds\|^2 \leq TE \int_0^T \|f(\cdot, s)\|^2 ds, \qquad (5.67)$$

and

$$E \sup_{0 \leq t \leq T} \|\int_0^t G_{t-s} M(\cdot, ds)\|^2 \leq K_T E \int_0^t Tr\, Q_s ds, \qquad (5.68)$$

for some $K_T > 0$, where $Tr\, Q_s = \sum_{j=1}^n \int q_{jj}(x, x; s) dx$.

Proof. In view of (5.61), we see that the matrix-norm $|\hat{K}(\xi, t)| \leq e^{-\alpha t}$. The first inequality (5.66) follows immediately from the L^2-isometry as mentioned above,

$$\|G_t h\|^2 = \|\hat{K}(\cdot, t)\hat{h}\|^2 \leq e^{-2\alpha t} \|\hat{h}\|^2 \leq \|h\|^2.$$

Next, by making use of the above result, we obtain

$$E \|\int_0^t G_{t-s} f_s ds\|^2 \leq TE[\int_0^t \|G_{t-s} f(\cdot, s)\|^2 ds$$

$$\leq T \int_0^t e^{-2\alpha(t-s)} \|f_s\|^2 ds,$$

which implies the inequality (5.67).

For the last inequality, let $\mu(\cdot, t) = \int_0^t G_{t-s} M(\cdot, ds)$ and let $\hat{\mu}(\cdot, t)$ denote its Fourier transform. Referring to (5.62) and (5.62), in terms of the eigenfunctions of $A(\xi)$, $\hat{\mu}$ can be written as

$$\hat{\mu}(\xi, t) = \sum_{j=1}^n \hat{\mu}_j(\xi, t) \phi_j(\xi),$$

where

$$\hat{\mu}_j(\xi, t) = \int_0^t e^{-\beta_j(\xi)(t-s)} \widehat{N}_j(\xi, ds),$$

and $\widehat{N}_j(\xi, t) = (\widehat{M}(\xi, t), \phi_j(\xi))$ is a martingale with parameter ξ. Recall that $\beta_j = \alpha + i\lambda_j$ so that

$$\|\hat{\mu}_j(\cdot, t)\|^2 = \int |\int_0^t e^{-\beta_j(\xi)(t-s)} \widehat{N}_j(\xi, ds)|^2 d\xi$$

$$= \int \{|\int_0^t e^{-\alpha(t-s)}[\cos \lambda_j(\xi)(t-s)]\widehat{N}_j(\xi, ds)|^2$$

$$+|\int_0^t e^{-\alpha(t-s)}[\sin \lambda_j(\xi)(t-s)]\widehat{N}_j(\xi, ds)|^2\}d\xi.$$

By means of this equation and similar estimates as in Lemma 3.3, we can get

$$E \sup_{0\leq t\leq T} \|\hat{\mu}_j(\cdot, t)\|^2 \leq 4 \int \{E \sup_{0\leq t\leq T} \{|\int_0^t e^{\alpha s}[\cos \lambda_j(\xi)s]\widehat{N}_j(\xi, ds)|^2$$

$$+|\int_0^t e^{\alpha s}[\sin \lambda_j(\xi)s]\widehat{N}_j(\xi, ds)|^2\}d\xi$$

$$\leq 4E |\int_0^T e^{2\alpha t} \int (\hat{q}(\xi, \xi; t)\phi_j(\xi), \phi_j(\xi))dt\, d\xi$$

$$\leq 4e^{2\alpha T} E \int_0^T \int (\hat{q}(\xi, \xi; t)\phi_j(\xi), \phi_j(\xi))dt\, d\xi,$$

where use was made of a submartingale inequality. Consequently,

$$E \sup_{0\leq t\leq T} \|\mu(\cdot, t)\|^2 \leq \sum_{j=1}^n E \sup_{0\leq t\leq T} \|\hat{\mu}_j(\cdot, t)\|^2$$

$$\leq 4e^{2\alpha T} \sum_{j=1}^n E \int_0^T (\hat{q}(\xi, \xi; t)\phi_j(\xi), \phi_j(\xi))dt d\xi = 4e^{2\alpha T} E \int_0^T Tr\, Q_t dt,$$

which verifies the inequality (5.68) with $K_T = 4e^{2\alpha T}$. $\qquad \square$

Theorem 6.2 Suppose that the conditions for Lemma 6.1 are fulfilled. Then, given $h \in \mathbf{H}$, the mild solution u of the Cauchy problem given by (5.64) belongs to $L^2(\Omega; \mathbf{C}([0, T]; \mathbf{H}))$, and there is $C_T > 0$ such that the energy inequality holds:

$$E \sup_{0\leq t\leq T} \|u(\cdot, t)\|^2 \leq C_T E \{\|h\|^2 + \int_0^T \|f(\cdot, s)\|^2 ds$$

$$+ \int_0^T \|\Sigma(\cdot, s)\|^2 ds\}, \tag{5.69}$$

where $\|\Sigma(\cdot, t)\|^2 = \sum_{j=1}^n \sum_{k=1}^m \|\sigma_{jk}(\cdot, t)\|^2$.

Proof. By the L^2-isometry as mentioned above, we will first show the Fourier transform $\hat{u}$ of the solution u belongs to the space $L^2(\Omega; \mathbf{C}([0,T]; \mathbf{H}))$. In view of (5.62) and (5.63), we have $\|u(\cdot,t)\|^2 = \sum_{j=1}^{n} \|\hat{v}_j(\cdot,t)\|^2$, where $\hat{v}_j(\cdot,t)$ can be shown to be continuous in H. It follows that $u(\cdot,t)$ is continuous in $\mathbf{H}$.

From (5.64), by applying the inequalities in Lemma 3.1, we can obtain

$$
E \sup_{0 \le t \le T} \|u(\cdot,t)\|^2 = 3E \sup_{0 \le t \le T} \{\|G_t h\|^2 + \| \int_0^t (G_{t-s} f)(\cdot,s) ds\|^2\}
$$
$$
+ \| \int_0^t (G_{t-s} M)(\cdot, ds)\|^2
$$
$$
\le 3\{\|g\|^2 + TE \int_0^T \|f_s\|^2 ds + K_T E \int_0^T Tr\, Q_s ds\}.
$$

Recall that $Tr\, Q_s = \sum_{j=1}^{n} \int q_{jj}(x,x,s) dx$, where, by (5.56),

$$
q_{jj}(x,x,s) = r(x,x) \sum_{k=1}^{m} \sigma_{jk}^2(x,s) \le r_0 \sum_{k=1}^{m} \sigma_{jk}^2(x,t).
$$

It follows that $E \sup_{0 \le t \le T} \|u(\cdot,t)\|^2 < \infty$, and the energy inequality (5.69) holds true. $\qquad \square$

Now let us consider a semilinear hyperbolic system for which the random fields f_j and σ_{jk} in (5.54) depend on the unknowns $u_1, \cdots, u_n$. The associated Cauchy problem reads

$$
\frac{\partial u_i}{\partial t} = \sum_{j=1}^{n} \sum_{k=1}^{d} a_{ij}^k \frac{\partial u_j}{\partial x_k} - \alpha u_i + f_i(u_1, \cdots, u_n, x, t)
$$
$$
+ \dot{M}_i(u_1, \cdots, u_n, x, t), \quad t \in (0,T), \tag{5.70}
$$
$$
u_i(x,0) = g_i(x), \quad i = 1, 2, \cdots, n; \ x \in \mathbf{R}^d,
$$

in which

$$
\dot{M}_i(u_1, \cdots, u_n, x, t) = \sum_{j=1}^{m} \sigma_{ij}(u_1, \cdots, u_n, x, t) \dot{W}_j(x, t). \tag{5.71}
$$

Here $f_i(u_1, \cdots, u_n, x, t)$ and $\sigma_{ij}(u_1, \cdots, u_n, x, t)$ are predictable random fields, and $W_1, \cdots, W_m$, are the same Wiener random fields as given before.

By using the matrix notation, the system (5.70) can be written as

$$
\frac{\partial u}{\partial t} = \sum_{k=1}^{d} A^k \frac{\partial u}{\partial x_k} - \alpha u + f(u, x, t)
$$
$$
+ \Sigma(u, x, t) \dot{W}(x, t), \quad t \in (0,T), \tag{5.72}
$$
$$
u(x,0) = g(x), \quad x \in \mathbf{R}^d.
$$

To prove the existence theorem for the above system, we suppose that, for $i = 1, \cdots, n; j = 1, \cdots, m$, $f_i(\rho, x, t)$ and $\sigma_{ij}(\rho, x, t)$ are predictable random fields with parameters $\rho \in \mathbf{R}^n$, $x \in \mathbf{R}^d$, such that the following conditions H hold:

(H.1) For $i = 1, \cdots, n$, let $f_i(\cdot, \cdot, t)$ be a predictable H-valued process and set $\bar{f}_i(\cdot, t) = f_i(0, \cdot, t)$. Suppose there exists positive constant b_1 such that

$$E \int_0^T \|\bar{f}(\cdot, t)\|^2 dt = E \sum_{i=1}^n \int_0^T \|\bar{f}_i(\cdot, t)\|^2 dt \leq b_1.$$

(H.2) For $i = 1, \cdots, n; j = 1, \cdots, m$, $\sigma_{ij}(\rho, x, t)$ is a predictable H-valued process and there exists a positive constant b_2 such that

$$|\Sigma(\rho, x, t)|^2 = \sum_{i=1}^n \sum_{j=1}^m |\sigma_{ij}(\rho, x, t)|^2 \leq b_2(1 + |\rho|^2) \quad \text{a.s.,}$$

for any $\rho \in \mathbf{R}^n$, $x \in \mathbf{R}^d$, and $t \in [0, T]$.

(H.3) There exist positive constants b_3 and b_4 such that

$$|f(\rho, x, t) - f(\rho', x, t)|^2 = \sum_{i=1}^n |f_i(\rho, x, t) - f_i(\rho', x, t)|^2 \leq b_3|\rho - \rho'|^2,$$

and

$$|\Sigma(\rho, x, t) - \Sigma(\rho', x, t)|^2 = \sum_{i=1}^n \sum_{j=1}^m |\sigma_{ij}(\rho, x, t) - \sigma_{ij}(\rho', x, t)|^2$$
$$\leq b_4|\rho - \rho'|^2 \quad \text{a.s.,}$$

for any $\rho, \rho' \in \mathbf{R}^n$, $x \in \mathbf{R}^d$ and $t \in [0, T]$.

(H.4) The independent identically distributed Wiener random fields $W_1(x, t), \cdots, W_m(x, t)$ are given as before with common covariance function $r(x, y)$ bounded by r_0 and $TrR = \int r(x, x) dx = r_1$ is finite.

With the aid of Lemma 6.1 and Theorem 6.2, we shall prove the following existence theorem for the semilinear hyperbolic system (5.70) or (5.72) under conditions H.

Theorem 6.3 Suppose that conditions (H.1)–(H.4) are satisfied. For $h \in \mathbf{H}$, the Cauchy problem for the hyperbolic system (5.70) has a unique (mild) solution $u \in L^2(\Omega; \mathbf{C}([0, T]; \mathbf{H}))$ such that

$$E \sup_{0 \leq t \leq T} \|u(\cdot, t)\|^2 \leq C_T,$$

for some constant $C_T > 0$.

Proof.   We rewrite the system (5.72) in the integral form

$$u_t = G_t h + \int_0^t G_{t-s} f_s(u) ds + \int_0^t G_{t-s} \Sigma_s(u) dW_s. \qquad (5.73)$$

To show the existence of a solution to this equation, introduce a Banach space X_T of continuous adapted **H**-valued processes $u(\cdot, t)$ in $[0, T]$ with norm $|||u|||_T$ defined by

$$|||u|||_T = \{E \sup_{0 \le t \le T} ||u_t||^2\}^{1/2}.$$

Let Γ be an operator in X_T defined by

$$\Gamma_t u = G_t h + \int_0^t G_{t-s} f_s(u) ds + \int_0^t G_{t-s} \Sigma_s(u) dW_s. \qquad (5.74)$$

We will first show that $\Gamma : X_T \to X_T$ is bounded. For $u \in X_T$, clearly we have

$$E \sup_{0 \le t \le T} ||\Gamma_t u||^2 \le 3E \sup_{0 \le t \le T} \{||G_t h||^2 + ||\int_0^t G_{t-s} f_s(u) ds||^2 \qquad (5.75)$$
$$+ ||\int_0^t G_{t-s} \Sigma_s(u) dW_s||^2\},$$

provided that the upper bound is finite. This is the case because, by Lemma 6.1, we have $||G_t h||^2 \le ||h||^2$,

$$E \sup_{0 \le t \le T} ||\int_0^t G_{t-s} f_s(u) ds||^2 \le T E \int_0^T ||f_s(u)||^2 ds,$$

and

$$E \sup_{0 \le t \le T} ||\int_0^t G_{t-s} \Sigma_s(u) dW_s||^2 \le K_T E \int_0^T Tr \, Q_s(u) ds.$$

Furthermore, by condition (H.1), we obtain

$$||f_s(u)||^2 \le 2(||\bar{f}_s||^2 + ||f_s(u) - f_s(0)||^2)$$

$$\le 2\{||\bar{f}_s||^2 + b_3(1 + ||u_t||^2)\},$$

so that

$$E \int_0^T ||f_s(u)||^2 ds \le 2E \int_0^T \{||\bar{f}_s||^2 + b_3(1 + ||u_s||^2)\} ds \qquad (5.76)$$
$$\le 2\{b_1 + b_3 T(1 + |||u|||_T^2)\}.$$

By invoking conditions (H.2) and (H.4), we get

$$E \int_0^T Tr\, Q_s(u)ds = E \int_0^T \int r(x,x)|\Sigma(u,x,s)|^2 dx ds$$
$$\leq E \int_0^T \int r(x,x)b_2[\,1 + |u(x,s)|^2\,]dx\, ds \tag{5.77}$$
$$\leq b_2 T(r_1 + r_0 |||u|||_T^2).$$

When (5.76) and (5.77) are used in (5.75), it gives

$$|||\Gamma u|||_T^2 \leq 3\{\, \|h\|^2 + 2[b_1 + b_3 T(1 + |||u|||_T^2) + 4b_2 T(r_1 + r_0 |||u|||_T^2)]\},$$

which verifies the boundedness of the map Γ.

To show it is a contraction map, let $u, v \in \mathbf{H}$ and consider

$$\|\Gamma_t u - \Gamma_t v\|^2 = \|\int_0^t G_{t-s}\delta f_s(u,v)ds$$
$$+ \int_0^t G_{t-s}\delta\Sigma_s(u,v)dW_s\|^2 \tag{5.78}$$
$$\leq 2\{\|\int_0^t G_{t-s}\delta f_s(u,v)ds\|^2 + \|\int_0^t G_{t-s}\delta\Sigma_s(u,v)dW_s\|^2\},$$

where we set $\delta f_s(u,,v) = [f_s(u) - f_s(v)]$ and $\delta\Sigma_s(u,v) = [\Sigma_s(u) - \Sigma_s(v)]$. By means of Lemma 6.1 and conditions H, we can deduce that

$$E \sup_{0 \leq t \leq T} \|\int_0^t G_{t-s}\delta f_s(u,v)ds\|^2 \leq T E \int_0^T \|f_s(u) - f_s(v)\|^2 ds$$
$$\leq b_3 T E \int_0^T \|u_s - v_s\|^2 ds \leq b_3 T^2 |||u - v|||_T^2,$$

and

$$E \sup_{0 \leq t \leq T} \|\int_0^t G_{t-s}\delta\Sigma_s(u,v)dW_s\|^2 \leq K_T E \int_0^T Tr\, \delta Q_s(u,v)ds$$
$$= K_T E \int_0^T \int r(x,x)|\Sigma(u,x,s) - \Sigma(v,x,s)|^2 dx\, ds$$
$$\leq K_T b_4 r_0 E \int_0^T \|u_s - v_s\|^2 ds \leq K_T b_4 r_0 T |||u - v|||_T^2.$$

By taking the above two upper bounds into account, the inequality (5.78) gives rise to

$$|||\Gamma u - \Gamma v|||_T^2 \leq 2T(b_3 T + K_T b_4 r_0) |||u - v|||_T^2,$$

which shows that Γ is a contraction map for small T. Again, by the principle of contraction mapping, the equation (5.73) has a unique solution $u \in$

$L^2(\Omega; \mathbf{C}([0,T]; \mathbf{H}))$ as claimed. Since $u = \Gamma u$, from the estimates in (5.75)–(5.77), we can show there exist constants C_3, C_4 such that

$$\|\|u\|\|_T^2 \le C_3 + C_4 \int_0^T \|\|u\|\|_t^2 dt,$$

which, by Gronwall's inequality, implies that $\|\|u\|\|_T^2 = \sup_{0 \le t \le T} E \|u(\cdot, t)\|^2 \le C_T$ for some $C_T > 0$. $\square$

As an example, consider the problem of the jet noise propagation in one dimension. A simple model consists of an acoustic wave propagating over the mean flow field due to a randomly fluctuating source. Let v_0, ρ and c denote the unperturbed flow velocity, fluid density and the local sound speed, respectively, which are assumed to be constants. Then the acoustic velocity $v(x,t)$ and pressure $p(x,t)$ for this model satisfy the following system:

$$\frac{\partial v}{\partial t} + v_0 \frac{\partial v}{\partial x} + \frac{1}{\rho} \frac{\partial p}{\partial x} = f_1(v, p, x, t) + \dot{M}_1(x, t),$$

$$\frac{\partial p}{\partial t} + v_0 \frac{\partial p}{\partial x} + c^2 \rho \frac{\partial v}{\partial x} = f_2(v, p, x, t) + \dot{M}_2(x, t), \quad t \in (0, T), \tag{5.79}$$

$$v(x, 0) = h_1(x), \; p(x, 0) = h_2(x), \quad x \in \mathbf{R}.$$

In the above equations, $f_i(v, p, x, t), i = 1, 2$, are predictable random fields which account for the fluctuation of the flow field; $\dot{M}_i(x, t), i = 1, 2$, are the random source terms, and $h_i(x), i = 1, 2$, are the initial data. In particular let $f_i(v, p, x, t)$ be linear in v and p so that

$$f_i(v, p, x, t) = \bar{f}_i(x, t) + \sum_{j=1}^{2} b_{ij}(x, t) u_j(x, t), \tag{5.80}$$

and, as before,

$$\dot{M}_i(x, t) = \sum_{j=1}^{2} \sigma_{ij}(x, t) \dot{W}_j(x, t), \quad i = 1, 2, \tag{5.81}$$

where we set $u_1 = v, u_2 = p$, and $\bar{f}_i(x, t), b_{ij}(x, t), \sigma_{ij}(x, t)$ are some given predictable random fields. With the notation $u = (v; p)$, the system (5.79) can be rewritten in the matrix form (5.72), with $d = 1, n = 2, \alpha = 0$, as follows

$$\frac{\partial u}{\partial t} = A \frac{\partial u}{\partial x} + f(u, x, t) + \Sigma(x, t) \dot{W}(x, t), \quad t \in (0, T), \tag{5.82}$$

$$u(x, 0) = h(x), \quad x \in \mathbf{R},$$

where

$$A = - \begin{bmatrix} v_0 & 1/\rho \\ c^2 \rho & v_0 \end{bmatrix}$$

has two distinct real eigenvalues $\lambda = v_0 \pm c$. Therefore the system (5.79) is strongly hyperbolic. It follows from Theorem 6.3 that the system has a unique solution under some suitable conditions.

Theorem 6.4 Let $f_i(v, p, x, t), \dot{M}_i(x, t), i = 1, 2$, be given by (5.80), (5.81) respectively. Suppose that $\bar{f}_i(\cdot, t)$ and $\sigma_{ij}(\cdot, t)$ are predictable processes in H such that

$$E\{\int_0^T \|\bar{f}_i(\cdot, s)\|^2 ds + \int_0^T \|\sigma_{ij}(\cdot, s)\|^2 ds\} < \infty, \quad i, j = 1, 2,$$

and $b_{ij}(x, t), i = 1, 2$, are bounded, continuous adapted random fields. If condition (H.4) holds, for $h_i \in H, i = 1, 2$, the Cauchy problem for the system (5.79) has a unique solution $(v; p) \in L^2(\Omega; \mathbf{C}([0, T]; H \times H))$ such that

$$E \sup_{0 \le t \le T} \{\|v(\cdot, t)\|^2 + \|p(\cdot, t)\|^2\} \le C_T,$$

for some constant $C_T > 0$. $\qquad\qquad\square$

From the assumptions, it is easy to check that the conditions (H.1)–(H.4) for Theorem 6.3 are satisfied and, consequently, the conclusion of Theorem 6.4 is true.

Chapter 6

Stochastic Evolution Equations in Hilbert Spaces

6.1 Introduction

So far we have studied two principal types of stochastic partial differential equations, namely the parabolic and hyperbolic equations. To provide a unified theory with a wider range of applications, we shall consider stochastic evolution equations in a Hilbert space setting. We first introduce the notion of Hilbert space-valued martingales in Section 6.2. Then stochastic integrals in Hilbert spaces are defined and an Itô's formula is given in Section 6.3 and Section 6.4, respectively. After a brief introduction to stochastic evolution equations in Section 6.5, we consider, in Section 6.6, the mild solutions by the semigroup approach. This subject was treated extensively in the book by Da Prato and Zabczyk [19] and the article by Walsh [84]. For further development, one is referred to this classic book. Next, in Section 6.7, the strong solutions of linear and nonlinear stochastic evolution equations are studied mainly under the so-called coercivity and monotonicity conditions. These conditions are commonly assumed in the study of deterministic partial differential equations [57] and were introduced to the stochastic counterpart by Bensoussan and Teman [4], [5], and Pardoux [67]. Their function analytic approach will be adopted to study the existence, uniqueness and regularity of strong solutions. We should also mention that a real analytical approach to strong solutions was developed by Krylov [47]. Finally the strong solutions of second-order stochastic evolution equations will be treated separately, in Section 6.8, due to the fact that they are neither coercive nor monotone. Some simple examples are given in various sections, and more interesting ones will be discussed in Chapter Eight.

6.2   Hilbert Space-Valued Martingales

Let H, K be real separable Hilbert spaces with inner products $(\cdot, \cdot)_H$, $(\cdot, \cdot)_K$ and norms $\| \cdot \|_H = (\cdot, \cdot)_H^{1/2}$ and $\| \cdot \|_K = (\cdot, \cdot)_K^{1/2}$, respectively. The subscripts will often be dropped when there is little chance of confusion. Suppose that Γ is a linear operator from K into H. Introduce the following classes of linear operators:

(1) $\mathcal{L}(K; H)$: the space of *bounded linear operators* $\Gamma : K \to H$ with the uniform operator norm $\|\Gamma\|_{\mathcal{L}}$.

(2) $\mathcal{L}_2(K; H)$: the space of *Hilbert-Schmidt operators* with norm defined by

$$\|\Gamma\|_{\mathcal{L}_2} = \{\sum_{k=1}^{\infty} \|\Gamma e_k\|^2\}^{1/2},$$

where $\{e_k\}$ is a complete orthonormal basis for K.

(3) $\mathcal{L}_1(K; H)$: the space of *nuclear (trace class) operators* with norm given by

$$\|\Gamma\|_{\mathcal{L}_1} = Tr\hat{\Gamma} = \sum_{k=1}^{\infty}(\hat{\Gamma}e_k, e_k),$$

where $\hat{\Gamma} = (\Gamma^*\Gamma)^{1/2}$ and Γ^* denotes the adjoint of Γ.

Clearly we have the inclusions: $\mathcal{L}_1(K; H) \subset \mathcal{L}_2(K; H) \subset \mathcal{L}(K; H)$. If $K = H$, we set $\mathcal{L}_k(H, H) = \mathcal{L}_k(H)$, for $k = 0, 1, 2$, with $\mathcal{L}_0(H; H) = \mathcal{L}(H)$. Notice that, If $\Gamma \in \mathcal{L}_1(H)$ is self-adjoint, $\|\Gamma\|_{\mathcal{L}_1} = Tr\ \Gamma$. The class of self-adjoint operators in $\mathcal{L}_1(K)$ will be denoted by $\hat{\mathcal{L}}_1(K)$. The following lemma will be found useful later on.

Lemma 2.1   Let $A : K \to H$ and $B : H \to H$. Then the following inequalities hold.

(1) $\|A\|_{\mathcal{L}} \leq \|A\|_{\mathcal{L}_2} \leq \|A\|_{\mathcal{L}_1}$.

(2) $\|BA\|_{\mathcal{L}_1} \leq \|A\|_{\mathcal{L}_2}\|B\|_{\mathcal{L}_2}$.

(3) $\|BA\|_{\mathcal{L}_j} \leq \|B\|_{\mathcal{L}}\|A\|_{\mathcal{L}_j}, \ j = 1, 2.$ $\square$

Let $(\Omega, \mathcal{F}, P)$ be a complete probability space and let $\{\mathcal{F}_t, t \in [0, T]\}$ be a filtration of increasing sub σ-fields of $\mathcal{F}$. In what follows we shall present some basic definitions and facts about the martingales and Wiener processes in Hilbert spaces. The proofs of the theorems can be founded in several books, such as [19], [63], [72] and [75].

Let M_t, $t \in [0, T]$, be a $\mathcal{F}_t$-adapted K-valued stochastic process defined on Ω such that $E\|M_t\| = E\|M_t\|_K < \infty$. It is said to be a K-*valued martingale* if

$$E\{M_t \mid \mathcal{F}_s\} = M_s, \quad a.s.,$$

for any $s, t \in [0, T]$ with $s \leq t$. Then the norm $\|M_t\|$ is a submartingale so that

$$E\{\|M_t\| \mid \mathcal{F}_s\} \geq \|M_s\| \quad a.s.$$

If there exists a sequence of stopping times $\{\tau_n\}$ with $\tau_n \uparrow \infty$ a.s. such that $M_{t \wedge \tau_n}$ is a K-valued martingale for each n, then M_t is said to be a K-valued local martingale.

If $E\|M_t\|^p < \infty$ with $p \geq 1$, M_t is said to be a K-*valued L^p-martingale*. For such a martingale, if there exists an integrable $\widehat{\mathcal{L}}_1(K)$-valued process $\widehat{Q}_t$ over $[0, T]$ such that

$$\langle (M, g), (M, h) \rangle_t = (\widehat{Q}_t g, h), \quad \text{a.s.} \ \forall\, g, h \in K, \tag{6.1}$$

then define $[M]_t = \widehat{Q}_t$ as the *covariation operator* of M_t. If it has a density Q_t such that

$$[M]_t = \int_0^t Q_s ds,$$

then Q_t is said to be the *local covariation operator* or the *local characteristic operator* of M_t. Let N_t be another continuous K-valued L^2-martingale. The covariation operator $\langle M, N \rangle_t = \widehat{Q}_t^{MN}$ and the *local covariation operator* Q_t^{MN} for M and N are defined similarly as follows,

$$\langle (M, g), (N, h) \rangle_t = (\widehat{Q}_t^{MN} g, h) = \int_0^t (Q_s^{MN} g, h) ds, \quad \text{a.s.} \ \forall\, g, h \in K. \tag{6.2}$$

Lemma 2.2 Let M_t be a continuous L^2-martingale in K with $M_0 = 0$ and covariation operator $\widehat{Q}_t \in \widehat{\mathcal{L}}_1(K)$. Then

$$E\|M_t\|^2 = E \ Tr\,[M]_t = E\,(Tr\,\widehat{Q}_t). \tag{6.3}$$

Proof. Let $\{e_k\}$ be a complete orthonormal basis for K. Then

$$E\|M_t\|^2 = \sum_{k=1}^{\infty} E\,(M_t, e_k)^2. \tag{6.4}$$

Since $m_t^k = (M_t, e_k)$ is a real continuous martingale with $m_0^k = 0$, by Itô's formula, we have

$$|m_t^k|^2 = 2 \int_0^t m_s^k dm_s^k + [m^k]_t,$$

so that

$$E\,|m_t^k|^2 = E[m^k]_t = E\,(\hat{Q}_t e_k, e_k).$$

When this equation is used in (6.4), we obtain

$$E\,\|M_t\|^2 = \sum_{k=1}^{\infty} E\,(\hat{Q}_t e_k, e_k) = E\,(Tr\,\hat{Q}_t),$$

as to be shown. $\square$

Let $\mathcal{M}_T^p(K)$ denotes the Banach space of continuous, K-valued L^p- martingales with norm given by

$$\|M\|_{p,T} = (E \sup_{0 \le t \le T} \|M_t\|^p)^{1/p}.$$

Theorem 2.3   Let $M_t, t \in [0,T]$, be a continuous K-valued, L^p-martingale. Then the following inequalities hold.

(1)   For $p \ge 1$ and for any $\lambda > 0$,

$$P\{ \sup_{0 \le t \le T} \|M_t\| \ge \lambda \} \le \lambda^{-p} E\,\|M_T\|^p. \tag{6.5}$$

(2)   For $p = 1$,

$$E\{ \sup_{0 \le t \le T} \|M_t\| \} \le 3E(Tr\,[\,M\,]_T)^{1/2}. \tag{6.6}$$

(3)   For $p > 1$,

$$E\{ \sup_{0 \le t \le T} \|M_t\|^p \} \le (\frac{p}{p-1})^p E\|M_T\|^p. \tag{6.7}$$

A proof of the inequality (6.6) can be found in (p. 6, [67]). Since $\|M_t\|$ is a positive submartingale, the Doob inequalities (6.5) and (6.7) are well known (see p. 21, [75]).

Theorem 2.4   Let M_t be a $\mathcal{M}_T^2(K)$-martingale with $M_0 = 0$, and let $\{e_k\}$ be a complete orthonormal basis for K. Then M_t has the following series expansion:

$$M_t = \sum_{k=1}^{\infty} m_t^k e_k, \tag{6.8}$$

where $m_t^k = (M_t, e_k)$ and the series converges in $\mathcal{M}_T^2(K)$.

Proof.   Let M_t^n denote the partial sum:

$$M_t^n = \sum_{k=1}^{n} m_t^k e_k, \tag{6.9}$$

where m_t^k is a real-valued martingale with covariation $\hat{q}_t^k = (\hat{Q}_t e_k, e_k)$. It suffices to show that $\{M_t^n\}$ is a Cauchy sequence of martingales in the space $\mathcal{M}_T^2(K)$. To this end, we see clearly that M_t^n is a continuous martingale and, by the maximal inequality (6.7),

$$\|M^n\|_{2,T}^2 = E \sup_{0 \leq t \leq T} \|M_t^n\|^2 \leq 4E \|M_T^n\|^2$$

$$= 4E \sum_{k=1}^n |m_T^k|^2 = E \sum_{k=1}^n (\hat{Q}_T e_k, e_k) \leq E(Tr \hat{Q}_T),$$

which, in view of Lemma 2.1, implies that $M^n \in \mathcal{M}_T^2(K)$. For $m > n$, let $M_t^{m,n} = (M_t^m - M_t^n)$, which is a continuous martingale in K. As before, we have

$$\|M^{m,n}\|_{2,T}^2 = E \sup_{0 \leq t \leq T} \|M_t^{m,n}\|^2$$

$$\leq 4E \sum_{k=n+1}^m |m_T^k|^2 = 4E \sum_{k=n+1}^m (\hat{Q}_T e_k, e_k) \to 0,$$

as $m, n \to \infty$, since $E(Tr \hat{Q}_T) = E \sum_{k=1}^\infty (\hat{Q}_T e_k, e_k) < \infty$. Therefore $\{M^n\}$ is a Cauchy sequence converging in $\mathcal{M}_T^2(K)$ to a limit $M^\infty = \sum_{k=1}^\infty m^k e_k$. Since $(M^\infty, e_k) = m^k = (M, e_k)$ for every $k \geq 1$, we can conclude that the series representation (6.8) for M is valid. $\qquad \square$

A stochastic process X_t in a Hilbert space K is said to be a *K-valued Gaussian process* if for any $g \in K$, (X_t, g) is a real-valued Gaussian process. Let $R \in \hat{\mathcal{L}}_1(K)$. A Wiener process in K with covariance operator R, or a *R-Wiener process* $\{W_t, t \geq 0\}$, is a centered, continuous K-valued Gaussian process with $W_0 = 0$ such that it has stationary, independent increments and the covariance:

$$E(W_t, g)(W_s, h) = (t \wedge s)(Rg, h), \quad \text{for any } s, t \in [0, \infty), \ g, h \in K. \quad (6.10)$$

It is easy to verify that W_t is a continuous $\mathcal{F}_t$-adapted L^2-martingale in K with local covariation operator $Q_t = R$ a.s. $\forall \ t \geq 0$.

Theorem 2.5 Let $\{e_k\}$ be the complete set of eigenfunctions of the trace-class covariance operator R such that $Re_k = \lambda_k e_k$, $k = 1, 2, \cdots$. Then a R-Wiener process W_t has the following series representation

$$W_t = \sum_{k=1}^\infty \sqrt{\lambda_k} b_t^k e_k, \quad (6.11)$$

where $\{b_t^k\}$ is a sequence of independent, identically distributed standard Brownian motions in one dimension given by $b_t^k = (1/\sqrt{\lambda_k})(W_t, e_k)$. Moreover the series converges uniformly on any finite interval $[0, T]$ with probability one.

Proof. Since W_t is a K-valued $\mathcal{F}_t$-martingale, by Lemma 2.1 and Theorem 2.2,

$$E \sup_{0 \leq t \leq T} \|W_t\|^2 \leq 4 E \|W_T\|^2 = 4 T (TrR).$$

Setting $m_t^k = \sqrt{\lambda_k} b_t^k = (W_t, e^k)$, we can apply Theorem 2.3 to show that the the series (6.11) converges in $\mathcal{M}_T^2(K)$ or in $L^2(\Omega; \mathbf{C}([0,T]; K))$. To show it converges in $\mathbf{C}([0,T]; K)$ almost surely, we invoke a theorem (p. 15, [42]) which says, if the terms of the series are mutually independent, symmetrically distributed random variables in a Banach space, then the convergence in probability implies the almost sure convergence. Clearly the terms $\sqrt{\lambda_k} b^k e_k = (W, e_k) e_k$ are symmetrically distributed random variables in the Banach space $\mathbf{C}([0,T]; K)$. It is easy to check that they are mutually independent by computing

$$E (W_t, e_j)(W_t, e_k) = t (Re_j, e_k) = t \lambda_j (e_j, e_k) = 0,$$

if $j \neq k$. Therefore the almost sure convergence of the series follows. □

So far we have considered the R-Wiener process W_t with a trace-class covariance operator. It is also of interest to consider the so-called *cylindrical Wiener process* B_t in K. Formally $B_t = R^{-1/2} W_t$ has the covariance operator I, the identity operator. In fact it makes sense to regard B_t as a generalized Wiener process with values in the dual space K_R' of K_R, where K_R denotes the completion of $(R^{1/2} K)$ with respect to norm $\| \cdot \|_R = \|R^{-1/2} \cdot \|$. Since the inclusion $i : K \hookrightarrow K_R'$ is Hilbert-Schmidt, the triple (i, K, K_R') forms a so-called *Abstract Wiener space* (p. 63, [53]). The cylindrical Wiener process B_t can be defined as a Wiener process in K_R'. In this case, K is known as a *reproducing kernel space* (p. 40, [19]).

Remarks:

(1) By means of an exponential estimate for a submartingale to be given in Lemma 7-6.2, in the proof of Theorem 2.5, one can easily show the a.s. convergence of the series (6.11) to W_t uniformly on $[0, T]$.

(2) $B_t, t \geq 0$, is a Gaussian process in K_R' with $E \langle B_t, g \rangle = 0$ and

$$E \langle B_t, g \rangle \langle B_s, h \rangle = (t \wedge s)(g, h)_K,$$

for any $g, h \in K_R$, where $\langle \cdot, \cdot \rangle$ denotes the duality product between K_R and K_R'.

(3) Let $\Pi_n : K_R' \to K$ be a projection operator such that $\Pi_n \to I$ strongly in K_R'. Then one can show that (p. 66, [53])

$$W_t = R^{1/2} B_t = \lim_{n \to \infty} R^{1/2} \Pi_n B_t$$

in the mean-square.

(4) The formal derivative $\dot{B}_t$ is a space-time white noise satisfying

$$E \langle \dot{B}_t, g \rangle = 0, \quad E \langle \dot{B}_t, g \rangle \langle \dot{B}_s, h \rangle = \delta(t - s)(g, h)_K,$$

for any $g, h \in K_R$, where $\delta(t)$ denotes the Dirac delta function. In fact $\dot{B}_t$ can be regarded as a Hida's white noise in the white noise distribution theory [54].

6.3 Stochastic Integrals in Hilbert Spaces

Let K, H be real separable Hilbert spaces and let $W_t, t \geq 0$, be a R-Wiener process in K defined in a complete probability space $(\Omega, \mathcal{F}, P)$ with a filtration $\mathcal{F}_t, t \geq 0$ of increasing sub σ-fields of $\mathcal{F}$. To define a stochastic integral with respect to the Wiener process W_t, let K_R be the Hilbert subspace of K introduced in the last section with norm $\| \cdot \|_R$ and the inner product

$$(g, h)_R = (R^{-1/2}g, R^{-1/2}h)_K = \sum_{k=1}^{\infty} \frac{1}{\lambda_k}(g, e_k)_K (h, e_k)_K, \quad g, h \in K. \quad (6.12)$$

Denote by $\mathcal{L}_R^2$ the space $\mathcal{L}_2(K_R, H)$ of H.S. operators. We claim that the following lemma holds.

Lemma 3.1 The class $\mathcal{L}_R^2$ of H.S. operators is a separable Hilbert space with norm

$$\|F\|_R = ((F, F))_R^{1/2} = [Tr(FRF^*)]^{1/2}, \quad (6.13)$$

induced by the inner product $((F, G))_R = Tr(GRF^*)$, for any $F, G \in \mathcal{L}_R^2$, where $*$ denotes the adjoint. $\qquad\square$

For the proof, one is referred to Theorem 1.3 in [53]. Now we turn to the stochastic integrals in H.

Let $\Phi_t(\omega) \in \mathcal{L}_R^2$, for $t \in [0, T]$, be a $\mathcal{F}_t$-adapted $\mathcal{L}_R^2$-valued process such that

$$E \int_0^T \|\Phi_s\|_R^2 ds < \infty. \quad (6.14)$$

We will first define a H-valued stochastic integral of the form

$$(\Phi \cdot W) = \int_0^T \Phi_s dW_s, \quad (6.15)$$

from which we then define

$$(\Phi \cdot W)_t = \int_0^T I_{[0,t]}(s)\Phi_s dW_s = \int_0^t \Phi_s dW_s, \quad t \leq T, \quad (6.16)$$

where $I_\Delta(\cdot)$ denotes the indicator function of the set Δ.

As usual we begin with a *simple operator* of the form:

$$\Phi_s^n = \sum_{k=1}^{n} I_{[t_{k-1}, t_k)}(s)\Phi_{k-1}, \qquad (6.17)$$

where $\pi = \{t_0, t_1, \cdots, t_k, \cdots, t_n\}$, with $0 = t_0 < t_1 < \cdots < t_k < \cdots < t_n = T$, is a partition of $[0, T]$, and Φ_k is a $\mathcal{F}_{t_k}$-adapted random operator in $\mathcal{L}_R^2$ such that

$$E \, \|\Phi_k\|_R^2 = E \, Tr(\Phi_k R\Phi_k^*) < \infty, \qquad (6.18)$$

for $k = 0, 1, \cdots, n$.

Let $\mathcal{S}_T(K_R, H)$ denote the set of all simple operators of the form (6.17). For $\Phi \in \mathcal{S}_T(K_R, H)$, define

$$(\Phi \cdot W) = \int_0^T \Phi_s dW_s = \sum_{k=1}^{n} \Phi_{k-1}(W_{t_k} - W_{t_{k-1}}), \qquad (6.19)$$

and, in this case, define the integral (6.16) directly as

$$(\Phi \cdot W)_t = \int_0^t \Phi_s dW_s = \sum_{k=1}^{n} \Phi_{k-1}(W_{t \wedge t_k} - W_{t \wedge t_{k-1}}). \qquad (6.20)$$

Lemma 3.2   For $\Phi \in \mathcal{S}_T(K_R, H)$, $X_t = (\Phi \cdot W)_t$ is a continuous H-valued L^2-martingale such that, for any $g, h \in H$, the following holds

$$E \, (X_t, g) = 0, \quad E \, (X_t, g)(X_s, h) = E \int_0^{t \wedge s} (Q_r g, h) dr, \qquad (6.21)$$

and

$$E \, \|X_t\|^2 = E \int_0^t Tr \, Q_r \, dr, \qquad (6.22)$$

where $(\cdot, \cdot) = (\cdot, \cdot)_H$ and $Q_t = (\Phi_t R\Phi_t^*)$, with $*$ denoting the adjoint.

Proof.   Let $\Phi = \Phi^n$ be the simple operator, where Φ^n is given by (6.17). Then, by definition (6.20),

$$(X_t, g) = \sum_{k=1}^{n}(g, \Phi_{k-1}\Delta W_{t \wedge t_k}), \qquad (6.23)$$

with $\Delta W_{t \wedge t_k} = (W_{t \wedge t_k} - W_{t \wedge t_{k-1}})$. It is easy to check that the sum (6.23) is a continuous martingale in H. Since

$$E \, (g, \Phi_{k-1}\Delta W_{t \wedge t_k}) = E \, E\{(\Phi_{k-1}^* g, \Delta W_{t \wedge t_k})|\mathcal{F}_{t_{k-1}}\} = 0,$$

we have $E\ (X_t, g) = 0$. For $j < k, t < s$,

$$E[(g, \Phi_{j-1}\Delta W_{t \wedge t_j})(h, \Phi_{k-1}\Delta W_{s \wedge t_k})]$$
$$= E\{(g, \Phi_{j-1}\Delta W_{t \wedge t_j})E[(h, \Phi_{k-1}\Delta W_{s \wedge t_k})|\mathcal{F}_{t_{k-1}}]\} = 0.$$

It follows that, for $t \le s$,

$$E(X_t, g)(X_s, h) = \sum_{t_k \le t} E[(g, \Phi_{k-1}\Delta W_{t_k})(h, \Phi_{k-1}\Delta W_{t_k})],$$

where

$$E[(g, \Phi_{k-1}\Delta W_{t_k})(h, \Phi_{k-1}\Delta W_{t_k})]$$
$$= E\ E\{(\Phi_{k-1}^* g, \Delta W_{t_k})(\Phi_{k-1}^* h, \Delta W_{t_k})|\mathcal{F}_{t_{k-1}}\}$$
$$= E(R(\Phi_{k-1}^* g, \Phi_{k-1}^* h)(t_k - t_{k-1}) = E(\Phi_{k-1}R\Phi_{k-1}^* g, h)(t_k - t_{k-1}).$$

Therefore

$$E\ (X_t, g)(X_s, h) = E \sum_{t_k \le t \wedge s} E(\Phi_{k-1}R\Phi_{k-1}^* g, h)(t_k - t_{k-1})$$
$$= E \int_0^{t \wedge s} (\Phi_r R\Phi_r^* g, h)dr,$$

which verifies (6.21). To show (6.22), take a complete orthonormal basis $\{f_k\}$ for H. For $g = h = f_k$, the equation (6.21) yields

$$E\ (X_t, f_k)^2 = E \int_0^t (\Phi_r R\Phi_r^* f_k, f_k)dr,$$

so that

$$E\ \|X_t\|^2 = \sum_{k=1}^\infty E\ (X_t, f_k)^2 = E \int_0^t \sum_{k=1}^\infty (\Phi_r R\Phi_r^* f_k, f_k)dr = E \int_0^t TrQ_r\, dr,$$

where the interchange of expectation, integration and summation can be justified by the monotone convergence theorem and the fact that

$$E \int_0^t TrQ_r\, dr \le T \sup_{1 \le k \le n} E\ (TrQ_k) < \infty,$$

by invoking the property (6.18). $\qquad\square$

To extend the integrand to a more general random operator in $\mathcal{L}_R^2$, we introduce the class $\mathcal{P}_T$. Let $\mathcal{B}_T$ be the σ-algebra generated by sets of the form $\{0\} \times B_0$ and $(s, t] \times B$ for $0 \le s < t < T$, $B_0 \in \mathcal{F}_0$, $B \in \mathcal{F}_s$. Then a $\mathcal{L}_R^2$-valued process Φ_t is said to be a *predictable operator* if the mapping $\Phi:$

$[0, T) \times \Omega \to \mathcal{L}_R^2$ is $\mathcal{B}_T$−measurable. Define $\mathcal{P}_T$ to be the space of predictable operators in $\mathcal{L}_R^2$ equipped with the norm

$$\|\Phi\|_{\mathcal{P}_T} = \{E \int_0^T Tr(\Phi_t R \Phi_t^*) dt\}^{1/2}. \tag{6.24}$$

In particular, it is known that $\Phi^n \in \mathcal{P}_T$. The following lemma is essential in extending a simple integrand to a predictable operator. A proof can be found in (p. 96, [19]).

Lemma 3.3 The set $\mathcal{P}_T$ is separable Hilbert space under the norm (6.24), and the set $\mathcal{S}_T(K_R, H)$ of all simple predictable operators is dense in $\mathcal{P}_T$. $\square$

In view of Lemma 3.2 and (6.24), for $\Phi^n \in \mathcal{S}_T(K_R, H)$, let $X_T^n = (\Phi^n \cdot W)$. Then we have

$$E \ \|X_T^n\|_H^2 = E \int_0^T Tr(\Phi_t^n R \Phi_t^{n*}) dt = E \int_0^T \|\Phi_t^n\|_R^2 \, dt. \tag{6.25}$$

For $\Phi \in \mathcal{P}_T$ satisfying

$$E \int_0^T Tr(\Phi_t R \Phi_t^*) dt < \infty, \tag{6.26}$$

by Lemma 3.3, there exists $\Phi^n \in \mathcal{S}_T(K_R, H)$ such that $\|\Phi - \Phi^n\|_{\mathcal{P}_T} \to 0$ as $n \to \infty$. The equation (6.25) shows that the stochastic integral is an isometric transformation from $\mathcal{S}_T(K_R, H) \subset \mathcal{P}_T$ into the space $\mathcal{M}_T^2$ of H-valued martingales. Define Itô's *stochastic integral of* Φ as the limit

$$(\Phi \cdot W) = \int_0^T \Phi_s dW_s = \lim_{n \to \infty} \int_0^T \Phi_s^n dW_s, \tag{6.27}$$

in $\mathcal{M}_T^2$, and then set

$$(\Phi \cdot W)_t = \int_0^T I_{[0,t]}(s) \Phi_s dW_s = \int_0^t \Phi_s dW_s, \tag{6.28}$$

which is a continuous H-valued martingale.

Remarks:

(1) In fact the Itô integral (6.27) can be defined as the limit of a sequence of integrals with simple predictable operators under a weaker condition [19]:

$$P\{\int_0^T Tr(\Phi_t R \Phi_t^*) dt < \infty\} = 1.$$

In this case the convergence in (6.27) is uniform in $t \in [0, T]$ with probability one, and X_t will be a continuous local martingale in H.

(2) As a special case, for $\Phi_t : H \to \mathbf{R}$, we shall write

$$(\Phi \cdot W)_t = \int_0^t (\Phi_s, dW_s). \tag{6.29}$$

As in the simple predictable case (Lemma 3.2), the Itô integral defined above enjoys the same properties.

Theorem 3.4 For $\Phi \in \mathcal{P}_T$, the stochastic integral $X_t = (\Phi \cdot W)_t$ defined by (6.28), $0 \le t \le T$, is a continuous L^2-martingale in H with the covariation operator

$$[(\Phi \cdot W)]_t = \int_0^t (\Phi_s R \Phi_s^*) ds, \tag{6.30}$$

and it has the following properties:

$$E\ (X_t, g) = 0, \quad E\ (X_t, g)(X_s, h) = E \int_0^{t \wedge s} (\Phi_r R \Phi_r^* g, h) dr, \tag{6.31}$$

for any $g, h \in H$. In particular,

$$E\ \|X_t\|^2 = E \int_0^t Tr(\Phi_s R \Phi_s^*) ds. \qquad \square \tag{6.32}$$

Here we skip the proof, which can be shown by applying Lemmas 3.2 and 3.3 via a sequence of simple predictable approximations.

Instead of stochastic integral in H with respect to a Wiener process W_t, as in the finite-dimensional case, we can also define a stochastic integral with respect a continuous martingale in K. For instance, if $f_t \in H$ is a bounded and predictable process over $[0, T]$, similar to (6.29), we can define $(f \cdot X)_t$ as the integral

$$(f \cdot X)_t = \int_0^t (f_s, dX_s) = \int_0^t (f_s, \Phi_s dW_s).$$

Then, by following the same procedure in defining the Iô integral, we can verify the following lemma which will be useful later.

Lemma 3.5 Let f_t and g_t be any H-valued processes which are bounded and predictable over $[0, T]$. Then the following holds.

$$E\ (f \cdot X)_t = 0,$$
$$E\ (f \cdot X)_t (g \cdot X)_t = E \int_0^t (\Phi_s R \Phi_s^* f_s, g_s) ds. \qquad \square \tag{6.33}$$

In the subsequent applications, we need to define stochastic integral of the form

$$(F_t \Phi \cdot W)_t = \int_0^t F(t, s) \Phi_s dW_s, \tag{6.34}$$

where $F : [0,T] \times [0,T] \to \mathcal{L}(H)$ is bounded and continuous for $s,t \in [0,T]$, and $\Phi_t \in \mathcal{P}_T$ satisfying the boundedness condition (6.26). One way to proceed goes as follows. For $t \in (0,T]$, instead of (6.34), define an integral of the form:

$$Y_t = \int_0^T F(t,s)\Phi_s dW_s, \tag{6.35}$$

as a process in H with $E \int_0^T \|Y_t\|^2 \, dt < \infty$. Then we let $(F_t\Phi \cdot W)_t = \int_0^T I_{[0,t)}(s)F(t,s)\Phi_s dW_s$. To define the integral (6.35), an obvious way is to proceed as before. Starting with a simple predictable approximation, let $\Phi^n \in \mathcal{S}_T(K_R, H)$ and let

$$F^n(t,s) = \sum_{k=1}^n I_{[t_{k-1},t_k)}(s)F(t,t_{k-1}).$$

Then we define

$$Y_t^n = \int_0^T F^n(t,s)\Phi_s^n dW_s = \sum_{k=1}^n F(t,t_{k-1})\Phi_{k-1}(W_{t_k} - W_{t_{k-1}}), \tag{6.36}$$

where, for $0 = t_0 < t_1 < \cdots < t_{n-1} < t_n = T$, Φ_k is a $\mathcal{F}_{t_k}$-adapted random operator from K into H. Since F is bounded, there is $C > 0$ such that

$$E \int_0^T Tr[F(t,t_{k-1})\Phi_{k-1}R\Phi_{k-1}^* F^*(t,t_{k-1})]dt$$
$$\leq CE \int_0^T Tr[\Phi_{k-1}R\Phi_{k-1}^*]dt \leq CT\,E\,\|\Phi_{k-1}\|_R^2 < \infty, \tag{6.37}$$

by condition (6.18). Clearly we have $E\ Y_t^n = 0$, and

$$E \int_0^T \|Y_t^n\|^2 dt = E \int_0^T \{\sum_{k=1}^n Tr[F(t,t_{k-1})\Phi_{k-1}R\Phi_{k-1}^* F^*(t,t_{k-1})]$$
$$\times (t_k - t_{k-1})\}dt \tag{6.38}$$
$$= E \int_0^T \int_0^T Tr[F(t,s)\Phi_s^n R\Phi_s^{n*} F^*(t,s)]dsdt.$$

In general, let $\Phi \in \mathcal{P}_T$ and $F(t,s) \in \mathcal{L}(H), s,t \in [0,T]$ is bounded and continuous. Then we have

$$E \int_0^T \int_0^T Tr[F(t,s)\Phi_s R\Phi_s^* F^*(t,s)]ds \leq \kappa^2 TE \int_0^T Tr(\Phi_s R\Phi_s^*)ds < \infty,$$

where $\kappa = \sup_{s,t} \|F(t,s)\|_{\mathcal{L}(H)}$. By Lemma 3.3, there exists $\Phi^n \in \mathcal{S}_T(K_R; H)$ such that

$$E \int_0^T \int_0^T \|F(t,s)\Phi_s - F^n(t,s)\Phi_s^n\|_R^2 dsdt \to 0$$

as $n \to \infty$. Correspondingly $\{Y^n\}$ is a Cauchy sequence in $L^2(\Omega \times (0, T); H)$ which converges to Y and denotes $Y(t) = \int_0^T F(t, s) \Phi_s^n dW_s$ as the stochastic integral in (6.35). Then we define the stochastic integral (6.34) by

$$(F_t \Phi \cdot W)_t = \int_0^T I_{[0,t]}(s) F(t, s) \Phi_s dW_s = \int_0^t F(t, s) \Phi_s dW_s. \qquad (6.39)$$

This integral will be called an *evolutional stochastic integral*. For the special case when $F(t, s) = F(t - s)$, we have

$$Y_t = \int_0^t F(t - s) \Phi_s dW_s,$$

which will be called a *convolutional stochastic integral*. The following lemma, which seems obvious, will be useful later. The proof is to be omitted.

Theorem 3.6   Let $F(t, s) \in \mathcal{L}(H)$ be bounded and continuous in $s, t \in [0, T]$ such that

$$\sup_{s,t \in [0,T]} \|F(t, s)\|_{\mathcal{L}} < \infty. \qquad (6.40)$$

Then the evolutional stochastic integral $Y_t = \int_0^t F(t, s) \Phi_s dW_s$ has a continuous version which is a $\mathcal{F}_t$-adapted process in H such that $EY_t = 0$ and

$$E(Y_t, Y_s) = \int_0^{t \wedge s} Tr[F(t, r) \Phi_s R \Phi_r^* F^*(s, r)] dr. \qquad \square \qquad (6.41)$$

Remark: Here, for simplicity, we assumed that $F(s, t)$ is uniformly continuous in $s, t \in [0, T]$. In fact the integral Y_t can also be defined under a weaker condition, such as $F(t, s)$ is a continuous for $t \neq s$ under a certain integrability condition over $[0, T] \times [0, T]$.

So far we have defined stochastic integrals with respect to a Wiener process. In a similar fashion, a generalization to stochastic integrals with respect to a continuous K-valued L^2-martingale M_t can be easily carried out. Suppose that M_t has a local characteristic operator Q_t. Let Φ_t be a predictable random operator in $\mathcal{L}(K; H)$ such that

$$E \int_0^T Tr(\Phi_t Q_t \Phi_t^*) dt < \infty. \qquad (6.42)$$

Then we can define the stochastic integral with respect to the martingale

$$Z_t = (\Phi \cdot M)_t = \int_0^t \Phi_s dM_s, \qquad (6.43)$$

analogously as in the case of a Wiener process.

Theorem 3.7   Let M_t be a continuous K-valued martingale with $M_0 = 0$ and local characteristic operator Q_t for $0 \leq t \leq T$. Suppose that Φ_t is a predictable process with values in $\mathcal{L}(K, H)$ such that the property (6.42) is satisfied. The stochastic integral Z_t given by (6.43) is a continuous H-valued L^2-martingale with mean $E\,Z_t = 0$ and covariation operator

$$[\,Z\,]_t = \int_0^t Tr(\Phi_s Q_s \Phi_s^*)ds. \qquad \square \qquad (6.44)$$

6.4   Itô's Formula

Let ξ be $\mathcal{F}_0$-random variable in H with $E\|\xi\|^2 < \infty$. Suppose that V_t is a predictable H-valued process integrable over $[0, T]$ and X_t is a continuous H-valued martingale with $X_0 = 0$ with local characteristic operator Θ_t such that

$$E\left\{\int_0^T \|V_t\|^2 dt + \int_0^T Tr\,\Theta_t dt\right\} < \infty. \qquad (6.45)$$

Let Y_t be a semimartingale given by

$$Y_t = \xi + \int_0^t V_s ds + X_t. \qquad (6.46)$$

For a functional $F(\cdot, \cdot) : H \times [0, T] \to \mathbf{R}$, let $\partial_t F(\eta, t) = \frac{\partial}{\partial t} F(\eta, t)$ and denote the first two partial (Fréchet) derivatives of F with respect to η by $F'(\eta, t) \in H$ and $F''(\eta, t) \in \mathcal{L}(H)$, respectively. F is said to be in *class* $\mathbf{C}_U^2$-*Itô functionals* if it satisfies the following properties:

(1) $F : H \times [0, T] \to \mathbf{R}$ and its partial derivatives $\partial_t F(\eta, t), F'(\eta, t)$ and $F''(\eta, t)$ are continuous in $H \times [0, T]$.

(2) F and its derivatives $\partial_t F$ and $F' \in H$ are bounded and uniformly continuous on bounded subsets of $H \times [0, T]$.

(3) For any $\Gamma \in \mathcal{L}_1(H)$, the map: $(\eta, t) \to Tr[F''(\eta, t)\Gamma]$ is bounded and uniformly continuous on bounded subsets of $H \times [0, T]$.

Under the above conditions we shall present the following Itô's formula without proof.

Theorem 4.1   Let Y_t, $t \in [0, T]$, be a semimartingale given by (6.46) satisfying condition (6.45) and let $F(t, \eta)$ be a $\mathbf{C}_U^2$–Itô functional defined as above.

Then the following formula holds a.s.

$$
\begin{aligned}
F(Y_t, t) = F(\xi, 0) &+ \int_0^t \{\partial_s F(Y_s, s) + (F'(Y_s, s), V_s)\} ds \\
&+ \int_0^t (F'(Y_s, s), dX_s) + \frac{1}{2} \int_0^t Tr[F''(Y_s, s)\Theta_s] ds. \quad \square
\end{aligned}
\tag{6.47}
$$

When F is independent of t, this is a special case of a more general Itô's formula for a possible jump process Y_t given in (Chapter 3, [63]), where the proof is based on a generalization of the finite-dimensional result. For a time-dependent F, it can be easily modified to give the formula (6.47). In particular, when X_t represents the stochastic integral (6.16), we obtain the following Itô's formula as a corollary of Theorem 4.1. A direct proof of this theorem is given in Theorem 4.17 [19].

Theorem 4.2 Let

$$
Y_t = \xi + \int_0^t V_s ds + \int_0^t \Phi_s dW_s,
\tag{6.48}
$$

and let F be a $\mathbf{C}_U^2$-Itô functional. Then the following formula holds.

$$
\begin{aligned}
F(Y_t, t) = F(\xi, 0) &+ \int_0^t \{\partial_s F(Y_s, s) + (F'(Y_s, s), V_s)\} ds \\
&+ \int_0^t (F'(Y_s, s), \Phi_s dW_s) + \frac{1}{2} \int_0^t Tr[F''(Y_s, s)\Phi_s R\Phi_s^\star] ds. \quad \square
\end{aligned}
\tag{6.49}
$$

Later on it will be necessary to generalize the formula under weaker conditions in connection with stochastic evolution equations. As an application of the Itô formula (6.49), we shall sketch a proof of the Burkholder-Davis-Gundy inequality which has been used several times before.

Lemma 4.3 Let $X_t = \int_0^t \Phi_s dW_s$ such that, for $p \geq 1$,

$$
E\{\int_0^T Tr[\Phi_s R\Phi_s^\star] ds]\}^p < \infty.
$$

Then there exists a constant $C_p > 0$ such that

$$
E \sup_{0 \leq s \leq t} \|X_s\|^{2p} \leq C_p E\{\int_0^t Tr[\Phi_s R\Phi_s^\star] ds]\}^p, \quad 0 < t \leq T.
\tag{6.50}
$$

Proof. Since X_t is a martingale, clearly (6.50) holds for $p = 1$. For $p > 1$, let $F(\eta) = \|\eta\|^{2p}$, $\eta \in H$. Then $\partial_t F = 0, F(0) = 0$, and

$$
F'(\eta) = 2p\|\eta\|^{2(p-1)}\eta,
$$
$$
F''(\eta) = 2p\|\eta\|^{2(p-1)}I + 4p(p-1)\|\eta\|^{2(p-2)}(\eta \otimes \eta),
$$

where I is the identity operator and $\otimes$ denotes the tensor product. Let $\xi_t = \sup_{0 \le s \le t} \|X_s\|^{2p}$. Notice that, for $\Gamma \in \mathcal{L}_1(H)$,

$$\begin{aligned} Tr[F''(\eta)\Gamma] &\le 2p\|\eta\|^{2(p-1)}Tr\Gamma + 4p(p-1)\|\eta\|^{2(p-2)}(\Gamma\eta, \eta) \\ &\le 2p(2p-1)\|\eta\|^{2(p-1)}Tr\Gamma, \end{aligned} \tag{6.51}$$

and, by the maximal inequality (6.7),

$$E(\sup_{0 \le s \le t} \|X_s\|^{2p}) \le c_p E\|X_t\|^{2p}, \tag{6.52}$$

for some constant $c_p > 0$. By applying the Itô formula (6.49) and making use of (6.51), we get

$$\begin{aligned} E\|X_t\|^{2p} &\le p\ E \int_0^t \|X_s\|^{2(p-1)}Tr[\Phi_s R\Phi_s^\star]ds \\ &\quad + 2p(p-1)E\int_0^t \|X_s\|^{2(p-2)}(\Phi_s R\Phi_s^\star X_s, X_s)ds. \\ &\le p(2p-1)E\{\sup_{0 \le s \le t} \|X_s\|^{2(p-1)} \int_0^t Tr(\Phi_s R\Phi_s^\star)ds\}. \end{aligned} \tag{6.53}$$

By Hölder's inequality with exponents $p > 1$ and $q = p/(p-1)$, the above yields

$$\begin{aligned} E\|X_t\|^{2p} &\le p(2p-1)\ \{E\sup_{0 \le s \le t} \|X_s\|^{2p}\}^{(p-1)/p} \\ &\quad \times \{E[\int_0^t Tr(\Phi_s R\Phi_s^\star)ds]^p\}^{1/p}. \end{aligned} \tag{6.54}$$

Now it follows from (6.52) and (6.54) that

$$\begin{aligned} E(\sup_{0 \le s \le t} \|X_s\|^{2p}) &\le c_p p(2p-1)\{E\sup_{0 \le s \le t} \|X_s\|^{2p}\}^{(p-1)/p} \\ &\quad \times \{E[\int_0^t Tr(\Phi_s R\Phi_s^\star)ds]^p\}^{1/p}, \end{aligned}$$

or

$$\{E(\sup_{0 \le s \le t} \|X_s\|^{2p})\}^{1/p} \le c_p\, p(2p-1)E[\int_0^t Tr(\Phi_s R\Phi_s^\star)ds]^p,$$

which implies the desired inequality (6.50) with $C_p = [p(2p-1)c_p]^p$. $\square$

6.5   Stochastic Evolution Equations

In the previous chapters, we have studied stochastic parabolic and hyperbolic partial differential equations. For a unified approach, they will be treated as two special cases of stochastic ordinary differential equations in a Hilbert or Banach space, known as stochastic evolution equations. This approach, though more technical mathematically, has the advantage of being conceptually simpler, and the general results can be applied to a wider range of problems. To fix the idea, let us revisit the parabolic Itô equation considered in Chapter Three:

$$
\begin{aligned}
\frac{\partial u}{\partial t} &= Au + f(Ju, x, t) + \sigma(Ju, x, t) \cdot \dot{W}(x, t), \\
u(x, 0) &= h(x), \quad x \in D, \\
u(\cdot, t)|_{\partial D} &= 0, \quad t \in (0, T),
\end{aligned}
\tag{6.55}
$$

where $A = (\kappa \Delta - \alpha)$, $Ju = (u; \partial_x u)$ and

$$
\sigma(Ju, , x, t)] \cdot \dot{W}(x, t) = \sum_{k=1}^{m} \sigma_k(Ju, x, t)] \dot{W}_k(x, t).
$$

As done previously, let $H = L^2(D)$. Write $u_t = u(\cdot, t)$ and $W_t^k = W_k(\cdot, t)$ with $W_t = (W_t^1, \cdots, W_t^m)$, where $W_t^{k'}s$ are independent R_k-Wiener processes in H. Introduce the m-product Hilbert space $K = (H)^m$. Then W_t is a R-Wiener process in K with $R = diag\{R_1, \cdots, R_m\}$ being a diagonal operator in K. We also set $F_t(u) = f(Ju, \cdot, t)$ and define the (multiplication) operator $\Sigma_t(u) : K \to H$ by

$$
[\Sigma_t(u)v](x) = \sigma(Ju, x, t) \cdot v(x) = \sum_{k=1}^{m} \sigma_k(Ju, x, t)v_k(x),
$$

for $v = (v_1 \cdots v_m) \in K$. In view of the homogeneous boundary condition in (6.55), we restrict A to the domain $\mathcal{D}(A) = H^2 \cap H_0^1$. Denote H_0^1 by V, H^{-1} by V'. Let $A : V \to V'$ be continuous and coercive, $F_t(\cdot) : V \to H$ and $\Sigma_t(\cdot) : V \to \mathcal{L}_R^2$ satisfy certain regularity conditions. The equation (6.55) can be interpreted as an Itô equation in V':

$$
u_t = h + \int_0^t Au_s ds + \int_0^t F_s(u_s) ds + \int_0^t \Sigma_s(u_s) dW_s,
\tag{6.56}
$$

where the last term is an Itô integral defined in Section 6.3.

Alternatively, let $g(t - s, x, y)$ denote the Green's function for the linear problem associated with equation (6.55). As in Chapter Three, let G_t denote the Green's operator defined by

$$
[G_t \phi](x) = \int_D g(t, x, y)\phi(y) dy, \quad \phi \in H, \, t > 0.
$$

It is easy to check, as a linear operator in H, $G_t : H \to H$ is bounded and strongly continuous in $[0, \infty)$. Moreover the family $\{G_t, t \geq 0\}$ forms a *semigroup of bounded linear operators* on H with the properties: $G_0 = I$ and $G_t G_s = G_{t+s}$ for any $t, s \geq 0$. The differential operator A is known as the *infinitesimal generator* of the semigroup G_t, which is often written as e^{tA} (see [70]). When written as an integral equation in terms of the Green's operator, it yields the stochastic integral equation in H:

$$u_t = G_t h + \int_0^t G_{t-s} F_s(u_s) ds + \int_0^t G_{t-s} \Sigma_s(u_s) dW_s, \tag{6.57}$$

where the last term is a convolutional stochastic integral with a Green operator G_t as defined in Section 6.3.

The two interpretations given by (6.56) and (6.57) lead to two different notions of solutions: namely the strong and mild solutions, as discussed previously. The formulation (6.57) is often referred to as a semigroup approach. We will generalize the above setup to a large class of problems.

In general let K, H be two real separable Hilbert spaces. Most results to be presented in this section hold when H is complex by a change of notation. Let $V \subset H$ be a reflexive Banach space. Identify H with its dual H' and denote the dual of V by V'. Then we have

$$V \subset H \cong H' \subset V'$$

where the inclusions are assumed to be dense and compact. The triad (V, H, V') is known as a *Gelfand triple*.

For $t \in [0, T]$, let $A_t(\omega)$ be a family of closed $\mathcal{F}_t$-adapted random linear operators in H. $F_t(\cdot, \omega) : V \to H$ and $\Sigma_t(\cdot, \omega) : V \to \mathcal{L}_R^2$, $\omega \in \Omega$, are $\mathcal{F}_t$-adapted random mappings. In K the R-Wiener process is denoted again by W_t, $t \geq 0$. Corresponding to equation (6.55), we consider

$$\begin{aligned} du_t &= A_t u_t dt + F_t(u_t) dt + \Sigma_t(u_t) dW_t, \quad t \in (0, T), \\ u_0 &= h \in H. \end{aligned} \tag{6.58}$$

As before, unless necessary, the dependence on ω will be omitted. The equation (6.56) is now generalized to

$$u_t = h + \int_0^t A_s u_s ds + \int_0^t F_s(u_s) ds + \int_0^t \Sigma_s(u_s) dW_s. \tag{6.59}$$

For $0 \leq s \leq t \leq T$, let $G(t, s)$ be the Green's operator or the propagator associated with A_t satisfying

$$\frac{dG(t, s)}{dt} = A_t G(t, s), \qquad G(s, s) = I, \tag{6.60}$$

where I is the identity operator in H. Corresponding to (6.57), when equation (6.58) is written formally as a stochastic integral equation, it yields

$$u_t = G(t, 0)h + \int_0^t G(t, s) F_s(u_s) ds + \int_0^t G(t, s) \Sigma_s(u_s) dW_s, \tag{6.61}$$

where $G(t, s)$ is a random Green's operator. However, since $G(t, s)$ is not $\mathcal{F}_s$-adapted, the evolutional stochastic integral in (6.61) is not defined. For this reason, in studying the mild solution, we shall assume that $A_t = A$ is a deterministic operator independent of t. In particular, if A_t, F_t and Σ_t are independent of t, the equation (6.58) is said to be *autonomous*.

The following sections will be concerned with the existence and uniqueness questions for equation (6.58). We will take up the mild solution first, because it is technically less complicated.

6.6   Mild Solutions

To consider the mild solutions, for the reason given above, it is necessary to assume that $A_t = A$ is non-random and independent of t with domain $\mathcal{D}(A)$ dense in H. Moreover, we will adopt the semigroup approach by assuming that A generates a strongly continuous C_0-semigroup G_t, for $t \geq 0$, which satisfies the properties: $\lim_{t \downarrow 0} G_t h = h$, $\forall\, h \in H$, and $\|G_t\|_{\mathcal{L}} \leq M \exp\{\alpha t\}$, for some positive constants M, α [70]. In addition, suppose that, for a.e. $(t; \omega) \in [0, T] \times \Omega$, $F_t(\cdot, \omega) : H \to H$ and $\Sigma_t(\cdot, \omega) : H \to \mathcal{L}(K, H)$ are $\mathcal{F}_t$-adapted and satisfy a certain integrability condition. We consider the stochastic equation:

$$
\begin{aligned}
du_t &= [Au_t + F_t(u_t)]dt + \Sigma_t(u_t)dW_t, \quad t \in (0, T), \\
u_0 &= h \in H.
\end{aligned}
\tag{6.62}
$$

Then the integral equation (6.61) for a mild solution takes the form:

$$
u_t = G_t h + \int_0^t G_{t-s} F_s(u_s)ds + \int_0^t G_{t-s} \Sigma_s(u_s)dW_s.
\tag{6.63}
$$

Let us first consider the convolutional stochastic integral

$$
X_t = \int_0^t G_{t-s} \Sigma_s dW_s,
\tag{6.64}
$$

where G_t is a C_0-semigroup and $\Sigma. \in \mathcal{P}_T$ is a predictable process in $\mathcal{L}_R^2$. Then we have

Lemma 6.1   Let $\Sigma \in \mathcal{P}_T$. Then the stochastic integral X_t given by (6.64) is a $\mathcal{F}_t$-adapted, mean-square continuous H-valued process over $[0, T]$ with mean $EX_t = 0$ and the covariance operator

$$
\Lambda_t = E \int_0^t (G_{t-s} \Sigma_s R \Sigma_s^* G_{t-s}^*)ds,
\tag{6.65}
$$

so that

$$E \|X_t\|^2 = E \int_0^t Tr(G_{t-s}\Sigma_s R\Sigma_s^* G_{t-s}^*)ds. \tag{6.66}$$

Proof. For any $\phi, \psi \in H$, we have

$$E (X_t, \phi) = E \int_0^t (\phi, G_{t-s}\Sigma_s dW_s) = E \int_0^t (G_{t-s}^*\phi, \Sigma_s dW_s),$$

and

$$E (X_t, \phi)(X_t, \psi) = E[\int_0^t (G_{t-s}^*\phi, \Sigma_s dW_s)][\int_0^t (G_{t-s}^*\psi, \Sigma_s dW_s)].$$

For a fixed t, let $f_s = G_{t-s}^*\phi$ and $g_s = G_{t-s}^*\psi$. It follows from Lemma 3.6 that

$$E (X_t, \phi) = 0,$$
$$E (X_t, \phi)(X_t, \psi) = (\Lambda_t\phi, \psi) = E \int_0^t (G_{t-s}\Sigma_s R\Sigma_s^* G_{t-s}^*\phi, \psi)ds, \tag{6.67}$$

which imply that $EX_t = 0$ and the equation (6.65) holds. Let $\{\varphi_k\}$ be any complete orthonormal basis of H. By taking $\phi = \psi = \varphi_k$ in the second equation of (6.67) and summing over k, we obtain (6.66).

To show the mean-square continuity, consider

$$E\|X_{t+\tau} - X_t\|^2 = E\| \int_t^{t+\tau} G_{t+\tau-s}\Sigma_s dW_s$$
$$+ \int_0^t (G_{t+\tau-s} - G_{t-s})\Sigma_s dW_s\|^2 \leq 2\{E\| \int_t^{t+\tau} G_{t+\tau-s}\Sigma_s dW_s\|^2 \tag{6.68}$$
$$+E\| \int_0^t (G_{t+\tau-s} - G_{t-s})\Sigma_s dW_s\|^2\} = 2(J_\tau + K_\tau).$$

Similar to (6.66), we can get

$$J_\tau = E \int_t^{t+\tau} \|(G_{t+\tau-s}\Sigma_s\|_R^2 ds \leq CE \int_t^{t+\tau} \|\Sigma_s\|_R^2 ds,$$

and

$$K_\tau = E \int_t^{t+\tau} \|G_{t-s}(G_\tau - I)\Sigma_s\|_R^2 ds \leq CE \int_t^{t+\tau} \|(G_\tau - I)\Sigma_s\|_R^2 ds,$$

where we recalled that $\|\Phi\|_R^2 = Tr(\Phi R\Phi^*)$ and made use of the bound: $\|G_t\|_{\mathcal{L}(H)}^2 \leq C$. Since $J_\tau \to 0$ and $K_\tau \to 0$ as $\tau \to 0$ by the dominated convergence theorem, the mean-square continuity of X follows from (6.68). $\square$

In what follows, we shall present two maximal inequalities that are essential in studying mild solutions. To proceed we need a couple of definitions. A C_0-semigroup G_t with infinitesimal generator A is said to be a *contraction semigroup* if $\|G_t h\| \leq \|h\|$ for $h \in H$ and $(A\varphi, \varphi) \leq 0$ for $\varphi \in \mathcal{D}(A)$. In particular, when $\{U_t, -\infty \leq t < \infty\}$ is a C_0-group of *unitary operators* in a Hilbert space H, we have $\|U_t h\| = \|h\|$ and $U_{-t} = U_t^{-1} = U_t^\star$, where U_t^{-1} and $U_t^\star$ denote the inverse and the adjoint of U_t, respectively [70].

Theorem 6.2　Suppose that A generates a contraction semigroup G_t on H and $\Sigma \in \mathcal{P}_T$ satisfies the condition:

$$E \int_0^T [Tr(\Sigma_s R \Sigma_s^\star)]^p ds < \infty, \tag{6.69}$$

for $p \geq 1$. Then there exists a constant $C_p > 0$ such that, for any $t \in [0, T]$,

$$E\{ \sup_{0 \leq r \leq t} \| \int_0^r G_{t-s} \Sigma_s dW_s \|^{2p} \} \leq C_p E\{ \int_0^t \|\Sigma_s\|_R^2 ds \}^p, \tag{6.70}$$

where $\|\Sigma_s\|_R^2 = Tr(\Sigma_s R \Sigma_s^\star)$. Moreover the stochastic integral X_t given by (6.64) has a continuous version.

Proof.　Since the C_0-semigroup $\{G_t, t \geq 0,\}$ is a contraction, by making use of a theorem of Sz.-Nagy and Foias (Theorem I.8.1, [76]), there exist a Hilbert space $\mathcal{H} \supset H$ and a unitary group of operators $\{U_t, -\infty \leq t \leq \infty\}$ on $\mathcal{H}$ such that $G_t h = PU_t h, \forall h \in H$, where $P : \mathcal{H} \to H$ is an orthogonal projection. Therefore, noticing $U_{t-s} = U_t U_s^\star$, the convolutional stochastic integral (6.64) can written as

$$X_t = \int_0^t PU_{t-s} \Sigma_s dW_s = PU_t Y_t, \tag{6.71}$$

where

$$Y_t = \int_0^t U_s^\star \Sigma_s dW_s, \tag{6.72}$$

is a well-defined stochastic integral in $\mathcal{H}$. Since

$$E \int_0^T [Tr_\mathcal{H}(U_s \Sigma_s R \Sigma_s^\star U_s^\star)]^p ds \leq E \int_0^T [Tr(\Sigma_s R \Sigma_s^\star)]^p ds < \infty, \tag{6.73}$$

it follows from Lemma 4.3 that

$$E\{ \sup_{0 \leq r \leq t} \|Y_r\|_\mathcal{H}^{2p} \} \leq C_p E\{ \int_0^t [Tr(\Sigma_s R \Sigma_s^\star)]^p ds. \tag{6.74}$$

In view of (6.71) and (6.74), we obtain

$$E\{ \sup_{0 \leq r \leq t} \int_0^r \|X_r\|^2 \} \leq E\{ \sup_{0 \leq r \leq t} \int_0^r \|Y_r\|_\mathcal{H}^2 \}$$
$$\leq C_p E\{ \int_0^t [Tr(\Sigma_s R \Sigma_s^\star)]^p ds, \tag{6.75}$$

which verifies the maximal inequality (6.70). The continuity of X_t follows from the fact that the stochastic integral Y_t has a continuous version in $\mathcal{H}$. $\square$

Remark: The maximal inequality for the convolutional stochastic integral has been established by several authors (see, e.g., [79], [46]). A simple proof given here based on expressing G_t as the product of the projector P and the unitary operator U_t was adopted from the paper [35].

The above theorem was extended to the case of a general C_0-semigroup G_t (see Proposition 7.3, [19]), where, by the *method of factorization*, the maximal inequality was proved with a slightly weaker upper bound.

Theorem 6.3   Let G_t be a strongly continuous semigroup on H and let the following condition hold:

$$E\{\int_0^t \|\Sigma_s\|_R^{2p} ds\} < \infty.$$

Then, for $p > 1$, $t \in [0, T]$, there exists constant $C_{p,T} > 0$ such that the following inequality holds

$$E\{\sup_{0 \le r \le t} \|\int_0^r G_{t-s}\Sigma_s dW_s\|^{2p}\} \le C_{p,T} E\{\int_0^t \|\Sigma_s\|_R^{2p} ds\}. \tag{6.76}$$

Moreover the process X_t in H has a continuous version. $\square$

Now consider the linear equation with an additive noise:

$$\begin{aligned} du_t &= [Au_t + f_t]dt + \Sigma_t dW_t, \quad t \in (0, T), \\ u_0 &= h \in H, \end{aligned} \tag{6.77}$$

where f_t is a $\mathcal{F}_t$-adapted, locally integrable process in H. Then the mild solution of (6.77) is given by

$$u_t = G_t h + \int_0^t G_{t-s} f_s ds + \int_0^t G_{t-s} \Sigma_s dW_s. \tag{6.78}$$

With the aid of the above theorems, we can prove the following result.

Theorem 6.4   Suppose that G_t is a strongly continuous semigroup on H, f_t is a $\mathcal{F}_t$-adapted, locally integrable process in H and $\Sigma \in \mathcal{P}_T$ such that, for $p \ge 2$,

$$E \int_0^T (\|f_s\|^p + \|\Sigma_s\|_R^p) ds < \infty. \tag{6.79}$$

Then, for $p = 2$, the mild solution of the linear equation (6.77) given by (6.78) is a mean-square continuous, $\mathcal{F}_t$-adapted process in H and there exists constant $C_1(T) > 0$ such that

$$\sup_{0 \le t \le T} E\|u_t\|^2 \le C_1(T)\{\|h\|^2 + E\int_0^T (\|f_s\|^2 + \|\Sigma_s\|_R^2)ds\}. \qquad (6.80)$$

For $p > 2$, the solution has a continuous sample path in H with the properties: $u \in L^p(\Omega; \mathbf{C}([0,T];H))$ and there exists a constant $C_p(T) > 0$ such that

$$E\{\sup_{0 \le t \le T} \|u_t\|^p\} \le C_p(T)\{\|h\|^p + E\int_0^T (\|f_s\|^p + \|\Sigma_s\|_R^p)ds\}. \qquad (6.81)$$

Proof. From (6.78), we have

$$\|u_t\|^p = 3^p\{\|G_t h\|^p + \|\int_0^t G_{t-s}f_s ds\|^p + \|\int_0^t G_{t-s}\Sigma_s dW_s\|^p\}$$
$$\le 3^p\{\alpha^p\|h\|^p + \alpha^p T^{(p-1)}\int_0^t \|f_s\|^p ds + \|X_t\|^p\}, \qquad (6.82)$$

for some $\alpha > 0$, where X_t is given by (6.64). For $p = 2$, it is easy to show that u_t belongs to $\mathbf{C}([0,T];L^2(\Omega;H))$ and the inequality (6.80) follows after taking an expectation of (6.82).

For $p > 2$, by making use of Theorem 4.3, the inequality (6.82) yields

$$E\{\sup_{0 \le r \le t} \|u_r\|^p\} \le 3^p\{\alpha^p\|h\|^p + \alpha^p T^{(p-1)}E\int_0^t \|f_s\|^p ds + E(\sup_{0 \le r \le t} \|X_r\|^p)\}$$
$$\le 3^p\{\alpha\|h\|^p + T^{(p-1)}E\int_0^T \|f_s\|^p ds + C_{p/2,T}E\int_0^t \|\Sigma_s\|_R^p ds\},$$

which implies the desired inequality (6.81). The continuity of the solution can be shown by making use of the Kolmogorov criterion. □

Remark: Here and hereafter, for conciseness, the initial datum $u_0 = h$ is assumed to be non-random. The theorem is still true if $u_0 = \xi \in H$ is a $\mathcal{F}_0$-adapted L^p-random variable such that $E\|\xi\|^p < \infty$. In this case, we simply replaced the term $\|h\|^p$ in (6.81) by $E\|\xi\|^p$.

We now turn to the nonlinear stochastic equation (6.62) with Lipschitz-continuous coefficients and a sublinear growth. To be precise, assume the following conditions A:

(A.1) A is the infinitesimal generator of a strongly continuous semigroup $G_t, t \ge 0$, in H.

(A.2) For $\phi \in H, t \in [0,T]$, $F_t(\phi, \omega)$ is a $\mathcal{F}_t$-adapted, locally integrable H-valued process, and $\Sigma_t(\phi, \omega)$ is a $\mathcal{L}_R^2$-valued process in $\mathcal{P}_T$.

(A.3) There exist positive constants b and c such that

$$E \int_0^T [\, \|F_s(0,\omega)\|^p + \|\Sigma_s(0,\omega)\|_R^p \,] ds \leq b,$$

with $p \geq 2$, and, for any $g \in H$,

$$\|F_t(g,\omega) - F_t(0,\omega)\|^2 + \|\Sigma_t(g,\omega) - \Sigma_t(0,\omega)\|_R^2 \leq c(1 + \|g\|^2),$$

for a.e. $(t,\omega) \in [0,T] \times \Omega$.

(A.4) There exists a constant $\kappa > 0$ such that, for any $g, h \in H$,

$$\|F_t(g,\omega) - F_t(h,\omega)\|^2 + \|\Sigma_t(g,\omega) - \Sigma_t(h,\omega)\|_R^2 \leq \kappa \|g - h\|^2,$$

for a.e. $(t,\omega) \in [0,T] \times \Omega$.

Under conditions A, we can generalize Theorem 6.4 to the nonlinear case as follows.

Theorem 6.5   Suppose that the conditions (A.1)–(A.4) are satisfied. Then, for $h \in H$ and $p \geq 2$, the nonlinear equation (6.62) has a unique mild solution u_t, $0 \leq t \leq T$ with $u \in \mathbf{C}([0,T]; L^p(\Omega; H))$ such that

$$E\{ \sup_{0 \leq t \leq T} \|u_t\|^p \} \leq K_p(T)\{ 1 + E \int_0^T [\, \|F_s(0)\|^p + \|\Sigma_s(0)\|_R^p \,] ds \}, \qquad (6.83)$$

for some constant $K_p(T) > 0$. For $p > 2$, the solution has continuous sample paths with $u \in L^p(\Omega; \mathbf{C}([0,T]; H))$.

Proof.   We will first take care of the uniqueness question. Suppose that u_t and $\tilde{u}_t$ are both mild solutions satisfying the integral equation (6.63). Set $v_t = u_t - \tilde{u}_t$. Then the following equation holds:

$$v_t = \int_0^t G_{t-s}[\, F_s(u_s) - F_s(\tilde{u}_s)\,] ds + \int_0^t G_{t-s}[\, \Sigma_s(u_s) - \Sigma_s(\tilde{u}_s)\,] dW_s. \quad (6.84)$$

Let $\alpha = \sup_{0 \leq t \leq T} \|G_t\|_{\mathcal{L}(H)}$. By making use of condition (A.4), we can deduce from (6.84) that

$$E\, \|v_t\|^2 \leq 2\alpha^2 E\{ \int_0^t [\, \|F_s(u_s) - F_s(\tilde{u}_s)\|^2 + \|\Sigma_s(u_s) - \Sigma_s(\tilde{u}_s)\|_R^2 \,] ds \}$$

$$\leq 2\alpha^2 \kappa \int_0^t E\, \|v_s\|^2 ds.$$

It then follows from the Gronwall inequality that

$$E\, \|v_t\|^2 = E\, \|u_t - \tilde{u}_t\|^2 = 0,$$

for all t in $[0, T]$. Therefore the solution is unique.

The existence proof is based on the standard contraction mapping principle. Since the proof of existence and the inequality (6.83) for $p = 2$ is relatively simple, we will prove only the second part of the theorem with $p > 2$. This will be done in steps.

(Step 1) For $p > 2$, let $\mathbf{X}_{p,T}$ denote the Banach space of $\mathcal{F}_t$-adapted, continuous processes in H with norm

$$\|u\|_{p,T} = \{ E \sup_{0 \leq t \leq T} \|u_t\|^p \}^{1/p}. \tag{6.85}$$

Define the mapping $\Gamma : \mathbf{X}_{p,T} \to \mathbf{X}_{p,T}$ as follows:

$$\mu_t = \Gamma_t(u) = G_t h + \int_0^t G_{t-s} F_s(u_s) ds + \int_0^t G_{t-s} \Sigma_s(u_s) dW_s. \tag{6.86}$$

To show this map is well defined, let $I_1(t) = G_t h$;

$$I_2(t) = \int_0^t G_{t-s} F_s(u_s) ds, \quad \text{and} \quad I_3(t) = \int_0^t G_{t-s} \Sigma_s(u_s) dW_s,$$

so that $\mu_t = \Gamma_t(u) = \sum_{i=1}^{3} I_i(t)$. Then

$$\|\Gamma u\|_{p,T} \leq \sum_{i=1}^{3} \|I_i\|_{p,T} = \sum_{i=1}^{3} \eta_i(T), \tag{6.87}$$

where $\eta_i(T) = \{ E \sup_{0 \leq t \leq T} \|I_i(t)\|^p \}^{1/p}, \quad i = 1, 2, 3.$

We shall estimate $\eta_i^p(T)$ separately. First we have

$$\eta_1^p(T) = \sup_{0 \leq t \leq T} \|G_t h\|^p \leq \alpha^p \|h\|^p, \tag{6.88}$$

and

$$\eta_2^p(T) = E(\sup_{0 \leq t \leq T} \| \int_0^t G_{t-s} F_s(u_s) ds \|^p) \leq \alpha^p E[\int_0^T \|F_s(u_s)\| ds]^p$$

$$\leq \alpha^p E\{ \int_0^T [\|F_s(u_s) - F_s(0)\| + \|F_s(0)\|] ds \}^p.$$

By making use of condition (A.3) and the Hölder inequality, we can deduce from the above that

$$\eta_2^p(T) \leq (2\alpha)^p T^{p-1} E\{ (2c)^{p/2} \int_0^T (1 + \|u_s\|_1^p) ds + \int_0^T \|F_s(0)\|^p ds \} \tag{6.89}$$

$$\leq (2\alpha)^p T^{p-1} \{ b + 2^p (2c)^{p/2} T (1 + \|u\|_{p,T}^p) \}.$$

Similarly, by Theorem 5.3 and (A.3), we have

$$\eta_3^p(T) \le C_{p/2,T} \{E \int_0^T \|\Sigma_s\|_R^2 ds\}^{p/2}$$

$$\le 2^p C_{p/2,T} T^{(p-2)/2} \{ b + (2c)^{p/2} T (1 + \|u\|_{p,T}^p) \}. \tag{6.90}$$

By means of Theorem 6.4, it can be shown that $\mu_t = \Gamma_t(u)$ is continuous in t. This fact together with (6.87), the estimates (6.88), (6.89) and (6.90) for I_i show that Γ maps $\mathbf{X}_{p,T}$ into itself.

(Step 2) We will show that Γ is a contraction mapping for a small $T = T_1$. To this end, for $u, \tilde{u} \in \mathbf{X}_{p,T}$, we define

$$J_1(t) = \int_0^t G_{t-s}[F_s(u_s) - F_s(\tilde{u}_s)]ds,$$

$$J_2(t) = \int_0^t G_{t-s}[\Sigma_s(u_s) - \Sigma_s(\tilde{u}_s)]dW_s.$$

Then

$$\|\Gamma(u) - \Gamma(\tilde{u})\|_{p,T} \le \|J_1\|_{p,T} + \|J_2\|_{p,T}. \tag{6.91}$$

By condition (A.4),

$$\|J_1(t)\| \le \alpha \int_0^t \|F_s(u_s) - F_s(\tilde{u}_s)\|ds,$$

so that

$$\|J_1\|_{p,T} \le \alpha \kappa^{1/2} T \|u - \tilde{u}\|_{p,T}. \tag{6.92}$$

By virtue of Theorem 6.3 and condition (A.4), we can get

$$\|J_2\|_{p,T}^p \le C_{p/2,T} \{E \int_0^t \|\Sigma_s(u_s) - \Sigma_s(\tilde{u}_s)\|_R^2 ds\}^{p/2}$$

$$\le C_{p/2,T} \kappa^{p/2} T^{(p-2)/2} E \int_0^T \|u_s - \tilde{u}_s\|^p ds \le C_{p/2,T} (\kappa T)^{p/2} \|u - \tilde{u}\|_{p,T}^p,$$

or

$$\|J_2\|_{p,T} \le (\kappa T)^{1/2} (C_{p/2,T})^{1/p} \|u - \tilde{u}\|_{p,T}. \tag{6.93}$$

By taking (6.91), (6.92) and (6.93) into account, we get

$$\|\Gamma(u) - \Gamma(\tilde{u})\|_{p,T} \le \rho(T)\|u - \tilde{u}\|_{p,T},$$

where $\rho(T) = \alpha\{\kappa^{1/2}T + (\kappa T)^{1/2}(C_{p/2,T})^{1/p}\}$. Let $T = T_1$ be sufficiently small so that $\rho(T_1) < 1$. Then Γ is a Lipschitz-continuous contraction mapping in $\mathbf{X}_{p,T}$ which has a fixed point u. This element u_t is the unique mild solution of (6.62). For $T > T_1$, we can extend the solution by continuation, from T_1 to T_2 and so on. This completes the existence and uniqueness proof.

(Step 3) To verify the inequality (6.83), we make a minor change of the estimates for η_2^p and η_3^p in (6.88) and (6.90) to get

$$\eta_2^p(T) \le (2\alpha)^p \, T^{p-1} \{ \, b + (2c)^{p/2} (T + \int_0^T \|u\|_{p,s}^p ds) \, \},$$

and

$$\eta_3(T) \le 2^p \, C_{p/2,T} \, T^{(p-2)/2} \{ \, b + (2c)^{p/2} (T + \int_0^T \|u\|_{p,s}^p ds) \, \}.$$

Therefore it follows from (6.87) and the previous estimates that

$$\|u\|_{p,T}^p = \|\Gamma u\|_{p,T}^p \le 3^p \sum_{i=1}^3 \eta_i^p(T)$$

$$\le C_1(1 + \|h\|^p) + C_2 \{ \int_0^T \|u\|_{p,s}^p ds + E \int_0^T [\, \|F_s(0)\|^p + \|\Sigma_s(0)\|_R^p \,] ds \, \},$$

for some constants C_1, C_2 depending on p, T. With the aid of Gronwall's lemma, the above inequality implies the desired inequality (6.83). $\quad\square$

Remarks:

(1) As in the linear case, the theorem holds when the initial datum $u_0 = \xi$ is a $\mathcal{F}_0$-measurable L^p- random variable in H with $\|h\|^p$ in (6.83) replaced by $E\|\xi\|^p$. Also here we assumed the Hilbert space H is real. The theorem holds for a complex Hilbert space as well.

(2) For simplicity we impose the standard Lipschitz-continuity and linear growth conditions on the nonlinear terms. As seen from concrete problems in the previous chapters, the Lipschitz condition may be relaxed to hold locally. But, then, the solution may also be local, unless there exists an *a priori* bound on the solution.

(3) If the semigroup is analytic and the coefficients are smooth, it is possible to study the regularity properties of a mild solution in Sobolev subspaces of H. More extensive treatment of mild solutions is given in [19].

To connect the general theorem with specific problems studied in Chapters Three and Four, we will give two examples. Further applications will be provided later.

(Example 6.1) Parabolic Itô Equation

As in Chapter Three, let $D \subset \mathbf{R}^d$ be a bounded domain with a smooth boundary ∂D. We assume that

(1) Let A be a second-order strongly elliptic operator in D given by

$$A = \sum_{j,k=1}^{d} \frac{\partial}{\partial x_j}\left[a_{jk}(x)\frac{\partial}{\partial x_k}\right] + \sum_{j=1}^{d} a_j(x)\frac{\partial}{\partial x_j} + a_0(x), \qquad (6.94)$$

with smooth coefficients $a's$ on $\bar{D}$ such that

$$\sum_{j,k=1}^{d} \int_{D} a_{jk}(x)\frac{\partial\varphi(x)}{\partial x_j}\frac{\partial\varphi(x)}{\partial x_k}dx$$

$$> \sum_{j=1}^{d} \int_{D} a_j(x)\frac{\partial\varphi(x)}{\partial x_j}\varphi(x)dx + \int_{D} a_0(x)\varphi^2(xdx),$$

for any $\mathbf{C}^1$-function φ on D.

(2) Let $f(r,x,t)$ and $\sigma(r,x,t)$ be real-valued continuous functions on $\mathbf{R} \times D \times [0,T]$. Suppose that there exist constants $b_1, b_2 > 0$ such that

$$|f(r,x,t)|^2 + |\sigma(r,x,t)|^2 \le b_1(1+|r|^2),$$

$$|f(0,x,t)| + |\sigma(0,x,t)| \le b_2,$$

for any $(r,x,t) \in \mathbf{R} \times D \times [0,T]$.

(3) There is a positive constant b_3 such that

$$|f(r,x,t) - f(s,x,t)| + |\sigma(r,x,t) - \sigma(s,x,t)| \le b_3|r-s|,$$

for any $r,s \in \mathbf{R}$, $x \in D$ and $t \in [0,T]$.

(4) $W_t(x)$ is a R-Wiener process in $L^2(D)$ with covariance function $r(x,y)$ which is bounded and continuous for $x,y \in D$.

Now consider the parabolic Itô equation

$$\frac{\partial u}{\partial t} = Au + f(u,x,t) + \sigma(u,x,t)\frac{\partial}{\partial t}W(x,t), \quad x \in D, t \in (0,T),$$

$$u(x,0) = h(x), \quad u|_{\partial D} = 0.$$
$$(6.95)$$

Let $H = L^2(D)$. Under condition (1), by the semigroup theory (p. 210, [70]), with the domain $\mathcal{D}(A) = H^2 \cap H_0^1$, A generates a contraction semigroup G_t in H. Setting $F_t(g) = f(g(\cdot),\cdot,t)$ and regarding $\Sigma_t(g) = \sigma(g(\cdot),\cdot,t)$ as a multiplication operator from $K = H$ into H, it is easy to check that conditions (1)–(3) imply that conditions (A1)–(A4) for Theorem 6.5 are satisfied. Moreover condition (4) means that $W_t = W(\cdot,t)$ is a R-Wiener process K with $TrR < \infty$. By applying Theorem 6.5, we conclude that the initial-boundary problem (6.94) has a unique continuous mild solution $u \in L^p(\Omega; \mathbf{C}([0, T]; H))$ for any $p > 2$, and there exists constant $K(p, T) > 0$ such that

$$E\{ \sup_{0 \le t \le T} (\int_D |u(x, t)|^2 dx)^{p/2} \} \le K_p(T)\{ 1 + \int_0^T [\int_D f^2(0, x, t) \, dx]^{p/2} dt$$

$$+ \int_0^T [\int_D r(x, x)\sigma^2(0, x, t) dx]^{p/2} \} dt.$$

(Example 6.2) Hyperbolic Itô Equation

Consider the initial-boundary value problem in domain D:

$$\frac{\partial^2 v}{\partial t^2} = Av + f(v, x, t) + \sigma(v, x, t) \cdot \frac{\partial}{\partial t} W(x, t),$$

$$v|_{\partial D} = 0, \tag{6.96}$$

$$v(x, 0) = g(x), \quad \frac{\partial v}{\partial t}(x, 0) = h(x).$$

Here we assume that A is given by (6.94) with $a_j = 0$ for $j \ne 0$. Then the equation (6.94) becomes

$$A = \sum_{j,k=1}^d \frac{\partial}{\partial x_j} [a_{jk}(x) \frac{\partial}{\partial x_k}] + a_0(x), \tag{6.97}$$

where $a_{jk} = a_{kj}$ and a_0 are smooth coefficients. Then A with domain $H^2 \cap H_0^1$ is a self-adjoint, strictly negative operator. Let $u = \begin{bmatrix} v \\ v' \end{bmatrix}$ with $v_0 = \begin{bmatrix} g \\ h \end{bmatrix}$,

$\mathcal{A} = \begin{bmatrix} 0 & 1 \\ A & 0 \end{bmatrix}$, $F_t(u) = \begin{bmatrix} 1 \\ f(v, \cdot, t) \end{bmatrix}$ and $\Sigma_t(u) = \begin{bmatrix} 1 \\ \sigma(v, \cdot, t) \end{bmatrix}$.

Introduce the spaces $H = L^2(D)$ and $\mathbb{H} = H_0^1 \times H$ with the energy norm $\|u\|_e = \{ \|v\|_1^2 + \|v'\|^2 \}^{1/2}$, where $\| \cdot \|_1$ and $\| \cdot \|$ denote the norms in $H^1(D)$ and $L^2(D)$, respectively. Let $\mathcal{D}(\mathcal{A}) = \mathcal{D}(A) \times H^1$. Then the equation (6.96) can be put in the standard form (6.62) with A replaced by $\mathcal{A}$. It is known that $\mathcal{A}$ generates a strongly continuous semigroup, in fact, a group in $\mathbb{H}$ (p. 423, [86]). Under the conditions (1)–(4) in Example 6.1, the assumptions (A1)–(A4) are satisfied. Therefore, by applying Theorem 6.5, we can conclude that, for $g \in H_0^1, h \in L^2(D)$, the problem (6.96) has a unique continuous mild solution $u \in L^p(\Omega; \mathbf{C}([0, T]; H_0^1))$ for any $p > 2$, and there exists constant $K_p(T) > 0$ such that

$$E\{ \sup_{0 \le t \le T} [\|u(\cdot, t)\|_1^2 + \|\frac{\partial u}{\partial t}(\cdot, t)\|^2] \} \le K_p(T) \{ 1 + \int_0^T [\int_D f^2(0, x, t) \, dx]^{p/2} dt$$

$$+ \int_0^T [\int_D r(x, x)\sigma^2(0, x, t) dx]^{p/2} \} dt.$$

6.7   Strong Solutions

As in Section 6.5, let K, H be two real separable Hilbert spaces and let $V \subset H$ be a reflexive Banach space with dual space V'. By identifying H with its dual H', we have

$$V \subset H \cong H' \subset V'$$

where the inclusions are assumed to be dense and compact. Following the previous notation, let the norms in V, H and V' be denoted by $\| \cdot \|_V, \| \cdot \|$ and $\| \cdot \|_{V'}$, respectively. The inner product in H and the duality scalar product between V and V' will be denoted by $(\cdot, \cdot)$ and $\langle \cdot, \cdot \rangle$.

Let $A_t(\omega) : V \to V'$, $F_t(\cdot, \omega) : V \to V'$ and $\Sigma_t(\cdot, \omega) : V \to \mathcal{L}_R^2$ a.e. $(t, \omega) \in \Omega_T = [0, T] \times \Omega$. As usual W_t is a K-valued R-Wiener process with $TrR < \infty$. Here we say that a $\mathcal{F}_t$-adapted V-valued process u_t is a *strong solution* of the equation (6.55) if $u \in L^2(\Omega_T; V)$ and, for any $\phi \in V$, the following equation

$$
\begin{aligned}
(u_t, \phi) = (h, \phi) &+ \int_0^t \langle A_s u_s, \phi \rangle ds + \int_0^t (F_s(u_s), \phi) ds \\
&+ \int_0^t (\phi, \Sigma_s(u_s) dW_s)
\end{aligned}
\tag{6.98}
$$

holds for each $t \in [0, T]$ a.s. Alternatively, u_t satisfies the Itô equation (6.56) in V'.

We shall impose the following conditions on A_t:

(B.1) Let A_t be a continuous family of closed random linear operator with domain $\mathcal{D}(A)$ (independent of t, ω) dense in H such that $A_t : V \to V'$ and, for any $v \in V$, $A_t v$ is an adapted continuous V'-valued process.

(B.2) For any $u, v \in V$ there exists $\alpha > 0$ such that

$$|\langle A_t u, v \rangle| \leq \alpha \|u\|_V \|v\|_V, \quad a.e. \ (t, \omega) \in \Omega_T.$$

(B.3) A_t satisfies the coercivity condition: There exist constants $\beta > 0$ and γ such that

$$\langle A_t v, v \rangle \leq -\beta \|v\|_V^2 + \gamma \|v\|^2, \quad \forall \ v \in V, \ a.e. \ (t, \omega) \in \Omega_T.$$

Before dealing with the equation (6.98), consider the linear problem:

$$
\begin{aligned}
du_t &= A_t u_t dt + f_t dt + \Phi_t dW_t, \ 0 < t < T, \\
u_0 &= h,
\end{aligned}
\tag{6.99}
$$

or

$$u_t = h + \int_0^t A_s u_s ds + \int_0^t f_s ds + M_t, \tag{6.100}$$

where f_t is an adapted, locally integrable process in $L^2(\Omega_T; H)$; Φ_t is a predictable operator in $\mathcal{P}_T$ and $M_t = \int_0^t \Phi_s dW_s$ is a V_0-valued martingale, where $V_0 \subset \mathcal{D}(A)$ is a Hilbert space dense in H. We shall first prove the existence of a unique solution to (6.99) by assuming that the martingale M_t is smooth.

Lemma 7.1 Suppose that the conditions (B1)–(B3) hold and f_t is a $\mathcal{F}_t$-adapted process in $L^2(\Omega_T; H)$. Assume that M_t is a V_0-valued process such that $N_t = A_t M_t$ is a L^2-martingale in H. Then, for $h \in H$, the equation (6.100) has a unique strong solution $u \in L^2(\Omega; \mathbf{C}([0, T]; H)) \cap L^2(\Omega_T; V)$ such that the energy equation holds:

$$\begin{aligned}
\|u_t\|^2 = \|h\|^2 &+ 2\int_0^t \langle A_s u_s, u_s \rangle ds + 2\int_0^t (f_s, u_s) ds \\
&+ 2\int_0^t (u_s, \Phi_s dW_s) + \int_0^t \|\Phi_s\|_R^2 ds.
\end{aligned} \tag{6.101}$$

Proof. Since the martingale in (6.100) is smooth, it is possible, by a simple transformation, to reduce the problem to an almost deterministic case. To proceed we let

$$v_t = u_t - M_t, \quad \text{and} \quad N_t = A_t M_t. \tag{6.102}$$

Then it follows from (6.100) that $v_t \in V$ and it satisfies

$$v_t = h + \int_0^t A_s v_s ds + \int_0^t (f_s + N_s) ds, \tag{6.103}$$

for almost every $(t, \omega) \in \Omega_T$. Since $\tilde{f}_t = (f_t + N_t) \in L^2((0, T); H)$ a.s., by appealing to a deterministic result in [58], the equation (6.103) has a unique solution $v \in \mathbf{C}([0, T]; H) \cap L^2((0, T); V)$ satisfying

$$\|v_t\|^2 = \|h\|^2 + 2\int_0^t \langle A_s v_s, v_s \rangle ds + 2\int_0^t (f_s + N_s, v_s) ds \quad a.s. \tag{6.104}$$

By condition (B.3) and the Gronwall inequality, we can deduce that

$$E\{\sup_{0 \le t \le T} \|v_t\|^2 + \int_0^T \|v_s\|_V^2 ds\} \le K,$$

for some $K > 0$. In view of (6.102) and the above inequality, we can assert that u belongs to $L^2(\Omega; \mathbf{C}([0, T]; H)) \cap L^2(\Omega_T; V)$ and it is the unique solution of (6.100). The energy equation (6.101) can be obtained from (6.101) by substituting $v = (u - M)$ back in (6.104) and making use of Itô's formula. $\square$

Now we will show that the statement of the lemma holds true when M_t is a H-valued martingale.

Theorem 7.2 Let conditions (B.1)–(B.3) hold true. Assume that f_t is a predictable H-valued process and $\Phi \in \mathcal{P}_T$ such that

$$E \int_0^T (\|f_s\|^2 + \|\Phi_s\|_R^2)ds < \infty. \qquad (6.105)$$

Then, for $h \in H$, the linear problem (6.99) has a unique strong solution $u \in L^2(\Omega; \mathbf{C}([0,T]; H)) \cap L^2(\Omega_T; V)$ such that the energy equation holds:

$$\begin{aligned}
\|u_t\|^2 = \|h\|^2 &+2 \int_0^t \langle A_s u_s, u_s \rangle ds + 2 \int_0^t (f_s, u_s)ds \\
&+2 \int_0^t (u_s, \Phi_s dW_s) + \int_0^t \|\Phi_s\|_R^2 ds.
\end{aligned} \qquad (6.106)$$

Proof. For uniqueness, suppose that u and $\tilde{u}$ are both strong solutions of (6.99). Let $v = (u - \tilde{u})$. Then v satisfies the homogeneous equation

$$dv_t = A_t v_t dt, \quad v_0 = 0,$$

which, as in the deterministic case, implies that

$$\|v_t\|^2 = 2 \int_0^t \langle A_s v_s, v_s \rangle ds \quad a.s.$$

By condition (B.3), the above yields

$$\|v_t\|^2 + 2\beta \int_0^t \|v_s\|_V^2 ds \leq 2\gamma \int_0^t \|v_s\|^2 ds.$$

It follows from Gronwall's inequality that

$$\|v_t\|^2 = \|u_t - \tilde{u}_t\|^2 = 0,$$

and

$$\int_0^T \|u_t - \tilde{u}_t\|_V^2 dt = 0, \quad a.s.$$

We will carry out the existence proof in several steps.

(Step 1) Approximate Solutions:

Let $\{\phi_k\}$, with $\phi_k \in V_0$, $k = 1, 2, \cdots$, be a complete orthonormal basis for H, which also spans V_0. Let $H_n = span\{\phi_1, \cdot, \phi_n\}$. Suppose that $P_n : H \to H_n$ is an orthogonal projection operator defined by $P_n g = \sum_{k=1}^n (g, \phi_k)\phi_k$ for

$g \in H$. Then $P_n g \in \mathcal{D}(A)$. Since $\Phi \in \mathcal{P}_T$, M_t is a L^2-martingale in H. Define $M_t^n = P_n M_t$ so that M_t^n is a L^2-martingale in $\mathcal{D}(A)$. In lieu of (6.100), for $n = 1, 2, \cdots$, consider the approximate equations:

$$u_t^n = h + \int_0^t A_s u_s^n ds + \int_0^t f_s ds + M_t^n. \tag{6.107}$$

Because $N_t^n = A_t M_t^n$ is a L^2-martingale in H, it follows from Lemma 7.1 that the approximate equation (6.107) has a unique solution u^n belongs to $L^2(\Omega; \mathbf{C}([0,T]; H)) \cap L^2(\Omega_T; V)$ such that it satisfies

$$\|u^n(\cdot, t)\|^2 = \|h\|^2 + 2\int_0^t (A_s u_s^n, u_s^n) ds + 2\int_0^t (f_s, u_s^n) ds$$
$$+ 2\int_0^t (u_s^n, dM_s^n) + \int_0^t \|\Phi_s^n\|_R^2 ds. \tag{6.108}$$

(Step 2) Convergent Sequence of Approximations:

Let $\mathbf{Y}_T$ denote the Banach space of $\mathcal{F}_t$-adapted, V-valued processes over [0,T], with continuous sample paths in H and norm $\|\cdot\|_T$ defined by

$$\|u\|_T = \{E[\sup_{0 \le t \le T} \|u_t\|^2 + \int_0^T \|u_s^n\|_V^2 ds]\}^{1/2}. \tag{6.109}$$

Let u^m be the solution of (6.107) with n replaced by m and put $u^{mn} = (u^m - u^n)$ and $M_t^{mn} = (M_t^m - M_t^n)$. Then it follows from (6.107) that u^{mn} satisfies the equation:

$$u_t^{mn} = \int_0^t A_s u_s^{mn} ds + M_t^{mn}. \tag{6.110}$$

By the coercivity condition (B.3) and (6.108), we obtain from (6.110) that

$$\|u_t^{mn}\|^2 + 2\beta \int_0^t \|u_s^{mn}\|_V^2 ds \le 2\gamma \int_0^t \|u_s^{mn}\|^2 ds$$
$$+ \int_0^t Tr\, Q_s^{mn} ds + 2\int_0^t (u_s^{mn}, dM_s^{mn}), \tag{6.111}$$

where Q_s^{mn} denotes the local covariation operator for M_s^{mn}. After taking the expectation, the above gives

$$E\|u_t^{mn}\|^2 + 2\beta E \int_0^t \|u_s^{mn}\|_V^2 ds$$
$$\le 2\gamma \int_0^t E\|u_s^{mn}\|^2 ds + E \int_0^t Tr\, Q_s^{mn} ds. \tag{6.112}$$

Therefore, by condition (6.105) and the Gronwall lemma, we can deduce, first, that

$$\sup_{0 \le t \le T} E\|u_t^{mn}\|^2 \le K_1 E \int_0^T Tr\, Q_s^{mn} ds, \tag{6.113}$$

so that (6.112) yields

$$2\beta E \int_0^t \|u_s^{mn}\|_V^2 ds \leq 2\gamma T \sup_{0 \leq t \leq T} E\|u_t^{mn}\|^2 + E \int_0^t Tr\, Q_s^{mn} ds$$

$$\leq (2\gamma K_1 T + 1)E \int_0^t Tr\, Q_s^{mn} ds,$$

$$\text{or} \quad E \int_0^T \|u_s^{mn}\|_V^2 ds \leq K_2 E \int_0^T Tr\, Q_s^{mn} ds, \qquad (6.114)$$

for some constants $K_1, K_2 > 0$. Moreover, it can be derived from (6.111) that

$$E \sup_{0 \leq r \leq t} \|u_r^{mn}\|^2 \leq 2\gamma \int_0^t E \sup_{0 \leq r \leq s} \|u_r^{mn}\|^2 ds$$
$$+ E \int_0^t Tr\, Q_s^{mn} ds + 2E \sup_{0 \leq r \leq t} |\int_0^r (u_s^{mn}, dM_s^{mn})|\, ds, \qquad (6.115)$$

where

$$\int_0^t Tr\, Q_s^{mn} ds = \int_0^t \|(P_m - P_n)\Phi_s\|_R^2\, ds. \qquad (6.116)$$

By a maximal inequality from Theorem 2.3,

$$E \sup_{0 \leq r \leq t} |\int_0^r (u_s^{mn}, dM_s^{mn})|ds \leq 3E\{\int_0^t (Q_s^{mn} u_s^{mn}, u_s^{mn})ds\}^{1/2}$$

$$\leq 3E\{\sup_{0 \leq r \leq t} \|u_r^{mn}\|\}\{\int_0^t Tr\, Q_s^{mn} ds\}^{1/2} \qquad (6.117)$$

$$\leq \frac{1}{4}E \sup_{0 \leq r \leq t} \|u_r^{mn}\|^2 + 9E \int_0^t Tr\, Q_s^{mn} ds.$$

By taking (6.115), (6.117) into account and invoking the Gronwall inequality again, we can find constant $K_3 > 0$ such that

$$E \sup_{0 \leq t \leq T} \|u_t^{mn}\|^2 \leq K_3 E \int_0^t Tr\, Q_s^{mn} ds. \qquad (6.118)$$

From (6.114) and (6.118), we obtain

$$\|u_{mn}\|_T^2 = \|u_m - u_n\|_T^2 \leq (K_2 + K_3)E \int_0^t Tr\, Q_s^{mn} ds.$$

In view of (6.116), we can deduce that $E \int_0^t Tr\, Q_s^{mn} ds \to 0$ as $m, n \to \infty$. Therefore $\{u^n\}$ is a Cauchy sequence in Y_T which converges to a limit u.

(Step 3) Strong Solution:

We will show that the limit u thus obtained is a strong solution. To this end, let $\phi \in V$ and apply the Itô formula to get

$$(u_t^n, \phi) = (h, \phi) + \int_0^t (A_s u_s^n, \phi)ds + \int_0^t (f_s, \phi)ds + (M_t^n, \phi). \qquad (6.119)$$

By taking the limit as $n \to \infty$, the equation (6.119) converges term-wise in $L^2(\Omega)$ to

$$(u_t, \phi) = (h, \phi) + \int_0^t \langle A_s u_s, \phi \rangle ds + \int_0^t (f_s, \phi)ds + (M_t, \phi), \qquad (6.120)$$

for any $\phi \in V$. Therefore u is the strong solution.

(Step 4) Energy Equation:

To verify the energy equation (6.106), we proceed by taking the term-wise limit in the approximate equation (6.108). It is easy to show that $\|u^n(\cdot, t)\|^2 \to \|u_t\|^2$ in the mean and $\int_0^t (f_s, u_s^n)ds \to \int_0^t (f_s, u_s)ds$ in mean-square. By the monotone convergence, the term $\int_0^t \|\Phi_s^n\|_R^2 ds$ converges in the mean to $\int_0^t \|\Phi_s\|_R^2 ds$. For the remaining two terms in (6.108), first consider

$$| \int_0^t \langle A_s u_s, u_s \rangle ds - \int_0^t (A_s u_s^n, u_s^n)ds |$$

$$\leq \int_0^t |\langle A_s(u_s - u_s^n), u_s \rangle| ds + \int_0^t |\langle A_s u_s^n, u_s - u_s^n \rangle| ds$$

$$\leq \alpha \int_0^t (\|u_s\|_V + \|u_s^n\|_V)\|u_s - u_s^n\|_V \, ds,$$

where use was made of condition (B.2). Therefore

$$E| \int_0^t \langle A_s u_s, u_s \rangle ds - \int_0^t \langle A_s u_s^n, u_s^n \rangle ds |^2$$

$$\leq 2\alpha^2 \{E \int_0^T (\|u_s\|_V^2 + \|u_s^n\|_V^2 \, ds \} \{E \int_0^T \|u_s - u_s^n\|_V^2 ds\},$$

which converges to zero as $n \to \infty$. Finally, for the stochastic integral term, we have

$$E| \int_0^t (u_s, dM_s) - \int_0^t (u_s^n, dM_s^n)| \leq E| \int_0^t (u_s - u_s^n, dM_s)| + E| \int_0^t (u_s^n, d\tilde{M}_s^n)|,$$

where $\tilde{M}_s^n = M_s - M_s^n$. Now, by a maximal inequality, we have

$$E \sup_{0 \leq t \leq T} | \int_0^t (u_s - u_s^n, dM_s)| \leq 3E\{ \int_0^T \|\Phi_s\|_R^2 \|(u_s - u_s^n)\|^2 \, ds \}^{1/2}$$

$$\leq 3\{E \sup_{0 \leq t \leq T} \|u_s - u_s^n\|^2\}^{1/2} \{E \int_0^T \|\Phi_s\|_R^2 ds\}^{1/2} \to 0,$$

as $n \to \infty$. Similarly,

$$E \sup_{0 \le t \le T} |\int_0^t (u_s^n, d\tilde{M}_s^n)| \le 3E\{\int_0^T \|(\Phi_s - \Phi_s^n)\|_R^2 \|u_s^n\|^2 ds\}^{1/2}$$

$$\le 3\{E \sup_{0 \le t \le T} \|u_s^n\|^2\}^{1/2}\{E \int_0^T \|\Phi_s - \Phi_s^n\|_R^2 ds\}^{1/2},$$

which also tends to zero as $n \to \infty$. The above three inequalities imply that $E \sup_{0 \le t \le T} |\int_0^t (u_s - u_s^n, dM_s)| \to 0$. Finally, by taking all of the above inequalities into account, we obtain the energy equation as given by (6.106). □

Remark: Here we assumed that f_t is a predicable H-valued process. It is not hard to show that the theorem holds for $f \in L^2(\Omega_T; V')$ as well.

As a consequence of this theorem, the following lemma will be found useful later.

Lemma 7.3 Assume that all conditions for Theorem 7.2 are satisfied. Then the following energy inequality holds:

$$E \sup_{0 \le t \le T} \|u_t\|^2 + E \int_0^T \|u_s\|_V^2 ds$$

$$\le C_T\{\|h\|^2 + E \int_0^T (\|f_s\|^2 + \|\Phi_s\|_R^2) ds\}, \tag{6.121}$$

for some constant $C_T > 0$.

Proof.   As expected, the inequality of interest follows from the energy equation (6.106). Since the proof is similar to that in the (Step 2) of Theorem 7.2, it will only be sketched. By the coercivity condition (B.3), we can deduce from (6.106) that

$$\|u_t\|^2 + \beta \int_0^t \|u_s\|_V^2 ds \le \|h\|^2 + (2\gamma + 1) \int_0^t \|u_s\|^2 ds$$

$$+ \int_0^t (\|f_s\|^2 + \|\Phi_s\|_R^2) ds + 2 \int_0^t (u_s, \Phi_s dW_s). \tag{6.122}$$

By taking the expectation over the above equation and making use of Gronwall's inequality, it can be shown that there exist positive constants C_1 and C_2 depending on T such that

$$\sup_{0 \le t \le T} E\|u_t\|^2 \le C_1\{\|h\|^2 + E \int_0^t (\|f_s\|^2 + \|\Phi_s\|_R^2) ds\}, \tag{6.123}$$

and

$$E \int_0^T \|u_s\|_V^2 ds \le C_2 \{ \|h\|^2 + E \int_0^t (\|f_s\|^2 + \|\Phi_s\|_R^2) ds \}. \qquad (6.124)$$

Returning to (6.122), we can get

$$E \sup_{0 \le t \le T} \|u_t\|^2 \le \|h\|^2 + (2\gamma + 1) E \int_0^T \|u_s\|^2 ds$$
$$+E \int_0^T (\|f_s\|^2 + \|\Phi_s\|_R^2) ds + 2E \sup_{0 \le t \le T} | \int_0^t (u_s, \Phi_s dW_s)|. \qquad (6.125)$$

With the aid of (6.123) and a maximal inequality, it is possible to derive from (6.125) that

$$E \sup_{0 \le t \le T} \|u_t\|^2 \le C_3 \{ \|h\|^2 + E \int_0^t (\|f_s\|^2 + \|\Phi_s\|_R^2) ds \}. \qquad (6.126)$$

Now the energy inequality follows from (6.124) and (6.126). $\qquad \square$

We will next consider the strong solution of the nonlinear equation (6.98) with regular nonlinear terms. More precisely let them satisfy the following linear growth and the Lipschitz conditions:

(C.1) For any $v \in H$, $F_t(v)$ and $\Sigma_t(v)$ are $\mathcal{F}_t$-adapted processes with values in H and $\mathcal{L}_R^2$, respectively. Suppose that there exist positive constants b, C such that

$$E \int_0^T \{ \|F_t(0)\|^2 + \|\Sigma_t(0)\|_R^2 \} dt \le b,$$

and

$$\|\hat{F}_t(v)\|^2 + \|\hat{\Sigma}_t(v)\|_R^2 \le C(1 + \|v\|^2), \quad a.e. \ (t, \omega) \in \Omega_T,$$

where we set

$$\hat{F}_t(v) = F_t(v) - F_t(0), \quad \hat{\Sigma}_t(v) = \Sigma_t(v) - \Sigma_t(0).$$

(C.2) There exists constant $\kappa > 0$, such that, for any $u, v \in H$, the Lipschitz continuity condition holds:

$$\|F_t(u) - F_t(v)\|^2 + \|\Sigma_t(u) - \Sigma_t(v)\|_R^2 \le \kappa \|u - v\|^2, \quad a.e. \ (t, \omega) \in \Omega_T.$$

Theorem 7.4 Let conditions (B.1)–(B.3), (C1) and (C2) hold true. Then, for $h \in H$, the nonlinear problem (6.58) has a unique strong solution $u \in$

$L^2(\Omega; \mathbf{C}([0,T]; H)) \cap L^2(\Omega_T; V)$ such that the energy equation holds:

$$
\begin{aligned}
\|u_t\|^2 = \|h\|^2 &+2 \int_0^t \langle A_s u_s, u_s \rangle ds + 2 \int_0^t (F_s(u_s), u_s) ds \\
&+2 \int_0^t (u_s, \Sigma_s(u_s) dW_s) + \int_0^t \|\Sigma_s(u_s)\|_R^2 \, ds.
\end{aligned} \tag{6.127}
$$

Proof.   The theorem can be proved by the contraction mapping principle. To this end, let the Y_T be the space introduced in the proof of Theorem 7.2 with norm $\|\cdot\|_T$ defined by (6.109). Given $u \in Y_T$, let μ be the solution of the equation

$$
\mu_t = h + \int_0^t A_s \mu_s ds + \int_0^t F_s(u_s) ds + \int_0^t \Sigma_s(u_s) dW_s. \tag{6.128}
$$

Then, by condition (C.1), we have

$$
\begin{aligned}
E \int_0^T & [\, \|F_s(u_s)\|^2 + \|\Sigma_s(u_s)\|_R^2 \,] ds \\
&\leq 2 \int_0^T \{\, [\, \|\hat{F}_s(u_s)\|^2 + \|\hat{\Sigma}_s(u_s)\|_R^2 \,] + [\, \|F_s(0)\|^2 + \|\Sigma_s(0)\|_R^2 \,] \,\} ds \\
&\leq 2 \, [C(T + E \int_0^T \|u_s\|_1^2 ds) + b].
\end{aligned}
$$

For a given $u \in Y_T$, by Theorem 7.2, the equation (6.128) has a unique solution $\mu \in Y_T$. Let Γ denote the solution map so that $\mu_t = \Gamma_t u$. By Lemma 7.3, the map $\Gamma: Y_T \to Y_T$ is bounded. To show the contraction mapping, for $u, \tilde{u} \in Y_T$, let $\mu = \Gamma u$ and $\tilde{\mu} = \Gamma \tilde{u}$. It follows from (6.128), by definition, that $\nu = (\mu - \tilde{\mu})$ satisfies

$$
\nu_t = \int_0^t A_s \nu_s ds + \int_0^t [F_s(u_s) - F_s(\tilde{u}_s)] ds + \int_0^t [\Sigma_s(u_s) - \Sigma_s(\tilde{u}_s)] dW_s. \tag{6.129}
$$

By making use of the energy equation (6.106) and taking conditions (B.3) and (C.2) into account, we can deduce that

$$
\begin{aligned}
\|\nu_t\|^2 + 2\beta \int_0^t \|\nu_s\|_V^2 ds &\leq 2\gamma \int_0^t \|\nu_s\|^2 ds \\
&+2 \int_0^t (\, [F_s(u_s) - F_s(\tilde{u}_s)\,], \nu_s) ds + \int_0^t \|\Sigma_s(u_s) - \Sigma_s(\tilde{u}_s)\|_R^2 ds \\
&+2 \int_0^t (\nu_s, [\Sigma_s(u_s) - \Sigma_s(\tilde{u}_s)] dW_s) \\
&\leq (2\gamma + 1) \int_0^t \|\nu_s\|^2 ds + 2\kappa \int_0^t \|u_s - \tilde{u}_s\|^2 ds \\
&+2 \int_0^t (\nu_s, [\Sigma_s(u_s) - \Sigma_s(\tilde{u}_s)] dW_s).
\end{aligned} \tag{6.130}
$$

Now, since the following estimates are similar to that in the proof of Lemma 7.3, we omit the details. First, by taking expectation of (6.130) and then by Gronwall's inequality, we can get

$$E\|\nu_t\|^2 \le C_1(T) \int_0^t \|u_s - \tilde{u}_s\|^2 ds,$$

and

$$E \int_0^T \|\nu_s\|_V^2 ds \le C_2(T) E \int_0^T \|u_s - \tilde{u}_s\|^2 ds, \qquad (6.131)$$

where $C_1 = 2\kappa \exp(2\gamma + 1)T$ and $C_2 = (C_1 T + 2\kappa)/2\beta$. Next, by taking the supremum of (6.130) over t and then the expectation, it can be shown that

$$E \sup_{0 \le t \le T} \|\nu_t\|^2 \le C_3(T) \int_0^T \|u_s - \tilde{u}_s\|^2 ds, \qquad (6.132)$$

where $C_3 = 2[C_1 T(2\gamma + 1) + 20\kappa]$. It follows from (6.131) and (6.132) that

$$\|\Gamma u - \Gamma \tilde{u}\|_T^2 = \|\nu\|_T^2 \le (C_2 + C_3) E \int_0^T \|u_s - \tilde{u}_s\|^2 ds$$
$$\le (C_2 + C_3) T \|u - \tilde{u}\|_T^2,$$

which shows that Γ is a contraction mapping in Y_T for small T. Therefore, by continuation, there exists a unique strong solution for any $T > 0$. The energy equation follows from the corresponding result in Theorem 7.2. □

In the above theorem, the nonlinear terms were assumed to be bounded from H into H. If they are only bounded in V, the problem becomes more complicated technically. A proof based on the contraction mapping principle fails unless the nonlinear terms are small in some sense. Instead, under certain reasonable assumptions, we will prove an existence theorem by a different approach. To be specific, assume that the following conditions hold:

(D.1) For any $v \in V$, $F_t(v)$ is a $\mathcal{F}_t$-adapted process in H, $\Sigma_t(v)$ is a predictable operator in $\mathcal{P}_T$, and there exist positive constants b, C such that

$$E \int_0^T \{\|F_t(0)\|^2 + \|\Sigma_t(0)\|_R^2\} dt \le b,$$

and, for any $N > 0$, there exists constant $C_N > 0$ such that

$$|(\hat{F}_t(v), v)| + \|\hat{\Sigma}_t(v)\|_R^2 \le C_N(1 + \|v\|_V^2), \quad a.e. \ (t, \omega) \in \Omega_T,$$

for any $v \in V$ with $\|v\|_V < N$, where

$$\hat{F}_t(v) = F_t(v) - F_t(0), \quad \hat{\Sigma}_t(v) = \Sigma_t(v) - \Sigma_t(0).$$

(D.2) For any $N > 0$, there exists constant $K_N > 0$ such that the local Lipschitz continuity condition holds:

$$\|F_t(u) - F_t(v)\|^2 + \|\Sigma_t(u) - \Sigma_t(v)\|_R^2 \le K_N \|u - v\|_V^2, \quad a.e. \ (t, \omega) \in \Omega_T,$$

for any $u, v \in V$ with $\|u\|_V < N$ and $\|v\|_V < N$.

(D.3) For any $v \in V$, there exist constants $\beta > 0, \lambda$ and C_1 such that the coercivity condition holds:

$$\langle A_t v, v \rangle + (\hat{F}_t(v), v) + \|\hat{\Sigma}_t(v)\|_R^2 \le -\beta \|v\|_V^2 + \lambda \|v\|^2 + C_1, \quad a.e. \ (t, \omega) \in \Omega_T.$$

(D.4) For any $u, v \in V$, the monotonicity condition holds:

$$2\langle A_t(u - v), u - v \rangle + 2(F_t(u) - F_t(v), u - v)$$
$$+ \|\Sigma_t(u) - \Sigma_t(v)\|_R^2 \le \delta \|u - v\|^2, \quad a.e. \ (t, \omega) \in \Omega_T,$$

for some constant δ.

Theorem 7.5 Let conditions (B.1)–(B.3) and (D1)–(D3) hold true. Then, for $h \in H$, the nonlinear problem (6.58) has a unique strong solution $u \in L^2(\Omega; \mathbf{C}([0, T]; H)) \cap L^2(\Omega_T; V)$. Moreover the energy equation holds:

$$\|u_t\|^2 = \|h\|^2 + 2 \int_0^t \langle A_s u_s, u_s \rangle ds + 2 \int_0^t (F_s(u_s), u_s) ds$$
$$+ \int_0^t \|\Sigma_s(u_s)\|_R^2 \, ds + 2 \int_0^t (u_s, \Sigma_s(u_s) dW_s). \tag{6.133}$$

Proof. For uniqueness, suppose that u and $\tilde{u}$ are strong solutions. Then, in view of (6.133) and condition (D.4), we can easily show that

$$E\|u_t - \tilde{u}_t\|^2 = 2E \int_0^t \langle A_s(u_s - \tilde{u}_s), u_s - \tilde{u}_s \rangle ds$$

$$+ E \int_0^t \{ 2(F_s(u_s) - F_s(u_s), u_s - \tilde{u}_s) + \|\Sigma_s(u_s) - \Sigma_s(\tilde{u}_s)\|_R^2 \} ds$$

$$\le \delta E \int_0^t \|u_s - \tilde{u}_s\|^2 ds,$$

which, with the aid of Gronwall's inequality, yields

$$E\|u_t - \tilde{u}_t\|^2 = 0, \quad \forall \ t \in [0, T],$$

which implies the uniqueness.

A detailed existence proof is rather lengthy. For clarity, we will sketch the proof in several steps to follow.

(Step 1) Approximate Solutions:

Let $\{v_k\}$ be a complete orthonormal basis for H with $v_k \in V$ and let $H_n = span\{v_1, \cdots, v_n\}$. Suppose that $P_n : H \to H_n$ is the orthogonal projector such that, for $h \in H$, $P_n h = \sum_{k=1}^n (h, v_k) v_k$. Extend P_n to a projection operator $P'_n : V' \to V_n'$ defined by $P'_n w = \sum_{k=1}^n \langle w, v_k \rangle v_k$ for $w \in V'$. Clearly we have $V_n = H_n = V_n'$. Let $\{e_k\}$ be the set of eigenfunctions of R and let $K_n = span\{e_1, \cdots, e_n\}$. Denote by Π_n the projection operator from K into K_n such that $\Pi_n \phi = \sum_{k=1}^n (\phi, e_k) e_k$. Introduce the following truncations:

$$A_t^n v = P'_n A_t v, \quad F_t^n(v) = P_n F_t(v), \quad \Sigma_t^n(v) = P_n \Sigma_t(v), \tag{6.134}$$

for $v \in V$. Consider the approximate equation to (6.98):

$$du^n(\cdot, t) = A_t^n u^n(\cdot, t)dt + F_t^n(u^n(\cdot, t))dt + \Sigma_t^n(u^n(\cdot, t))dW_t^n,$$
$$u_0^n = h^n, \tag{6.135}$$

for $t \in (0, T)$, where $W_t^n = \Pi_n W_t$ and $h^n = P_n h$ for $h \in H$.

Notice that the above equation can be regarded as an Itô equation in $\mathbf{R}^n$. It can be shown that, under conditions (D1)–(D4), the mollified coefficients F^n and Σ^n are locally bounded and Lipschitz continuous and monotone. Therefore, by an existence theorem in finite dimension, it has a unique solution $u^n(\cdot, t)$ in V_n in any finite time interval $[0, T]$. Moreover it satisfies the property: $u^n \in L^2(\Omega_T; V) \cap L^2(\Omega; \mathbf{C}([0, T]; H)) \subset Y_T$.

(Step 2) Boundedness of Approximate Solutions:

To show the sequence $\{u^n\}$ is bounded in the Banach space Y_T, it follows from the energy equation for (6.135) that:

$$\|u^n(\cdot, t)\|^2 = \|h^n\|^2 + 2\int_0^t (A_s^n u_s^n, u_s^n)ds + 2\int_0^t (F_s^n(u_s^n), u_s^n)ds$$
$$+ \int_0^t \|\Sigma_s^n(u_s^n)\|_R^2 ds + 2\int_0^t (u_s^n, \Sigma_s^n(u_s^n)\Pi_n dW_s). \tag{6.136}$$

By the simple inequality: $2(F_s(v), v) + \|\Sigma_s(v)\|_R^2 \leq 2[(\hat{F}_s(v), v) + \|\hat{\Sigma}_s(v)\|_R^2] + 2[(F_s(0), v) + \|\Sigma_s(0)\|_R^2]$, and by invoking condition (D.3), we can obtain from (6.136) that

$$\|u^n(\cdot, t)\|^2 + 2\beta \int_0^t \|u_s^n\|_V^2 ds \leq \|h^n\|^2 + 2C_1 T$$
$$+ (2\lambda + 1)\int_0^t \|u_s^n\|^2 ds + 2\int_0^t [\|F_s(0)\|^2 + \|\Sigma_s(0)\|_R^2]ds \tag{6.137}$$
$$+ 2\int_0^t (u_s^n, \Sigma_s^n(u_s^n)\Pi_n dW_s).$$

By using similar estimates to derive the inequalities (6.131) and (6.132), we can deduce from (6.137) that

$$E \int_0^T \|u_s^n\|_1^2 ds \le K_1, \text{ and } E \sup_{0 \le t \le T} \|u^n(\cdot, t)\|_1^2 \le K_2, \tag{6.138}$$

for some positive constants K_1, K_2.

(Step 3) Weak Limits:

From (6.138) it follows that there exists a subsequence, denoted by u^k, which converges weakly to u in $L^2(\Omega_T; V)$. Moreover, by virtue of (6.138) and condition (D.1), we can further deduce that $F^n(u^n) \dashrightarrow f$ in $L^2(\Omega_T; H)$, $\Sigma^n(u^n)\Pi_n \dashrightarrow \Phi$ in $L^2(\Omega_T; \mathcal{L}_R^2)$ and $u_T^n \dashrightarrow \xi$ in $L^2(\Omega; H)$, where $\dashrightarrow$ denotes the weak convergence .

For technical reasons, we extend the time interval from $[0, T]$ to $(-\varepsilon, T + \varepsilon)$ with $\varepsilon > 0$, and set the terms in the equation equal to zero for $t \ne [0, T]$. Let $\theta \in \mathbf{C}^1(-\varepsilon, T + \varepsilon)$ with $\theta(0) = 1$ and, for $v_k \in V$, define $v_t^k = \theta(t)v_k$. Apply the Itô formula to (u_t^n, v_t^k) to get

$$\begin{aligned}
(u_T^n, v_T^k) = (h^n, v_k) &+ \int_0^T (u_s^n, \dot{v}_s^k)ds + \int_0^T \langle A_s u_s^n, v_s^k \rangle ds \\
&+ \int_0^T (F_s^n(u_s^n), v_s^k)ds + \int_0^T (v_s^k, \Sigma_s^n(u_s^n)\Pi_n dW_s),
\end{aligned} \tag{6.139}$$

where $\dot{v}_s^k = [\frac{d}{ds}\theta(s)]v_k$. By letting $n \to \infty$ in (6.139), in view of the above weak convergence results, we can obtain

$$\begin{aligned}
(u_T, v_T^k) = (h, v_k) &+ \int_0^T (u_s, \dot{v}_s^k)ds + \int_0^T \langle A_s u_s, v_s^k \rangle ds \\
&+ \int_0^T (f_s, v_s^k)ds + \int_0^T (v_s^k, \Phi_s dW_s).
\end{aligned} \tag{6.140}$$

To justify the limit of the stochastic integral term, let $\Phi^n = \Sigma^n(u^n)\Pi_n$ in $L^2(\Omega_T; \mathcal{L}_R^2)$. The fact $\Phi^n \dashrightarrow \Phi$ in $\mathcal{P}_T$ means that, for any $Q \in \mathcal{P}_T$, $(\Phi^n, Q)_{\mathcal{P}_T} \to (\Phi, Q)_{\mathcal{P}_T}$, where the inner product $(\Phi, Q)_{\mathcal{P}_T} = E \int_0^T \|Q_s^\star \Phi_s\|_R^2 ds$. Now the limit of the stochastic integral follows from the inequality

$$|(\Phi^n - \Phi), Q)_{\mathcal{P}_T}| \le |(\Sigma^n(u^n) - \Phi, Q)_{\mathcal{P}_T}| + |(\Sigma^n(u^n), Q - Q\Pi_n)_{\mathcal{P}_T}|,$$

which goes to zero as $n \to \infty$.

Now choose $\theta_m \in \mathbf{C}^1(-\varepsilon, T + \varepsilon)$ with $\theta_m(0) = 1$, for $m = 1, 2, \cdots$, such that $\theta_m \dashrightarrow \chi_t$ and $\dot{\theta}_m \dashrightarrow \delta_t$, where $\chi_t(s) = 1$, for $s \le t$ and 0 otherwise, and $\delta_t(s) = \delta(t - s)$ is the Dirac $\delta-$ function. Let $v_t^{k,m} = \theta_m(t)v_k$. By replacing

v_t^k with $v_t^{k,m}$ in (6.139) and taking the limits as $m \to \infty$, it yields

$$
\begin{aligned}
(u_t, v_k) = (h, v_k) &+ \int_0^t \langle A_s u_s, v_k \rangle ds + \int_0^t (f_s, v_k) ds \\
&+ \int_0^t (v_k, \Phi_s dW_s), \quad \text{for } 0 < t < T,
\end{aligned}
\tag{6.141}
$$

with $(u_T, v_k) = (\xi, v_k)$, for any $v_k \in V$. Since V is dense in H, the above two equations hold with any $v \in V$ in place of v_k. Therefore u is a strong solution of the linear Itô equation:

$$
u_t = h + \int_0^t A_s u_s ds + \int_0^t f_s ds + \int_0^t \Phi_s dW_s, \quad u_T = \xi,
\tag{6.142}
$$

and, by Theorem 7.2, it satisfies the energy equation:

$$
\begin{aligned}
\|u_t\|^2 = \|h\|^2 &+ 2 \int_0^t \langle A_s u_s, u_s \rangle ds + 2 \int_0^t (f_s, u_s) ds \\
&+ 2 \int_0^t (u_s, \Phi_s dW_s) + \int_0^t \|\Phi_s\|_R^2 ds.
\end{aligned}
\tag{6.143}
$$

(Step 4) Strong Solution and Energy Equation:

We will show that $f_s = F_s(u_s)$ and $\Phi_s = \Sigma_s(u_s)$ so that u_t is a strong solution of the nonlinear problem (6.58). To this end, for any $v \in L^2(\Omega_T; V)$, we let

$$
\begin{aligned}
Z_n = 2E &\int_0^T e^{-\delta s} [\langle A_s(u_s^n - v_s), u_s^n - v_s \rangle - \delta \|u_s^n - v_s\|^2] ds \\
&+ 2E \int_0^T e^{-\delta s} (F_s^n(u_s^n) - F_s^n(v_s), u_s^n - v_s) ds \\
&+ E \int_0^T e^{-\delta s} \|\Sigma_s^n(u_s^n) - \Sigma_s^n(v_s)\|_R^2 ds \le 0,
\end{aligned}
\tag{6.144}
$$

by the monotonicity condition (D.4). In what follows, to simplify the computations, we set $\delta = 0$, (otherwise, replace A_t by $(A_t - \delta)$). Rewrite Z_n as

$$
Z_n = Z_n' + Z_n'' \le 0,
\tag{6.145}
$$

where

$$
\begin{aligned}
Z_n' = 2E &\int_0^T \langle A_s u_s^n, u_s^n \rangle ds + 2E \int_0^T (F_s^n(u_s^n), u_s^n) ds \\
&+ E \int_0^T \|\Sigma_s^n(u_s^n)\|_R^2 ds,
\end{aligned}
\tag{6.146}
$$

$$Z_n'' = 2E \int_0^T [\, \langle A_s v_s, v_s \rangle - \langle A_s u_s^n, v_s \rangle - \langle A_s v_s, u_s^n \rangle \,] ds$$

$$+2E \int_0^T [\, (F_s^n(v_s), v_s) - (F_s^n(u_s^n), v_s) - (F_s^n(v_s), u_s^n) \,] ds \qquad (6.147)$$

$$+E \int_0^T [\, \|\Sigma_s^n(v_s)\|_R^2 - 2(\Sigma_s^n(u_s^n), \Sigma_s^n(v_s))_R \,] ds,$$

and $(\cdot, \cdot)_R$ denotes the inner product in $\mathcal{L}_R^2$. In view of (6.136) and (6.146), we get

$$Z_n' = E\|u_T^n\|^2 - \|h^n\|^2 \geq E\|u_T^n\|^2 - \|h\|^2,$$

which, with the aid of Fatou's lemma and the equation (6.142), implies that

$$\liminf_{n \to \infty} Z_n' \geq E \ \|u_T\|^2 - \|h\|^2 = 2E \int_0^T \langle A_s u_s, u_s \rangle ds$$

$$+2E \int_0^T (f_s, u_s) ds + E \int_0^T \|\Phi_s\|_R^2 ds. \qquad (6.148)$$

In the meantime, noting that Z_n'' has termwise limits, we can deduce from (6.144)–(6.148) that

$$2E \int_0^T \langle A_s(u_s - v_s), u_s - v_s \rangle + 2E \int_0^T (F_s(u_s) - f_s, u_s - v_s) ds$$

$$+E \int_0^T \|\Sigma_s(u_s) - \Phi_s\|_R^2 ds \leq \liminf_{n \to \infty} Z_n \leq 0. \qquad (6.149)$$

Hence, by setting $u_s = v_s$, the above gives

$$E \int_0^T \|\Sigma_s(u_s) - \Phi_s\|_R^2 ds = 0,$$

or $\Phi_s = \Sigma_s(u_s)$ a.s. Then it follows from (6.149) that

$$2E \int_0^T \{\langle A_s(u_s - v_s), u_s - v_s \rangle + (F_s(v_s) - f_s, u_s - v_s)\} ds \leq 0. \qquad (6.150)$$

Let $v = u - \alpha w$, where $w \in L^2(\Omega_T; V)$ and $\alpha > 0$. Then the above inequality becomes

$$\alpha E \int_0^T \langle A_s(w_s, w_s) ds + E \int_0^T (F_s(u_s - \alpha w_s) - f_s, w_s) ds \leq 0.$$

Letting $\alpha \downarrow 0$, we get

$$E \int_0^T (F_s(u_s) - f_s, w_s) ds \leq 0, \quad \forall \ w \in L^2(\Omega_T; V),$$

which implies that $f_s = F_s(u_s)$ a.s. Therefore the weak limit u is in fact a strong solution.

The energy equation follows immediately from (6.143) after setting $f_s = F_s(u_s)$ and $\Phi_s = \Sigma_s(u_s)$. Finally, from the equations (6.58) for u and (6.135) for u^n together with the associated energy equations, one can show that $E \sup_{0 \le t \le T} \|u_t - u^n(\cdot, t)\| \to 0$ as $n \to \infty$ and, hence, $u \in L^2(\Omega; \mathbf{C}([0, T]; H)) \cap L^2(\Omega_T; V)$. $\qquad\square$

Remarks: Here the proof, based on functional analytic techniques, was adopted from the works of Bensoussan and Temam [4] and Pardoux [67]. If the coercivity and monotonicity conditions are not met, there may exist only a local solution. We wish to point out the obvious fact: Since A_t is a linear operator-valued process, in contrast with the equation (6.62), the semigroup approach fails to apply to the problem (6.58) treated in this section.

(**Example 7.1**) Consider the following initial-boundary value problem for a parabolic equation with random coefficients:

$$\frac{\partial u}{\partial t} = \sum_{j,k=1}^{d} \frac{\partial}{\partial x_j}[a_{jk}(x,t)\frac{\partial u}{\partial x_k}] + \sum_{k=1}^{d} b_k(x,t)\frac{\partial u}{\partial x_k} + c(x,t)u$$
$$+ f(x,t) + \sigma(x,t)\dot{W}(x,t), \quad x \in D, \; t \in (0,T), \tag{6.151}$$
$$u|_{\partial D} = 0, \quad u(x,0) = h(x),$$

where a_{jk}, b_k, c, f and σ are all given adpated random fields such that all of them are bounded and continuous in $D \times [0, T]$ a.s. and there exist positive constants α, β such that, for any $\xi \in \mathbf{R}^d$,

$$\alpha|\xi|^2 \le \sum_{j,k=1}^{d} a_{jk}(x,t)\xi_j\xi_k \le \beta|\xi|^2, \quad \forall (x,t) \in D \times [0,T], \quad \text{a.s.} \tag{6.152}$$

As usual, $W(x,t)$ is a Wiener random field in $L^2(D)$ with a bounded covariance function $r(x,y)$. Let $H = K = L^2(D), V = H_0^1$ and $V' = H^{-1}$ and define

$$A_t\phi = \sum_{j,k=1}^{d} \frac{\partial}{\partial x_j}[a_{jk}(\cdot,t)\frac{\partial \phi}{\partial x_k}] + \sum_{k=1}^{d} b_k(\cdot,t)\frac{\partial \phi}{\partial x_k} + c(\cdot,t)\phi, \quad \phi \in V. \tag{6.153}$$

We set $f_t = f(\cdot, t)$ and $M_t = \int_0^t \sigma(\cdot, s)dW(\cdot, s)$. Then the equation (6.151) takes the form (6.100). By an integration by parts, we get

$$\langle A_t\phi, \psi\rangle = -\int_D [\sum_{j,k=1}^{d} \{a_{jk}(x,t)\frac{\partial \phi}{\partial x_k}(x)\frac{\partial \psi}{\partial x_j}(x)]dx$$
$$+ \int_D [\sum_{k=1}^{d} b_k(x,t)\frac{\partial \phi}{\partial x_k}(x) + c(x,t)\phi(x)]\psi(x)dx, \tag{6.154}$$

for any $\phi, \psi \in V$. By condition (6.152) and the boundedness of the coefficients, we can verify from (6.154) that conditions (B1)–(B3) are satisfied. Therefore, by Theorem 7.2, if $h \in H$ and

$$E \int_0^T \{ \int_D |f(x,t)|^2 dx + \int_D r(x,x)\sigma^2(x,t)dx \} dt < \infty,$$

the equation (6.151) has a unique solution $u \in L^2(\Omega; \mathbf{C}([0,T]; H)) \cap L^2(\Omega_T; H_0^1)$.

(**Example 7.2**) Let us consider the randomly convective reaction-diffusion equation in domain $D \subset \mathbf{R}^3$ with a cubic nonlinearity:

$$\frac{\partial u}{\partial t} = (\kappa\Delta - \alpha)u - \gamma u^3 + f(x,t) + \sum_{k=1}^3 \frac{\partial u}{\partial x_k}\frac{\partial}{\partial t}W_k(x,t), \qquad (6.155)$$

$$u|_{\partial D} = 0, \quad u(x,0) = h(x),$$

where κ, α, γ are given positive constants, $f(\cdot, t)$ is a predictable process in H and $W_k(x,t), k = i, \cdots, d$, are independent Wiener random fields with continuous covariance functions $r_k(x,y)$ such that

$$E \int_0^T \int_D |f(x,t)|^2 dxdt < \infty, \quad \sup_{x,y\in D} |r_k(x,y)| \le r_0, \ \ k = 1,2,3. \quad (6.156)$$

Let the spaces H and V and the function f_t be defined as in Example 7.1. Here we define $K = (H)^3$ to be the triple Cartesian product of H. Let $A_t = (\kappa\Delta - \alpha)$ and $F_t(u) = -\gamma u^3 + f_t$. For $Z = (Z^1, Z^2, Z^3) \in K$, define $[\Sigma_t(u)]Z = \sum_{k=1}^3 (\frac{\partial}{\partial x_k}u)Z^k$, and set $W_t = (W_t^1, W_t^2, W_t^3)$. Then the equation (6.155) is of the form (6.58). The conditions (B1)–(B3) are clearly met. To check condition (D.1), noticing (6.156), we have $F_t(0) = f_t$, $\Sigma_t(0) = 0$ and

$$|(\hat{F}_t(v), v)| + \|\hat{\Sigma}_t(v)\|_R^2 = \gamma|v|_4^4 + \sum_{j=1}^3 \int_D r_j(x,x)|\frac{\partial v}{\partial x_j}(x)|^2 dx$$

$$\le \gamma|v|_4^4 + r_0\|\nabla v\|^2 \le \gamma|v|_4^4 + C_1\|v\|_1^2,$$

for some constant $C_1 > 0$. To check condition (D.2), for $u, v \in H^1$, we have

$$\|F_t(u) - F_t(v)\|^2 + \|\Sigma_t(u) - \Sigma_t(v)\|_R^2 \le \gamma\|u^3 - v^3\| + C_1\|u - v\|_1^2,$$

where

$$\|u^3 - v^3\|^2 = \|(u^2 + uv + v^2)(u - v)\|^2 \le 8\{\|u^2(u-v)\|^2 + \|v^2(u-v)\|^2\}.$$

Making use of the Sobolev imbedding theorem (p. 97, [1]), we can get: the L^4-norm $|u|_4 \le C_1\|u\|_1$ and $\|u^2 v\|^2 \le C_2\|u\|_1^4\|v\|_1^2$, for some constants $C_1, C_2 > 0$. Hence there exists $C_3 > 0$ such that the above inequality yields

$$\|u^3 - v^3\|^2 \le C_3(\|u\|_1^4 + \|v\|_1^4)\|u - v\|_1^2,$$

which satisfies condition (D.2). The condition (D.3) is also true because

$$\langle Av, v \rangle + (\hat{F}_t(v), v) + \|\hat{\Sigma}_t(v)\|_R^2 = \kappa \langle \Delta v, v \rangle - \alpha \|v\|^2 - \gamma(v^3, v) + \|\Sigma(v)\|_R^2$$
$$\leq -(\kappa - r_0)\|\nabla v\|^2 - \alpha\|v\|^2.$$

So condition (D.3) holds, provided that $\kappa > r_0$. This condition also implies the monotonicity condition (D.4), since

$$2\langle A(u - v), u - v \rangle + 2(F_t(u) - F_t(v), u - v) + \|\Sigma_t(u) - \Sigma_t(v)\|_R^2$$
$$\leq -(2\kappa - r_0)\|\nabla(u - v)\|^2 - 2\alpha\|u - v\|^2.$$

Consequently, by applying Theorem 7.5, we can conclude that, if the condition (6.156) holds with $r_0 < 2\kappa$, then, given $h \in H$, the equation (6.155) has a unique solution $u \in L^2(\Omega; \mathbf{C}([0, T]; H)) \cap L^2(\Omega_T; H_0^1)$. Moreover it can be shown from the energy equation that $u \in L^4(\Omega_T; H)$.

6.8 Stochastic Evolution Equations of the Second Order

Let us consider the Cauchy problem in V':

$$\frac{d^2 u_t}{dt^2} = A_t u_t + F_t(u_t, v_t) + \Sigma_t(u_t, v_t)\dot{W}_t, \quad t \in (0, T),$$
$$u_0 = g, \quad \frac{du_0}{dt} = h,$$
(6.157)

where $v_t = \frac{d}{dt}u_t$; $A_t, F_t(\cdot)$ and $\Sigma_t(\cdot)$ are defined similarly as in the previous section, and $g \in V, h \in H$ are the initial data. We rewrite the equation (6.157) as the first-order system

$$\begin{aligned} du_t &= v_t dt, \\ dv_t &= [A_t u_t + F_t(u_t, v_t)]dt + \Sigma_t(u_t, v_t)dW_t, \\ u_0 &= g, \quad v_0 = h, \end{aligned}$$
(6.158)

or, in the integral form

$$\begin{aligned} u_t &= g + \int_0^t v_s ds, \\ v_t &= h + \int_0^t [A_s u_s + F_s(u_s, v_s)]ds + \int_0^t \Sigma_s(u_s, v_s)dW_s. \end{aligned}$$
(6.159)

In Section 6.6 we studied the mild solution of such a first-order stochastic evolution equation. Here we will consider the strong solution of the system. To establish an *a priori* estimate for the nonlinear problem, we first consider the linear case:

$$\begin{aligned} du_t &= v_t dt, \\ dv_t &= [A_t u_t + f_t]dt + dM_t, \\ u_0 &= g, \quad v_0 = h, \end{aligned}$$
(6.160)

where f_t is a predictable H-valued process and M_t is a H-valued martingale with local characteristic operator Q_t. In addition to conditions (B.1)–(B.3), we impose the following conditions on A_t:

(B.3') There exists $\beta > 0$ such that, for any $v \in V$,

$$\langle A_t v, v \rangle \leq -\beta \|v\|_V^2 \quad a.e.(t, \omega) \in \Omega_T.$$

(B.4) Let $A'_t = \frac{d}{dt} A_t : V \to V'$ and there is a constant $\alpha_1 > 0$ such that, for any $v \in V$,

$$|\langle A'_t v, v \rangle| \leq \alpha_1 \|v\|_V^2, \quad a.e.(t, \omega) \in \Omega_T.$$

Introduce the Hilbert space $\mathbb{H} = V \times H$ and $\mathbf{C}([0, T]; \mathbb{H}) = \mathbf{C}([0, T]; V) \times \mathbf{C}([0, T]; H)$. Let $V_0 \subset \mathcal{D}(A)$ be a Hilbert space dense in H. The lemma given below is analogous to Lemma 7.1.

Lemma 8.1   Suppose that the conditions (B.1), (B.2) and (B.3') hold and f_t is a $\mathcal{F}_t$-predictable process in $L^2(\Omega_T; H)$. Assume that M_t is a continuous V_0-valued martingale such that $A_t M_t$ is a continuous $L^2(\Omega)$-process in H. Then, for $g \in V$ and $h \in H$, the system (6.160) has a unique strong solution $(u; v) \in L^2(\Omega; \mathbf{C}([0, T]; \mathbb{H}))$ such that the energy equation holds:

$$\|v_t\|^2 - \langle A_t u_t, u_t \rangle + \langle A_0 g, g \rangle = \|h\|^2 + 2 \int_0^t \langle A'_s u_s, u_s \rangle ds$$
$$+ 2 \int_0^t (f_s, v_s) ds + 2 \int_0^t (v_s, dM_s) + \int_0^t Tr\, Q_s ds. \tag{6.161}$$

Proof.   The proof is similar to that of Lemma 7.1 and will only be sketched. Define $N_t = \int_0^t M_s ds$. Let $\mu_t = u_t - N_t$ and $\nu_t = v_t - M_t$. Then it follows from (6.161) that μ_t and ν_t satisfy

$$d\mu_t = \nu_t dt, \quad d\nu_t = A_t \mu_t dt + \tilde{f}_t dt,$$
$$u_0 = g, \quad v_0 = h, \tag{6.162}$$

where, by assumption on M_t, $\mu \in \mathbf{C}([0, T]; V)$, $\nu \in \mathbf{C}([0, T]; H)$ and $\tilde{f}_t = (f_t + A_t N_t) \in L^2((0, T); H)$. By appealing to the deterministic theory [58] for the corresponding equation, we can deduce that the system (6.160) has a unique solution: $u \in \mathbf{C}([0, T]; V)$ and $v \in L^2((0, T); H)$ a.s. such that the energy equation holds:

$$\|\nu_t\|^2 - \langle A_t \mu_t, \mu_t \rangle + \langle A_0 g, g \rangle = \|h\|^2 + 2 \int_0^t \langle A'_s \mu_s, \mu_s \rangle ds$$
$$+ 2 \int_0^t (f_s, \nu_s) ds + 2 \int_0^t (A_s N_s, \nu_s) ds. \tag{6.163}$$

Therefore u and v satisfy the system (6.160) with the same degree of regularity as that of μ and ν, correspondingly. The energy equation (6.161) can be derived from (6.163) by expressing μ_s and ν_s in terms of u_s and v_s and then applying the Itô formula. □

With the aid of this lemma, we will prove the existence theorem for the linear equation. To this end we introduce the energy function $\mathbf{e}(u, v)$ on $\mathbb{H}$ defined by

$$\mathbf{e}(u, v) = \|u\|_V^2 + \|v\|^2, \tag{6.164}$$

and the energy norm $\|(u; v)\|_e = \{\mathbf{e}(u, v)\}^{1/2}$.

We shall seek a solution of the system (6.160) with a bounded energy. Since the idea of proof is the same as that of Theorem 7.2, some details will be omitted.

Theorem 8.2   Let conditions (B.1), (B2), (B.3') and (B.4) hold true. Assume that f_t is a predictable H-valued process and M_t is a continuous L^2-martingale in H with local covariation operator Q_t such that

$$E\left\{\int_0^T \{\|f_s\|^2 + Tr\, Q_s\}ds < \infty. \tag{6.165}$$

Then, for $g \in V$, $h \in H$, the linear problem (6.160) has a unique strong solution $(u; v)$ with $u \in L^2(\Omega; \mathbf{C}([0, T]; V))$ and $v \in L^2(\Omega; \mathbf{C}([0, T]; H))$ such that the energy equation (6.161) holds true.

Proof.   By means of the energy equation (6.161), the uniqueness question can be easily disposed of. For the existence proof, we are going to take similar steps as done in Theorem 7.2.

(Step 1)   Approximate Solutions:

Let $\{M_t^n\}$ be a sequence of continuous martingales in V_0 such that $A_t M_t^n$ is a continuous L^2-process in H and $E \sup_{0 \le t \le T} \|M_t - M_t^n\|^2 \to 0$ as $n \to \infty$. Consider the approximate system of (6.160):

$$\begin{aligned} du_t^n &= v_t^n dt, \quad dv_t^n = [A_t u_t^n + f_t]dt + dM_t^n, \\ u_0^n &= g, \quad v_0^n = h. \end{aligned} \tag{6.166}$$

By Lemma 8.1, the above system has a unique solution $(u^n; v^n) \in L^2(\Omega; \mathbf{C}([0, T]; \mathbb{H}))$ such that the energy equation holds:

$$\|v_t^n\|^2 - \langle A_t u_t^n, u_t^n \rangle + \langle A_0 g, g \rangle = \|h\|^2 + 2\int_0^t \langle A_s' u_s^n, u_s^n \rangle ds$$
$$+2\int_0^t (f_s, v_s^n)ds + 2\int_0^t (v_s^n, dM_s) + \int_0^t Tr\, Q_s ds. \tag{6.167}$$

(Step 2)   Convergence of Approximation Sequences:

Let $\mathbf{Z}_T$ be the Banach space of adapted continuous processes in the space $L^2(\Omega; \mathbf{C}([0,T]; \mathbb{H}))$ with norm $\| \cdot \|_T$ defined by

$$\|(u;v)\|_T = \{E \sup_{0 \le t \le T} \mathbf{e}(u_t, v_t)\}^{1/2}, \tag{6.168}$$

where $\mathbf{e}(\cdot, \cdot)$ denotes the energy function given by (6.164).

Let $(u^m; v^m)$ and $(u^n; v^n)$ be solutions. Set $(u^{mn}; v^{mn}) = (u^m - u^n; v^m - v^n)$ and $M_t^{mn} = (M_t^m - M_t^n)$. Then, in view of (6.160), the following equations hold:

$$\begin{aligned} u_t^{mn} &= \int_0^t v_s^{mn} ds, \\ v_t^{mn} &= \int_0^t A_t u_s^{mn} ds + M_t^{mn}. \end{aligned} \tag{6.169}$$

The corresponding energy equation reads

$$\begin{aligned} \|v_t^{mn}\|^2 - \langle A_t u_t^{mn}, u_t^{mn} \rangle &= 2 \int_0^t \langle A_s' u_s^{mn}, u_s^{mn} \rangle ds \\ &+ 2 \int_0^t (v_s^{mn}, dM_s^{mn}) + \int_0^t Tr\, Q_s^{mn} ds, \end{aligned} \tag{6.170}$$

where Q_t^{mn} denotes the local covariation operator for M_t^{mn}. By making use of conditions (B.3'), (B.4) and some similar estimates as in Theorem 7.2, we can show that there exists $C > 0$ such that

$$E \sup_{0 \le t \le T} \{\|u_t^{mn}\|_V^2 + \|v_t^{mn}\|^2\} \le CE \int_0^T Tr\, Q_s^{mn} ds \to 0 \tag{6.171}$$

as $m, n \to \infty$. Hence $\{(u^n; v^n)\}$ is a Cauchy sequence in $\mathbf{Z}_T$ converging to $(u; v) \in L^2(\Omega; \mathbf{C}([0,T]; \mathbb{H}))$.

(Step 3)   Strong Solution and Energy Equation:

For any $\phi \in H$ and $\psi \in V$, by using the Itô formula, we have

$$\begin{aligned} (u_t^n, \phi) &= (g, \phi) + \int_0^t (v_s^n, \phi) ds, \\ (v_t^n, \psi) &= (h, \psi) + \int_0^t (A_s u_s^n, \psi) ds + \int_0^t (f_s, \psi) ds + (M_t^n, \psi). \end{aligned}$$

By taking the termwise limit, the above equations converge in $L^2(\Omega)$ to

$$\begin{aligned} (u_t, \phi) &= (g, \phi) + \int_0^t (v_s, \phi) ds, \\ (v_t, \psi) &= (h, \psi) + \int_0^t (A_s u_s, \psi) ds + \int_0^t (f_s, \psi) ds + (M_t, \psi), \end{aligned}$$

which implies that $(u; v)$ is a strong solution.

Finally The energy equation can be obtained by taking the limits in the approximation equation (6.167) in a similar fashion as in the proof of Theorem 7.2. □

As a corollary of the theorem, we can easily obtain the following energy inequality.

Lemma 8.3 Assume the conditions of Theorem 8.2 are satisfied. Then the following energy inequality holds:

$$
\begin{aligned}
&E \sup_{0 \leq t \leq T} \{\|u_t\|_V^2 + \|v_t\|^2\} \\
&\leq C_T \{\|g\|_V^2 + \|h\|^2 + E \int_0^T (\|f_s\|^2 + Tr\, Q_s) ds\},
\end{aligned}
\tag{6.172}
$$

for some constant $C_T > 0$. □

Now we turn to the nonlinear system (6.158) and impose the following conditions on the nonlinear terms:

(E.1) For any $u \in V, v \in H$, let $F_t(u, v)$ and $\Sigma_t(u, v)$ be $\mathcal{F}_t$–adapted processes with values in H and $\mathcal{L}_R^2$, respectively. Suppose there exist positive constants b, C_1 such that

$$
E \int_0^T [\|F_t(0, 0)\|^2 + \|\Sigma_t(0, 0)\|_R^2] dt \leq b,
$$

and

$$
\|\hat{F}_t(u, v)\|^2 + \|\hat{\Sigma}_t(u, v)\|_R^2 \leq C_1(1 + \|u\|_V^2 + \|v\|^2), \quad a.e. \ (t, \omega) \in \Omega_T,
$$

where $\hat{F}_t(u, v) = F_t(u, v) - F_t(0, 0)$ and $\hat{\Sigma}_t(u, v) = \Sigma_t(u, v) - \Sigma_t(0, 0)$.

(E.2) For any $u, u' \in V$ and $v, v' \in H$, there is a constant $C_2 > 0$ such that the Lipschitz condition holds:

$$
\begin{aligned}
&\|F_t(u, v) - F_t(u', v')\|^2 + \|\Sigma_t(u, v) - \Sigma_t(u', v')\|_R^2 \\
&\leq C_2(\|u - u'\|_V^2 + \|v - v'\|^2), \quad a.e. \ (t, \omega) \in \Omega_T.
\end{aligned}
$$

Theorem 8.4 Let conditions (B.1), (B.2), (B.3'), (B.4) and (E.1), (E.2) hold true. Then, for $g \in V, h \in H$, the nonlinear problem (6.158) has a unique strong solution $(u; v)$ with $u \in L^2(\Omega; \mathbf{C}([0, T]; V))$ and $v \in L^2(\Omega; \mathbf{C}([0, T]; H))$ such that the energy equation holds:

$$\|v_t\|^2 - \langle A_t u_t, u_t \rangle = \|h\|^2 - \langle A_0 g, g \rangle + 2 \int_0^t \langle A_s' u_s, u_s \rangle ds$$

$$+ 2 \int_0^t (F_s(u_s, v_s), v_s) ds + 2 \int_0^t (v_s, \Sigma_s(u_s, v_s) dW_s) \tag{6.173}$$

$$+ \int_0^t \|\Sigma_s(u_s, v_s)\|_R^2 ds.$$

Proof. We will prove the theorem by the contraction mapping principle. In the proof, for brevity, we will set $F_t(0,0) = 0$ and $\Sigma_t(0,0) = 0$ so that $\hat{F}_t = F_t$ and $\hat{\Sigma}_t = \Sigma_t$. As seen from the proof of Theorem 6.5, such terms do not play an essential role in the proof. Given $u \in L^2(\Omega; \mathbf{C}([0,T]; V))$ and $v \in L^2(\Omega; \mathbf{C}([0,T]; H))$, consider the following system

$$\mu_t = g + \int_0^t v_s ds,$$
$$\nu_t = h + \int_0^t A_s \mu_s ds + \int_0^t F_s(u_s, v_s) ds + \int_0^t \Sigma_s(u_s, v_s) dW_s. \tag{6.174}$$

Let $f_t = F_t(u_t, v_t)$ and $M_t = \int_0^t \Sigma_s(u_s, v_s) dW_s$ with local characteristic operator Q_t. Then, by condition (E.1), we have

$$E \int_0^T \{ \|f_t\|^2 + \mathrm{Tr}\, Q_t \} dt = E \int_0^T \{ \|F_t(u_t, v_t)]\|^2 + \|\Sigma_t(u_t, v_t)\|_R^2 \} dt$$
$$\leq C_1 T \{ 1 + E \sup_{0 \leq t \leq T} [\|u_t\|_V^2 + \|v_t\|^2] \} < \infty.$$

Therefore, for $(u; v) \in L^2(\Omega; \mathbf{C}([0,T]; \mathbb{H}))$, we can apply Theorem 8.2 to assert that the system (6.174) has a unique solution $(\mu; \nu) = \Gamma(u, v) \in \mathbf{Z}_T$, and that the solution map $\Gamma : \mathbf{Z}_T \to \mathbf{Z}_T$ is well defined.

To show the map Γ is a contraction for small T, we let $u, \tilde{u} \in L^2(\Omega; \mathbf{C}([0,T]; V))$ and $v, \tilde{v} \in L^2(\Omega; \mathbf{C}([0,T]; H))$. Let $(\tilde{\mu}; \tilde{\nu}) = \Gamma(\tilde{u}, \tilde{v})$. Set $\xi = \mu - \tilde{\mu}$ and $\eta = \nu - \tilde{\nu}$. Then ξ and η satisfy the system:

$$\xi_t = \int_0^t \eta_s ds,$$
$$\eta_t = \int_0^t A_s \xi_s ds + \int_0^t [F_s(u_s, v_s) - F_s(\tilde{u}_s, \tilde{v}_s)] ds \tag{6.175}$$
$$+ \int_0^t [\Sigma_s(u_s, v_s) - \Sigma_s(\tilde{u}_s, \tilde{v}_s)] dW_s.$$

By applying Lemma 8.3 and taking condition (E.2) into account, we can deduce from (6.175) that

$$
\begin{aligned}
E \sup_{0 \le t \le T} \{\|\xi_t\|_V^2 + \|\eta_t\|^2\} &\le CE \int_0^T \{\, \|F_s(u_s, v_s) - F_s(\tilde{u}_s, \tilde{v}_s)\|^2 \\
&\quad + \|\Sigma_s(u_s, v_s) - \Sigma_s(\tilde{u}_s, \tilde{v}_s)\|_R^2 \,\} ds \\
&\le C_2 T \, E \sup_{0 \le t \le T} \{\|u_t - \tilde{u}_t\|_V^2 + \|v_t - \tilde{v}_t\|^2\},
\end{aligned} \tag{6.176}
$$

for some $C_2 > 0$. Hence $\Gamma : \mathbf{Z}_T \to \mathbf{Z}_T$ is a contraction mapping for small T, and the unique fixed point $(u; v)$ is the solution of the system (6.174) which can be continued to any $T > 0$. By making use of the equation (6.168) in Theorem 8.2, the energy equation (6.173) can be verified. □

Under the linear growth and the uniform Lipschitz continuity conditions, it was shown that the solution exists in any finite time interval. If the nonlinear terms satisfy these condition only locally, such as the case of polynomial nonlinearity, the solution may explode in finite time [12]. To obtain a global solution, it is necessary to impose some additional condition on the nonlinear terms to prevent the solution from explosion. The next theorem is a statement about this fact. Before stating the theorem, we make the following assumptions:

(F.1) For $u \in V$, $v \in H$, let $F_t(u, v)$ and $\Sigma_t(u, v)$ be $\mathcal{F}_t$-predictable processes with values in H and $\mathcal{L}_R^2$, respectively. For any $u \in V, v \in H$ and $N > 0$ with $\mathbf{e}(u, v) \le N^2$, there exist positive constants b, C_N such that

$$
E \int_0^T [\|F_t(0, 0)\|^2 + \|\Sigma_t(0, 0)\|_R^2] dt \le b,
$$

$$
\|\hat{F}_t(u, v)\|^2 + \|\hat{\Sigma}_t(u, v)\|_R^2 \le C_N, \quad a.e. \ (t, \omega) \in \Omega_T,
$$

where $\hat{F}_t(u, v) = F_t(u, v) - F_t(0, 0)$ and $\hat{\Sigma}_t(u, v) = \Sigma_t(u, v) - \Sigma_t(0, 0)$.

(F.2) For any $u, u' \in V$ and $v, v' \in H$ with $\mathbf{e}(u, v) \le N^2$ and $\mathbf{e}(u', v') \le N^2$, there is a constant $K_N > 0$ such that the Lipschitz condition holds:

$$
\begin{aligned}
&\|F_t(u, v) - F_t(u', v')\|^2 + \|\Sigma_t(u, v) - \Sigma_t(u', v')\|_R^2 \\
&\le K_N(\|u - u'\|_V^2 + \|v - v'\|^2), \quad a.e. \ (t, \omega) \in \Omega_T.
\end{aligned}
$$

(F.3) Fix $T > 0$. For any $\phi. \in \mathbf{C}([0, T]; V) \cap \mathbf{C}^1([0, T]; H)$, there is a constant $\gamma > 0$ such that

$$
\int_0^t \{2(\hat{F}_s(\phi_s, \dot{\phi}_s), \dot{\phi}_s) + \|\hat{\Sigma}_s(\phi_s, \dot{\phi}_s)\|_R^2\} ds \le \gamma\{1 + \int_0^t \mathbf{e}(\phi_s, \dot{\phi}_s) \, ds)\},
$$

for $t \in [0, T]$, where we set $\dot{\phi}_t = \frac{d}{dt}\phi_t$.

Theorem 8.5   Let conditions (B.1), (B2), (B.3'),(B.4) and (F.1), (F.2) be fulfilled. Then, for $g \in V, h \in H$, the nonlinear problem (6.158) has a unique continuous local solution $(u_t; v_t)$ with $u_t \in V$ and $v_t \in H$ for $t < \tau$, where τ is a stopping time. If, in addition, condition (F.3) is also satisfied, then the solution exists in any finite time interval $[0, T]$ with $u \in L^2(\Omega; \mathbf{C}([0, T]; V))$ and $v \in L^2(\Omega; \mathbf{C}([0, T]; H))$. Moreover the energy equation (6.173) holds true as well.

Proof.   As before, for the sake of brevity, we set $F_t(0, 0) = 0$ and $\Sigma_t(0, 0) = 0$. First we will show the existence of a local solution by truncation. For $u \in V, v \in H$ and $N > 0$, introduce the truncation operators π_N and χ_N given by

$$\pi_N(u) = \frac{Nu}{(N \vee \|u\|_1)}, \quad \chi_N(v) = \frac{Nv}{(N \vee \|v\|)}, \tag{6.177}$$

and define

$$F_t^N(u, v) = F_t[\pi_N(u), \chi_N(v)], \quad \Sigma_t^N(u, v) = \Sigma_t[\pi_N(u), \chi_N(v)]. \tag{6.178}$$

Consider the truncated system:

$$\begin{aligned}
du_t^N &= v_t^N dt, \\
dv_t^N &= [A_t u_t^N + F_t^N(u_t^N, v_t^N)]dt + \Sigma_t^N(u_t^N, v_t^N)dW_t, \\
u_0^N &= g, \quad v_0^N = h,
\end{aligned} \tag{6.179}$$

Since $\pi_N(u)$ and $\chi_N(v)$ are bounded and Lipschitz continuous, it then follows from condition (F.2) and (6.178) that F_t^N and Σ_t^N satisfy the conditions (E.1)and (E.2). Therefore, by applying Theorem 8.4, the system has a unique strong solution $(u_t^N; v_t^N)$ over $[0, T]$ with the depicted regularity.

Introduce a stopping time $\tau_N = \inf\{t \in (0, T] : \mathbf{e}(u_t^N, v_t^N) > N^2\}$ and put it equal to T if the set is empty. Let $(u^N(\cdot, t); v^N(\cdot, t))$ be the solution of equation (6.179). Then, for $t < \tau_N$, $(u_t; v_t) = (u_t^N; v_t^N)$ is the solution of equation (6.158). Since the sequence $\{\tau_N\}$ increases with N, the limit $\tau = \lim_{N \uparrow \infty} \tau_N$ exists a.s. For $t < \tau < T$, we have $t < \tau_n$ for some $n > 0$, and define $(u_t; v_t) = (u_t^n; v_t^n)$. Then (u;v) is the desired local strong solution.

To obtain a global solution under condition (F.3), we make use of the local version of the energy equation (6.173):

$$\begin{aligned}
&\|v_{t \wedge \tau_N}\|^2 - \langle A_{t \wedge \tau_N} u_{t \wedge \tau_N}, u_{t \wedge \tau_N} \rangle = \|h\|^2 - \langle A_0 g, g \rangle \\
&+ 2 \int_0^{t \wedge \tau_N} \langle A_s' u_s, u_s \rangle ds + 2 \int_0^{t \wedge \tau_N} (F_s(u_s, v_s), v_s) ds \\
&+ 2 \int_0^{t \wedge \tau_N} (u_s, \Sigma_s(u_s, v_s) dW_s) + \int_0^{t \wedge \tau_N} \|\Sigma_s(u_s, v_s)\|_R^2 ds.
\end{aligned} \tag{6.180}$$

By taking the expectation over the above equation and making use of conditions (B3'), (B.4) and (F.3), we can deduce that

$$E\{\|v_{t\wedge\tau_N}\|^2 + \beta\|u_{t\wedge\tau_N}\|_V^2\} \le \|h\|^2 + \beta\|g\|_V^2$$
$$+2\alpha_1 E \int_0^t \|u_{s\wedge\tau_N}\|_V^2 ds + \gamma\{1 + E \int_0^t e(u_{s\wedge\tau_N}, v_{s\wedge\tau_N})ds\},$$

which implies that there exist positive constants C_0 and λ such that

$$E\mathbf{e}(u_{t\wedge\tau_N}, v_{t\wedge\tau_N}) \le C_0 + \lambda \int_0^t E\mathbf{e}(u_{s\wedge\tau_N}, v_{s\wedge\tau_N})ds.$$

Therefore, by means of the Gronwall inequality, we get

$$E\mathbf{e}(u_{T\wedge\tau_N}, v_{T\wedge\tau_N}) \le C_0 e^{\lambda T}. \tag{6.181}$$

On the other hand, we have

$$E\mathbf{e}(u_{T\wedge\tau_N}, v_{T\wedge\tau_N}) \ge E\left\{\mathbf{I}(\tau_N \le T)\mathbf{e}(u_{T\wedge\tau_N}, v_{T\wedge\tau_N})\right\}$$
$$\ge N^2 P\{\tau_N \le T\}, \tag{6.182}$$

where $\mathbf{I}(\cdot)$ denotes the indicator function. In view of (6.181) and (6.182), we get

$$P\{\tau_N \le T\} \le C_0 e^{\lambda T}/N^2,$$

which, by invoking the Borel-Cantelli lemma, implies that $P\{\tau \le T\} = 0$. This shows that the strong solution exists in any finite time interval $[0, T]$. Finally the energy equation now follows from (6.180) by letting $N \to \infty$. □

(**Example 8.1**) Consider the Sine-Gordon equation in domain D in $\mathbf{R}^d$ perturbed by a white noise:

$$\frac{\partial^2 u}{\partial t^2} = (\triangle - \alpha)u + \lambda \sin u - \nu\frac{\partial u}{\partial t} + f(x, t) + \sigma(x, t)\frac{\partial}{\partial t}W(x, t),$$

$$u|_{\partial D} = 0, \tag{6.183}$$

$$u(x, 0) = g(x), \quad \frac{\partial u}{\partial t}(x, 0) = h(x),$$

where $\alpha > 0, \nu$ and λ are given constants, $f(\cdot, t)$ and $\sigma(\cdot, t)$ are continuous adapted processes in $H = L^2(D)$, and $W(\cdot, t)$ is a Wiener random field in $K = H$ with covariance function $r(x, y)$. Assume that

$$E\{\int_0^T \int_D |f(x, t)|^2 dx dt + \int_0^T \int_D r(x, x)\sigma_2(x, t)dx dt\} \le b,$$

for some $b > 0$. Let $A = (\triangle - \alpha), F_t(u, v) = (\lambda \sin u - \nu v) + f(\cdot, t)$ and $\Sigma_t = \sigma(\cdot, t)$. Then it is easy to check that, for $V = H_0^1$, all conditions in Theorem

8.4 are satisfied. Hence, for $g \in H_0^1, h \in H$, the equation (6.183) has a unique strong solution $u \in L^2(\Omega; \mathbf{C}([0,T]; H_0^1))$ with $\frac{\partial}{\partial t} u \in L^2(\Omega; \mathbf{C}([0,T]; H))$.

(**Example 8.2**)　　Let D be a domain in $\mathbf{R}^d$ for $d \leq 3$. Consider the stochastic wave equation in D with a cubic nonlinear term:

$$\frac{\partial^2 u}{\partial t^2} = (\triangle - \alpha)u + \lambda u^3 + \nu \frac{\partial u}{\partial t} + f(x,t) + \sigma(x,t) \frac{\partial}{\partial t} W(x,t),$$

$$u|_{\partial D} = 0, \tag{6.184}$$

$$u(x,0) = g(x), \quad \frac{\partial u}{\partial t}(x,0) = h(x),$$

where all the coefficients of the equation are given as in the previous example. Let $A = (\triangle - \alpha), F_t(u,v) = (\lambda u^3 - \nu v) + f(\cdot, t)$. Making use of the estimates for the cubic term u^3 in Example 7.2, it is easy to show that conditions (F.1) and (F.2) are satisfied. Therefore, by the first part of Theorem 8.5, the problem (6.184) has a unique local solution. To check condition (F.3), for any $\phi. \in \mathbf{C}([0,T]; V) \cap \mathbf{C}^1([0,T]; H)$, we have

$$\int_0^t \{2(\hat{F}_s(\phi_s, \dot{\phi}_s), \dot{\phi}_s) + \|\hat{\Sigma}_s(\phi_s, \dot{\phi}_s)\|_R^2\} ds$$

$$= \int_0^t \{2\lambda(\phi_s^3, \dot{\phi}_s) ds + 2\nu \int_0^t (\dot{\phi}_s, \dot{\phi}_s) ds + \int_0^t \|\sigma_s\|_R^2\} ds$$

$$\leq \frac{1}{2}\lambda(|\phi_t|_4^4 - |\phi_0|_4^4) + 2\nu \int_0^t \|\dot{\phi}_s\|^2 ds + b$$

$$\leq |\lambda| \, |\phi_0|_4^4 + b + 2|\nu| \int_0^t \mathbf{e}(\phi_s, \dot{\phi}_s) ds, \quad \forall \, t \in [0,T],$$

provided that $\lambda < 0$. In this case, (B.3) holds true. Thus, by Theorem 8.5, the equation (6.184) has a unique strong solution $u \in L^2(\Omega; \mathbf{C}([0,T]; V))$ with $\frac{\partial}{\partial t} u \in L^2(\Omega; \mathbf{C}([0,T]; H))$ for any $T > 0$. In fact, from the energy inequality, we can show $u \in L^4(\Omega_T; H)$.

Chapter 7

Asymptotic Behavior of Solutions

7.1 Introduction

In the previous chapter, we have discussed the existence and uniqueness questions for stochastic evolution equations. Now some properties of the solutions will be studied. In particular we shall consider some problems concerning the asymptotic properties of solutions, such as boundedness, asymptotic stability, invariant measures and small random perturbations. For stochastic equations in finite dimensions, the asymptotic problems have been studied extensively for many years by numerous authors. For the stability and related questions, a comprehensive treatment of the subject is given in the classic book by Khasminskii [43]. In particular his systematic development in the stability analysis based on the method of Lyapunov functions has a natural generalization to an infinite-dimensional setting [9]. As to be seen, the method of Lyapunov functionals will play an important role in the subsequent asymptotic analysis. For stochastic processes in finite dimensions, the small random perturbation and the related large deviations problems are treated in the well-known books by Freidlin and Wentzell [28], Deuschel and Stroock [23], and Varadhan [80]. For stochastic partial differential equations, there are relatively fewer papers in asymptotic results. An up-to-date discussion of some stability results for stochastic evolution equations in Hilbert spaces can be found in a recent book by Liu [59], and the subject of invariant measures is treated by Da Prato and Zabczyk [20] in detail.

This chapter consists of seven sections. In Section 7.2, a generalized Itô's formula and the Lyapunov functional are introduced. By means of the Lyapunov functionals, the boundedness of solutions and the asymptotic stability of the null solution to some stochastic evolution equations will be treated in Section 7.3 and Section 7.4, respectively. The question on the existence of invariant measures is to be examined in Section 7.5. Then, in Section 7.6, we shall take up the small random perturbation problems. Finally the large deviations problem will be discussed briefly in Section 7.7.

7.2   Itô's Formula and Lyapunov Functionals

Consider the linear equation:

$$u_t = u_0 + \int_0^t A_s u_s ds + \int_0^t f_s ds + M_t. \tag{7.1}$$

In particular, for $u_0 = h$ and $M_t = \int_0^t \Phi_s dW_s$, this equation yields equation (6.100) for strong solutions. We shall generalize the Itô formula (6.47) to the one for the strong solution u_t of (7.1). Recall that, in Section 6.4, we defined an Itô functional $F : H \times [0,T] \to \mathbf{R}$ to be a function $F(v,t)$ with derivatives $\partial_t F, F', F''$ satisfying conditions (1)–(3) in Theorem 6-4.1. Due to the fact $A_s u_s \in V'$, the previous Itô formula is no longer valid. To extend this formula to the present case, we need to impose stronger conditions on F. To this end, let $U \subset H$ be an open set and let $U \times [0,T] = U_T$. Here a functional $F : U_T \to \mathbf{R}$ is said to be a *strong Itô functional* if it satisfies

(1) $\Phi : U_T \to \mathbf{R}$ is locally bounded and continuous such that its first two partial derivatives $\partial_t \Phi(v,t), \Phi'(v,t)$ and $\Phi''(v,t)$ exist for each $(v,t) \in U_T$.

(2) The derivatives $\partial_t \Phi$ and $\Phi' \in H$ are locally bounded and continuous in U_T.

(3) For any $\Gamma \in \mathcal{L}_1(H)$, the map: $(v,t) \to Tr[\Phi''(v,t)\Gamma]$ is locally bounded and continuous in $(v,t) \in U_T$.

(4) $\Phi'(\cdot,t) : V \cap U \to V$ is such that $\langle \Phi'(\cdot,t), v' \rangle$ is continuous in $t \in [0,T]$ for any $v' \in V'$ and

$$\|\Phi'(v,t)\|_V \le \kappa(1 + \|v\|_V), \quad (v,t) \in (V \cap U) \times [0,T],$$

for some $\kappa > 0$.

The following theorem gives a generalized Itô formula for the strong solution of equation (7.1).

Theorem 2.1   Suppose that A_t satisfies conditions (B1)–(B3) given in Section 6.7, $f \in L^2((0,T);H)$ is an integrable, adapted process and M_t is a continuous L^2-martingale in H with local characteristic operator Q_t. Let u_t be the strong solution of (7.1) with $u_0 \in L^2(\Omega;H)$ being $\mathcal{F}_0$-measurable. For any strong Itô functional F on U_T, the following formula holds

$$\Phi(u_t, t) = \Phi(u_0, 0) + \int_0^t \partial_s \Phi(u_s, s)\, ds + \int_0^t \langle A_s u_s, \Phi'(u_s, s) \rangle ds$$
$$+ \int_0^t (f_s, \Phi'(u_s, s)) ds + \int_0^t (\Phi'(u_s, s), dM_s) \tag{7.2}$$
$$+ \frac{1}{2} \int_0^t Tr[\Phi''(u_s, s) Q_s] ds.$$

Proof. The idea of proof is based on a sequence of approximations to the equation (7.1) by regularizing its coefficients so that Theorem 6-4.2 is applicable. Then the theorem is proved by a limiting process. The proof is long and tedious [67, 68]. To show some technical aspect of the problem, we will sketch a proof for the special case when M_t is a continuous L^2-martingale in V.

By invoking Theorem 6-7.2, the equation (7.1) has a strong solution $u \in L^2(\Omega_T; V) \cap L^2(\Omega; \mathbf{C}([0, T], H))$. Let $\hat{u} = (u - M)$ and $v_t = (A_t u_t + f_t)$. Then, $d\hat{u} = vdt$, and for $M \in L^2(\Omega_T; V)$, we have $\hat{u} \in L^2(\Omega_T; V)$ and $v_t \in L^2(\Omega_T; V')$.

Let $Y_T = L^2(\Omega_T; V) \cap L^2(\Omega; \mathbf{C}([0, T], H))$ with norm $\|u\|_T$ defined as in Theorem 6-7.2. Suppose that $\{\hat{u}^n\}$ is a sequence in Y_T with $\hat{u}_0^n = u_0$ such that $\hat{u}^n \in \mathbf{C}^1((0, T); H)$ a.s., $\hat{u}^n \to \hat{u}$ in $L^2(\Omega_T; V)$ and $\frac{d}{dt} \hat{u}^n = v^n \to v$ in $L^2(\Omega_T; V')$. Let $u^n = \hat{u}^n + M$. Then, by Theorem 6-4.1, the following Itô formula holds

$$\Phi(\hat{u}_t^n + M_t, t) = \Phi(u_0, 0) + \int_0^t \partial_s \Phi(\hat{u}_s^n + M_s, s) ds$$
$$+ \int_0^t (v_s^n, \Phi'(\hat{u}_s^n + M_s, s)) ds + \int_0^t (\Phi'(\hat{u}_s^n + M_s, s), dM_s) \tag{7.3}$$
$$+ \frac{1}{2} \int_0^t Tr[\Phi''(\hat{u}_s^n + M_s, s) Q_s] ds.$$

To construct such a sequence $\hat{u}^n$, let $A_0 : V \to V'$ be a constant closed, coercive linear operator with domain $\mathcal{D}(A)$. Let $\hat{u}^n = u^n - M^n$, where $AM^n \in L^2(\Omega_T; H)$ such that $M^n \to M$ in $L^2(\Omega_T; V)$ and $u^n \to u$ in Y_T. In particular, we define u^n by the equation:

$$du^n(\cdot, t) = A_0 u^n(\cdot, t) dt + f_t^n dt + dM_t^n, \tag{7.4}$$
$$u_0^n = P_n u_0,$$

where $f_t^n = P_n'(v_t - A_0 u)$; $P_n : H \to V$ and $P_n' : V' \to V$ are two projectors such that $P_n u_0 \to u_0$ in H and $P_n' v_t \to v_t$ in $L^2(\Omega_T; V')$. By Theorem 6-7.2, we can assert that the problem (7.4) has a unique strong solution $u^n \in Y_T$ and, furthermore, $u^n \to u$ in Y_T. Hence we have $\hat{u}^n \to \hat{u}$ in Y_T. To check

$\hat{u}^n \in \mathbf{C}^1((0,T);H)$, in view of (7.4), $\hat{u}^n$ satisfies

$$d\hat{u}_t^n = A_0\hat{u}_t^n dt + (f_t^n + A_0 M_t^n)dt,$$
$$\hat{u}_0^n = P_n u_0,$$

which, by a regularity property of PDEs [57], has a solution $\hat{u}^n \in \mathbf{C}^1((0,T);H)$ a.s. Now we claim that, with the aid of the above convergence results and the conditions on Φ, the Itô formula (7.2) can be verified by taking the limit in (7.3).

For the general case, we can approximate M_t in H by a sequence M_t^n of V-valued martingales such that $M^n \to M$ in $L^2(\Omega_T;H)$, and prove the Itô formula (7.2) by a limiting process. However this will not be carried out here.
□

Let us consider the strong solution of the equation considered in Section 6.7:

$$du_t = A_t u_t dt + F_t(u_t)dt + \Sigma_t(u_t)dW_t, \quad t \geq 0, \tag{7.5}$$

where the coefficients A_t, F_t and Σ_t are assumed to be non-random or deterministic. It follows from Theorem 2.1 that the following holds.

Theorem 2.2 Let the conditions for Theorem 6-7.5 are satisfied so that equation (7.5) has a strong solution u_t over [0,T]. Suppose that $\Phi : H \times [0,T]$ is a strong Itô functional. Then we have

$$\Phi(u_t,t) = \Phi(u_0,0) + \int_0^t \mathcal{L}_s\Phi(u_s,s)ds + \int_0^t (\Phi'(u_s,s),\Sigma_s(u_s)dW_s), \tag{7.6}$$

where

$$\mathcal{L}_s\Phi(v,s) = \frac{\partial}{\partial s}\Phi(v,s) + \frac{1}{2}Tr[\Phi''(v,s)\Sigma_s(v)R\Sigma_s^\star(v)]$$
$$+\langle A_s v, \Phi'(v,s)\rangle + (F_s(v),\Phi'(v,s)). \tag{7.7}$$
□

Let $U \subset H$ be a neighborhood of the origin. A strong Itô functional $\Phi : U \times \mathbf{R}^+ \to \mathbf{R}$ is said to be a *Lyapunov functional* for the equation (7.5), if

(1) $\Phi(0,t) = 0$ for all $t \geq 0$, and, for any $\epsilon > 0$, there is $\delta > 0$ such that

$$\inf_{t \geq 0, \|h\| \geq \epsilon} \Phi(h,t) \geq \delta.$$

(2) For any $t \geq 0$ and $v \in U \cap V$,

$$\mathcal{L}_t\Phi(v,t) \leq 0. \tag{7.8}$$

In particular, if $\Phi = \Psi(v)$ is independent of t, we have $\mathcal{L}_t\Psi(v) = \mathcal{L}\Psi(v)$ with

$$\mathcal{L}\Psi(v) = \frac{1}{2}Tr[\Psi''(v)\Sigma_s(v)R\Sigma_s^*(v)] + \langle A_s v, \Psi'(v)\rangle + (F_s(v), \Psi'(v)). \quad (7.9)$$

Remark: For second-order stochastic evolution equations treated in Section 6.8, Itô's formula is not valid in general due to the fact that the solutions are insufficiently regular. However it does hold for a quadratic functional, such as the energy equation given in Theorem 6-8.4. The energy-related functionals will play an important role in studying the asymptotic behavior of solutions to second-order equations, such as a stochastic hyperbolic equation.

7.3 Boundedness of Solutions

Let u_t^h be a strong solution of the equation (7.5) with $u_0^h = h$. Here and throughout the remaining chapter, W_t denotes a K-valued Wiener process with a finite-trace covariance operator R. The solution is said to be *non-explosive* if

$$\lim_{r\to\infty} P\{\sup_{0\leq t\leq T} \|u_t^h\| > r\} = 0,$$

for any $T > 0$. If the above holds for $T = \infty$, the solution is said to be *ultimately bounded* .

Lemma 3.1 Let $\Phi : U \times \mathbf{R}^+ \to \mathbf{R}^+$ be a Lyapunov functional and let u_t^h denote the strong solution of (7.5). For $r > 0$, let $B_r = \{h \in H : \|h\| \leq r\}$ such that $B_r \subset U$. Define

$$\tau = \inf\{t > 0 : u_t^h \in B_r^c, h \in B_r\},$$

with $B_r^c = H \setminus B_r$. We put $\tau = T$ if the set is empty. Then the process $\phi_t = \Phi(u_{t\wedge\tau}^h, t\wedge\tau)$ is a local $\mathcal{F}_t$-supermartingale and the following Chebyshev inequality holds

$$P\{\sup_{0\leq t\leq T} \|u_t^h\| > r\} \leq \frac{\Phi(h,0)}{\Phi_r}, \quad (7.10)$$

where

$$\Phi_r = \inf_{0\leq t\leq T, h\in U\cap B_r^c} \Phi(h,t).$$

Proof. By Itô's formula and the properties of a Lyapunov functional, we have

$$\Phi(u_{t\wedge\tau}^h, t \wedge \tau) = \Phi(h,0) + \int_0^{t\wedge\tau} \mathcal{L}_s\Phi(u_s^h, s)ds + \int_0^{t\wedge\tau} (\Phi'(u_s^h, s), \Sigma_s(u_s^h))dW_s)$$

$$\leq \Phi(h,0) + \int_0^{t\wedge\tau} (\Phi'(u_s^h, s), \Sigma_s(u_s^h)dW_s),$$

so that $\phi_t = \Phi(u_{t\wedge\tau}^h, t \wedge \tau)$ is a supermartingale and

$$E\phi_t \leq E\phi_0 = \Phi(h,0).$$

But

$$E\phi_T = E\Phi(u_{T\wedge\tau}^h, T \wedge \tau) \geq E\{\Phi(u_\tau^h, \tau); \tau \leq T\}$$

$$\geq \inf_{0\leq t\leq T, \|h\|=r} \Phi(h,t)P\{\tau \leq T\}$$

$$\geq \Phi_r P\{\sup_{0\leq t\leq T} \|u_t^h\| > r\},$$

which implies (7.10).   □

As a consequence of this lemma, the following two boundedness theorems can be proved easily.

Theorem 3.2 Suppose that there exists a Lyapunov functional $\Phi : H \times \mathbf{R}^+ \to \mathbf{R}^+$ such that

$$\Phi_r = \inf_{t\geq 0, \|h\|\geq r} \Phi(h,t) \to \infty, \quad \text{as } r \to \infty.$$

Then the solution u_t^h is ultimately bounded.

Proof. By Lemma 3.1, we can deduce that

$$P\{\sup_{t\geq 0} \|u_t^h\| > r\} = \lim_{T\to\infty} P\{\sup_{0\leq t\leq T} \|u_t^h\| > r\} \leq \frac{\Phi(h,0)}{\Phi_r}.$$

Therefore we have

$$\lim_{r\to\infty} P\{\sup_{t\geq 0} \|u_t^h\| > r\} = 0$$

as to be shown.   □

Theorem 3.3 If there exist an Itô functional $\Psi : H \times \mathbf{R}^+ \to \mathbf{R}^+$ and a constant $\alpha > 0$ such that

$$\mathcal{L}_t\Psi \leq \alpha\Psi(v,t) \quad \text{for any} \quad v \in V,$$

and the infimum $\inf_{t\geq 0, \|h\|\geq r} \Psi(h,t) = \Psi_r$ exists such that $\lim_{r\to\infty} \Psi_r = \infty$, then the solution u_t^h does not explode in finite time.

Proof. Let $\Phi(v,t) = e^{-\alpha t}\Psi(v,t)$. Then it is easy to check that

$$\mathcal{L}_t\Phi(v,t) \leq 0,$$

so that Φ is a Lyapunov functional. Hence, by Lemma 3.1,

$$P\{\sup_{0 \leq t \leq T} \|u_t^h\| > r\} \leq \frac{\Phi(h,0)}{\Phi_r} = \frac{\Psi(h,0)}{\Psi_r} \to 0$$

as $r \to \infty$, for any $T > 0$. $\qquad\square$

(**Example 3.1**) Consider the reaction-diffusion equation in $D \subset \mathbf{R}^3$:

$$\frac{\partial u}{\partial t} = \kappa\Delta u + f(u) + \sum_{k=1}^{3} \frac{\partial u}{\partial x_k}\frac{\partial}{\partial t}W_k(x,t), \qquad (7.11)$$
$$u|_{\partial D} = 0, \quad u(x,0) = h(x),$$

where $f(\cdot) : H = L^2(D) \to L^p(D)$ is a nonlinear function with $p > 2$, and $W_k(x,t)$ are Wiener random fields with bounded covariance functions $r_{jk}(x,y)$ such that

$$\sum_{j,k=1}^{3} r_{jk}(x,x)\xi_j\xi_k \leq r_0|\xi|^2, \ \forall \ \xi \in \mathbf{R}^3, \qquad (7.12)$$

for some $r_0 > 0$. Suppose that, by Theorem 6-7.5, the problem (7.11) has a strong solution $u(\cdot,t) \in V$ with $V = L^{p+1}(D) \cap H_0^1 = H_0^1$ due to the Sobolev imbedding $H_0^1 \subset L^{p+1}(D)$, for $D \in \mathbf{R}^3$ and $p \leq 3$. Consider the reduced ordinary differential equation

$$\frac{dv}{dt} = f(v).$$

Assume there exist a $\mathbf{C}^2$-function $\phi(v) \geq 0$ and a constant α such that $\phi(v) = 0$ implies $v = 0$, and it satisfies

$$f(v)\phi'(v) \leq \alpha\phi(v), \ \phi''(v) \geq 0, \text{ and } \phi(v) \to \infty \text{ as } v \to \infty. \qquad (7.13)$$

Define

$$\Phi(v) = \int_D \phi[v(x)]dx, \ v \in V. \qquad (7.14)$$

Let $\Sigma(v) = \nabla v$ and $R = [r_{jk}]_{3\times 3}$. Then we have

$$\mathcal{L}_t\Phi(v) = \kappa(\Delta v, \Phi'(v)) + (f(v), \Phi'(v)) + \frac{1}{2}Tr[\Phi''(v)\Sigma(v)R\Sigma^*(v)]$$

$$= -\kappa\int_D \phi''[v(x)]|\nabla v(x)|^2 dx + \int_D f[v(x)]\phi'[v(x)]dx$$

$$+ \frac{1}{2}\int_D \phi''[v(x)]\sum_{j,k=1}^{3} r_{jk}(x,x)\frac{\partial v(x)}{\partial x_j}\frac{\partial v(x)}{\partial x_k}dx.$$

In view of (7.12) and (7.13), the above yields

$$\mathcal{L}_t \Phi(v) \leq -(\kappa - r_0/2) \int_D \phi''[v(x)]|\nabla v(x)|^2 dx + \alpha\Phi(v). \qquad (7.15)$$

Therefore, if $\kappa \geq r_0/2$, by Theorem 3.3, the solution does not explode in finite time. From Theorem 3.2, if $\kappa \geq r_0/2$ and $\alpha \leq 0$, the solution is ultimately bounded.

7.4 Stability of Null Solution

To study the stability problems, we assume throughout this section that, for given $u_0 = h \in H$, the initial-value problem (7.5) has a global strong solution $u^h \in L^2(\Omega \times (0, T); V) \cap L^2(\Omega; \mathbf{C}((0, T); H))$ for any $T > 0$. Suppose that the equation has an equilibrium solution $\hat{u}$. For simplicity suppose $F_t(0) = 0$ and $\Sigma_t(0) = 0$ so that $\hat{u} \equiv 0$ is such a solution for all $t > 0$. In particular we shall study the stability of the null solution.

For definitions of stability, we say that the null solution $u \equiv 0$ of (7.16) is *stable in probability* in H if for any $\epsilon_1, \epsilon_2 > 0$, there is $\delta > 0$ such that if $\|h\| < \delta$, then

$$P\{\sup_{t>0} \|u_t^h\| > \epsilon_1\} < \epsilon_2.$$

The null solution is said to be *asymptotically p-stable* in H for $p \geq 1$ if there exists $\delta > 0$ such that if $\|h\| < \delta$, then

$$\lim_{t\to\infty} E\|u_t^h\|^p = 0.$$

If, in addition, there are positive constants $K(\delta), \nu$ and T such that

$$E\|u_t^h\|^p \leq K(\delta)e^{-\nu t} \quad t > T,$$

then the null solution is *exponentially p-stable* in H. For $p = 2$, they are known as *asymptotically stable in mean-square* and *exponentially stable in mean-square*, respectively.

The null solution is said to be *a.s.(almost surely) asymptotically stable* in H if there exists $\delta > 0$ such that if $\|h\| < \delta$, then

$$P\{\lim_{t\to\infty} \|u_t^h\| = 0\} = 1,$$

and it is *a.s.(almost surely) exponentially stable* in H if there exist positive constants $\delta, K(\delta), \nu$ and a random variable $T(\omega) > 0$ such that if $\|h\| < \delta$, then

$$\|u_t^h\| \leq K(\delta)e^{-\nu t} \quad \forall \ t > T, \ a.s.$$

In view of the above definitions, it is clear that the exponential stability implies the asymptotic stability. If an asymptotic stability condition holds for any $\delta > 0$, then we say that it is *asymptotically globally stable* in H.

Theorem 4.1   Suppose there exists a Lyapunov functional $\Phi : H \times \mathbf{R}^+ \to \mathbf{R}^+$. Then the null solution of (7.5) is stable in probability.

Proof.   Let $r > 0$ such that $B_r = \{h \in H : \|h\| \leq r\} \subset U$. For any $\epsilon_1, \epsilon_2 > 0$, by Lemma 3.1, we have

$$P\{ \sup_{0 \leq t \leq T} \|u_t^h\| > \epsilon_1\} \leq P\{ \sup_{0 \leq t \leq T} \|u_t^h\| > (r \wedge \epsilon_1)\} \leq \frac{\Phi(h,0)}{\Phi_{(r \wedge \epsilon_1)}}.$$

As $T \to \infty$, we get

$$P\{\sup_{t \geq 0} \|u_t^h\| > \epsilon_1\} \leq \frac{\Phi(h,0)}{\Phi_{(r \wedge \epsilon_1)}}.$$

Since $\Phi(h,0)$ is continuous with $\Phi(0,0) = 0$, given $\epsilon_2 > 0$, there exists $\delta > 0$ such that the right-hand side of the above equation becomes less than ϵ_2 if $\|h\| < \delta$. $\square$

Theorem 4.2   Let the equation (7.5) have a global strong solution u_t^h such that $E\|u_t^h\|^p < \infty$ for any $t > 0$. Suppose there exists a Lyapunov functional $\Phi(h,t)$ on $H \times \mathbf{R}^+$ such that

$$\mathcal{L}_t \Phi(v,t) \leq -\alpha\Phi(v,t), \quad \forall\; v \in V, \; t > 0, \tag{7.16}$$

and

$$\beta\|h\|^p \leq \Phi(h,t) \leq \gamma\|h\|^p, \quad \forall\; h \in H, \; t > 0, \tag{7.17}$$

for some positive constants α, β, γ. Then the null solution is exponentially p-stable.

Proof.   For any $\delta > 0$, let $h \in B_\delta$. By Itô's formula, we have

$$e^{\alpha t}\Phi(u_t^h, t) = \Phi(h,0) + \int_0^t e^{\alpha s}[\mathcal{L}_s\Phi(u_s^h, s) + \alpha\Phi(u_s^h, s)]ds$$

$$+ \int_0^t e^{\alpha s}(\Phi'(u_s^h, s), \Sigma_s(u_s^h)dW_s),$$

so that, by condition (7.10),

$$Ee^{\alpha t}\Phi(u_t^h, t) = \Phi(h,0) + \int_0^t e^{\alpha s}E[\mathcal{L}_s\Phi(u_s^h, s) + \alpha\Phi(u_s^h, s)]\, ds$$

$$\leq \Phi(h,0).$$

In view of condition (7.11), the above implies

$$E\|u_t^h\|^p \le \frac{\Phi(h,0)}{\beta}e^{-\alpha t}, \quad \forall\, t > 0. \qquad\qquad \Box$$

Theorem 4.3   Suppose there exists a Lyapunov functional $\Phi : \mathbf{R}^+ \times H \to \mathbf{R}^+$ such that

$$\mathcal{L}_t\Phi(v,t) \le -\alpha\Phi(v,t), \quad \forall\, v \in V,\ t > 0, \qquad\qquad (7.18)$$

for some constant $\alpha > 0$. Then the null solution is a.s. asymptotically stable. If, in addition, there exist positive constants β, p, such that

$$\Phi(h,t) \ge \beta\|h\|^p, \quad \forall\, h \in H,\ t > 0, \qquad\qquad (7.19)$$

then the null solution is a.s. exponentially stable.

Proof.   For $h \in H$, let

$$\Psi(u_t^h,t) = e^{\alpha t}\Phi(u_t^h,t),$$

and let $\tau_n = \inf_{t>0}\{\|u_t^h\| > n\}$. By Itô's formula and condition (7.18), we have

$$\Psi(u_{t\wedge\tau_n}^h, t \wedge \tau_n) = \Phi(h,0) + \int_0^{t\wedge\tau_n}(\mathcal{L}_s + \alpha)\Psi(u_s^h,s)ds$$
$$+ \int_0^{t\wedge\tau_n}(\Psi'(u_s^h,s), \Sigma_s(u_s^h)dW_s).$$

It is easy to check that $\psi_t^n = \Psi(u_{t\wedge\tau_n}^h, t \wedge \tau_n)$ is a positive supermartingale. By Theorem 4.2 and a supermartingale convergence theorem (p. 65, [72]), it converges almost surely to a positive finite limit $\psi(\omega)$ as $t \to \infty$. That is, there exists $\Omega_0 \subset \Omega$ with $P\{\Omega_0\} = 1$ such that

$$\lim_{t\to\infty}\Psi(u_t^h,t) = \lim_{t\to\infty}\psi_t^n(\omega) = \psi(\omega), \quad \forall\, \omega \in \Omega_0.$$

It follows that, for each $\omega \in \Omega_0$, there exists $T(\omega) > 0$ such that

$$\Phi(u_t^h,t) \le 2\psi(\omega)e^{-\alpha t}, \quad \forall\, t > T(\omega), \qquad\qquad (7.20)$$

which implies that the null solution is a.s. asymptotically stable. If condition (7.19) also holds, the inequality (7.20) yields

$$\beta\|u_t^h\|^p \le 2\psi^h(\omega)e^{-\alpha t},$$

or

$$\|u_t^h\| \le K^h e^{-\nu t}, \quad \forall\, t > T,\ \text{a.s.},$$

with $K^h = (2\psi/\beta)^{1/p}$ and $\nu = \alpha/p$. Hence the null solution is a.s. exponentially stable. □

(**Example 4.1**)   Suppose that the coefficients of the equation (7.5) satisfy the coercivity condition as in Theorem 6-7.5:

$$2\langle A_t v, v \rangle + 2(F_t(v), v) + \|\Sigma_t(v)\|_R^2 \le \beta\|v\|^2 - \alpha\|v\|_V^2, \quad v \in V, \qquad (7.21)$$

for some constants α, β with $\alpha > 0$. Then the following stability result holds.

Theorem 4.4   Let the coercivity condition (7.21) be satisfied. If $\lambda = \inf_{v \in V} \dfrac{\|v\|_V^2}{\|v\|^2} > \beta/\alpha$, then the null solution of (7.5) is a.s. exponentially stable.

Proof.   Let $\Phi(h) = \|h\|^2$. Then, by condition (7.21),

$$\mathcal{L}\Phi(v) = 2\langle A_t v, v \rangle + 2(F_t(v), v) + \|\Sigma_t(v)\|_R^2$$
$$\le \beta\|v\|^2 - \alpha\|v\|_V^2 \le -(\alpha\frac{\|v\|_V^2}{\|v\|^2} - \beta)\|v\|^2$$
$$\le -(\alpha\lambda - \beta)\|v\|^2,$$

or

$$\mathcal{L}\Phi(v) \le -\kappa\Phi(v), \quad \text{for } v \in V,$$

with $\kappa = \alpha\lambda - \beta > 0$. Therefore, by Theorem 4.2, the null solution is a.s. exponentially stable. □

(**Example 4.2**)   Let us reconsider (Example 3.1) under the same assumptions. Suppose $f(0) = 0$ so that $u \equiv 0$ is a solution of (7.11). Let Φ be the Lyapunov functional given by (7.14). In view of (7.15), if $\kappa \ge r_0/2$ and $\alpha < 0$, then

$$\mathcal{L}\Phi(v) \le -|\alpha|\Phi(v), \quad \text{for } v \in V.$$

By invoking Theorem 4.2, the null solution is a.s. asymptotically stable.

7.5   Invariant Measures

Consider the autonomous version of the equation (7.5):

$$\begin{aligned} du_t &= Au_t dt + F(u_t)dt + \Sigma(u_t)dW_t, \quad t \ge 0, \\ u_0 &= \xi, \end{aligned} \qquad (7.22)$$

where $\xi \in L^2(\Omega; H)$ is a $\mathcal{F}_0$-measurable random variable. Suppose that the equation has a unique global strong solution $u_t, t \ge 0$, as depicted in Theorem

6-7.5. Then, as in finite dimensions, it is easy to show that the solution u_t is a time-homogeneous, continuous Markov process in H with the transition probability function $P(h, t, \cdot)$ defined by the conditional probability

$$P(h, t - s, B) = P\{u_t \in B | u_s = h\},$$

for $0 \leq s < t$, $h \in H$ and $B \in \mathcal{B}(H)$, the Borel field of H. For any bounded continuous function $\Phi \in \mathbf{C}_b(H)$, the transition operator P_t is defined by

$$(P_t\Phi)(h) = \int_H \Phi(g)P(h, t, dg) = E\{\Phi(u_t) | u_0 = h\}. \tag{7.23}$$

Recall that a probability measure μ on $(H, \mathcal{B}(H))$ is said to be an *invariant measure* for the given transition probability function if it satisfies the equation

$$\mu(B) = \int_H P(h, t, B)\mu(dh),$$

or, equivalently, the following holds:

$$\int_H (P_t\Phi)(h)\mu(dh) = \int_H \Phi(h)\mu(dh), \tag{7.24}$$

for any bounded continuous function $\Phi \in \mathbf{C}_b(H)$.

For $\Phi \in \mathbf{C}_b(H)$, if $P_t\Phi(h)$ is bounded and continuous for $t > 0$, $h \in H$, then the transition probability $P(h, t, \cdot)$ is said to possess the *Feller property*.

Let $\mathcal{M}_1(H)$ denote the space of probability measures on $\mathcal{B}(H)$. A sequence of probability measures $\{\mu_n\}$ is said to *converge weakly* to μ in $\mathcal{M}_1(H)$, or simply, $\mu_n \rightharpoonup \mu$, if

$$\lim_{n \to \infty} \int_H \Phi(h)\mu_n(dh) = \int_H \Phi(h)\mu(dh)$$

for any $\Phi \in \mathbf{C}_b(H)$.

Lemma 5.1   Let $\mu_t = \mathcal{L}\{u_t\}$ denote the probability measure for the solution u_t of (7.22) that has the Feller property. If $\mu_t \rightharpoonup \mu$ as $t \to \infty$, then μ is an invariant measure for the solution process.

Proof.   Let $\nu = \mathcal{L}\{\xi\}$ be the initial distribution. For $t > 0$ and $\Phi \in \mathbf{C}_b(H)$, since $\mu_t \rightharpoonup \mu$, we have

$$\int_H \Phi(h)\mu_t(dh) = \int_H (P_t\Phi)(h)\nu(dh) \to \int_H \Phi(h)\mu(dh),$$

as $t \to \infty$. Hence

$$\int_H (P_{t+s}\Phi)(h)\nu(dh) \to \int_H \Phi(h)\mu(dh), \tag{7.25}$$

for any $s > 0$. On the other hand,

$$\int_H (P_{t+s}\Phi)(h)\nu(dh) = \int_H [P_t(P_s\Phi)](h)\nu(dh) \rightarrow \int_H (P_s\Phi)(h)\mu(dh), \quad (7.26)$$

as $t \rightarrow \infty$ with fixed s, since $P_s\Phi \in \mathbf{C}_b(H)$ by the Feller property. In view of (7.25) and (7.26), we have

$$\int_H (P_s\Phi)d\mu = \int_H \Phi d\mu$$

for any $s > 0$, or μ is an invariant measure by (7.24). $\quad\square$

Lemma 5.2 Let $\mu_t^h(\cdot) = P(h, t, \cdot)$ be the distribution of u_t^h given $\xi = h \in H$. If $\mu_t^h \rightharpoonup \mu$ for all $h \in H$, then μ is the unique invariant measure for the solution process of equation (7.22).

Proof. By the weak convergence, for $\Phi \in \mathbf{C}_b(H)$,

$$(P_t\Phi)(h) = \int_H \Phi(g)\mu_t^h(dg) \rightarrow \int_H \Phi(g)\mu(dg).$$

Now, for any invariant measure $\nu \in \mathcal{M}_1(H)$, we have

$$\int_H \Phi(h)\nu(dh) = \lim_{t\to\infty} \int_H (P_t\Phi)(g)\nu(dg) = \int_H \Phi(g)\mu(dg),$$

or μ is the unique invariant measure. $\quad\square$

Theorem 5.3 Let μ be an invariant measure for the equation (7.22) with the initial distribution $\mathcal{L}\{\xi\} = \mu$. Then the solution $\{u_t^\xi, t > 0\}$ is a stationary process in H.

Proof. Let $\Phi_i, i = 1, 2, \cdots, n$ be bounded measurable functions and let $0 \le t_1 < t_2 < \cdots < t_n$. We must show that, for any $\tau > 0$,

$$M_n = E\{\Phi_1(u_{t_1+\tau}^\xi)\Phi_2(u_{t_2+\tau}^\xi)\cdots\Phi_n(u_{t_n+\tau}^\xi)\}$$

is independent of τ, for any $n \ge 1$. This can be verified by induction. For $n = 1$, due to the invariant μ,

$$M_1 = E\Phi_1(u_{t_1+\tau}^\xi) = \int_H (P_{t_1+\tau}\Phi_1)d\mu$$

$$= \int_H P_\tau(P_{t_1}\Phi_1)d\mu = \int_H (P_{t_1}\Phi_1)d\mu = E\Phi_1(u_{t_1}^\xi).$$

Now suppose that

$$M_{n-1} = E\{\Phi_1(u_{t_1+\tau}^\xi)\Phi_2(u_{t_2+\tau}^\xi)\cdots\Phi_{n-1}(u_{t_{n-1}+\tau}^\xi)\}$$

is independent of τ. Then, making use of the Markov property,

$$M_n = E\{[\Phi_1(u^\xi_{t_1+\tau}) \cdots \Phi_{n-1}(u^\xi_{t_{n-1}+\tau})] E[\Phi_n(u^\xi_{t_n+\tau})|\mathcal{F}_{t_{n-1}+\tau}]\}$$
$$= E\{[\Phi_1(u^\xi_{t_1+\tau}) \cdots \Phi_{n-2}(u^\xi_{t_{n-2}+\tau})][\Phi_{n-1}(u^\xi_{t_{n-1}+\tau}) P_{t_n-t_{n-1}} \Phi_n(u^\xi_{t_{n-1}+\tau})]\}$$
$$= E\{[\Phi_1(u^\xi_{t_1}) \cdots \Phi_{n-2}(u^\xi_{t_{n-2}+\tau})][\Phi_{n-1}(u^\xi_{t_{n-1}}) P_{t_n-t_{n-1}} \Phi_n(u^\xi_{t_{n-1}})]\},$$

so that M_n is also independent of τ. □

As a special case of (7.22), consider the linear equation with an additive noise:

$$\begin{aligned} du_t &= Au_t dt + dW_t, \quad t > 0, \\ u_0 &= \xi. \end{aligned} \tag{7.27}$$

Since A is coercive, it generates a strongly continuous semigroup G_t on H. It is known that the solution of (7.24) is given by

$$u_t = G_t\xi + \int_0^t G_{t-s} dW_s, \tag{7.28}$$

which has mean $Eu_t = G_t(E\xi)$ and covariance operator

$$\Gamma_t = \int_0^t (G_s R G_s^*) ds. \tag{7.29}$$

Theorem 5.4 Suppose that G_t is a contraction semigroup in H such that

(1) There are positive constants C, α such that

$$\|G_t h\| = C\|h\| e^{-\alpha t} \quad \forall\, t > 0, h \in H.$$

(2) $\int_0^\infty Tr(G_s R G_s^*) ds = \gamma < \infty.$

Then there exists an unique invariant measure μ for the equation (7.27) and $\mu \in \mathcal{N}(0, \Gamma)$, the Gaussian measure in H with mean zero and covariance operator $\Gamma = \int_0^\infty (G_s R G_s^*) ds.$

Proof. Fix $\xi = h \in H$. We know that the solution u_t^h given by (7.28) is a Gaussian process with mean $m_t^h = G_t h$ and covariance Γ_t defined by (7.29). Denote the corresponding Gaussian measure by μ_t^h. Let $\Psi_t^h(\lambda), \lambda \in H$, be the characteristic functional of u_t^h given by

$$\Psi_t^h(\lambda) = E\exp\{i(u_t^h, \lambda)\} = \exp\{i(m_t^h, \lambda) - \frac{1}{2}(\Gamma_t \lambda, \lambda)\}, \quad \lambda \in H. \tag{7.30}$$

To show $\mu_t^h \rightharpoonup \mu$, it suffices to prove that

$$\lim_{t\to\infty} \Psi_t^h(\lambda) = \Psi(\lambda) = \exp\{-\frac{1}{2}(\Gamma\lambda, \lambda)\}, \}$$

uniformly in λ on any bounded set $\Lambda \subset H$. By condition (1), it is easily seen that

$$|(m_t^h, \lambda)| \leq C\,\|h\|\,\|\lambda\|e^{-\alpha t} \to 0,$$

uniformly in $\lambda \in \Lambda$ for any $h \in H$, as $t \to \infty$. By condition (2), we see that $\int_t^\infty Tr(G_s R G_s^*)ds \to 0$ as $t \to \infty$. Therefore we have

$$|(\Gamma\lambda, \lambda) - (\Gamma_t\lambda, \lambda))| = ([\Gamma - \Gamma_t]\lambda, \lambda) = \left(\int_t^\infty (G_s R G_s^*)ds\lambda, \lambda\right)$$

$$\leq \|\lambda\|^2 \int_t^\infty Tr(G_s R G_s^*)ds \to 0$$

uniformly in $\lambda \in \Lambda$ as $t \to \infty$. Now, in view of the above results, we deduce that

$$|\Psi_t^h(\lambda) - \Psi(\lambda)| \leq |\exp\{i(m_t^h, \lambda)\} - \exp\{-\frac{1}{2}([\Gamma - \Gamma_t]\lambda, \lambda)\}|$$

$$\leq |\exp^{i(m_t^h, \lambda)} - 1| + |1 - \exp\{-\frac{1}{2}([\Gamma - \Gamma_t]\lambda, \lambda)\}|$$

$$\leq |(m_t^h, \lambda)| + \frac{1}{2}|[\Gamma - \Gamma_t]\lambda, \lambda)\}| \to 0$$

uniformly on Λ for each $h \in H$. Hence the corresponding probability distributions $\mu_t^h \rightharpoonup \mu \in \mathcal{N}(0, \Gamma)$ independent of h, and the invariant measure μ is unique by Lemma 5.2. $\qquad\square$

Remark: If the initial distribution of ξ is μ, then, by Theorem 5.3, the solution u_t^ξ is a stationary Gaussian process, known as the Ornstein-Uhlenbeck process.

For $N > 0$, let $K_N = \{v \in V : \|v\|_V \leq N\}$ and $A_N = \{h \in H \setminus K_N\}$. Recall that the inclusion $V \hookrightarrow H$ is compact so that K_N is a compact set in H. We will prove the following theorem.

Theorem 5.5 Suppose that the transition probability function $P(h, t, \cdot)$ for the solution process u_t of the equation (7.22) has the Feller property and satisfies the following condition:
For some $h \in H$, there exists a sequence $T_n \uparrow \infty$ such that

$$\lim_{N \to \infty} \frac{1}{T_n} \int_0^{T_n} P(h, t, A_N)dt = 0, \tag{7.31}$$

uniformly in n. Then the equation has an invariant measure μ on $(H, \mathcal{B}(H))$.

Proof. For any $B \in \mathcal{B}(H)$, define

$$\mu_n(B) = \frac{1}{T_n} \int_0^{T_n} P(h, t, B)dt. \tag{7.32}$$

Then $\{\mu_n\}$ is a family of probability measures on $(H, \mathcal{B}(H))$. By conditions (7.31) and (7.32), for any $\epsilon > 0$, there exists a compact set $K_N \subset H$ such that

$$\mu_n\{H \setminus K_N\} = \mu_n\{A_N\} < \epsilon, \ \forall \ n \geq 1.$$

Therefore, by the Prokhorov theorem [7], the family $\{\mu_n\}$ is weakly compact. It follows that there is a subsequence $\{\mu_{n_k}\}$ such that $\mu_{n_k} \rightharpoonup \mu$ in H.

To show μ being an invariant measure, let $\Phi \in \mathbf{C}_b(H)$. By the Feller property, $P_t \Phi \in \mathbf{C}_b(H)$. In view of the weak convergence, we have

$$\int_H P_t \Phi d\mu = \lim_{k \to \infty} \int_H P_t \Phi d\mu_{n_k} = \lim_{k \to \infty} \frac{1}{T_{n_k}} \int_0^{T_{n_k}} \int_H P_t \Phi(g) P(h, s, dg) ds$$

$$= \lim_{k \to \infty} \frac{1}{T_{n_k}} \int_0^{T_{n_k}} (P_{t+s} \Phi)(h) ds,$$

where use was made of the Fubini theorem and the Markov property. Now, for any fixed $t > 0$ and $h \in H$, we can write

$$\frac{1}{T_{n_k}} \int_0^{T_{n_k}} (P_{t+s} \Phi)(h) ds = \frac{1}{T_{n_k}} \{ \int_0^{T_{n_k}} (P_s \Phi)(h) ds + \int_{T_{n_k}}^{T_{n_k}+t} (P_s \Phi)(h) ds$$

$$- \int_0^t (P_s \Phi)(h) ds \}.$$

Since $\|P_t \Phi\| \leq \|\Phi\|$, we can deduce from the last two equations that

$$\int_H P_t \Phi d\mu = \lim_{k \to \infty} \frac{1}{T_{n_k}} \int_0^{T_{n_k}} (P_s \Phi)(h) ds$$

$$= \lim_{k \to \infty} \int_H \Phi d\mu_{n_k} = \int_H \Phi d\mu,$$

by the Fubini theorem and the weak convergence. Hence μ is an invariant measure as claimed. □

The following theorem is the main result of this section.

Theorem 5.6　Suppose that the coefficients of the equation (7.22), as a special case of equation (6.56), satisfy all the conditions of Theorem 6-7.5 except condition (D.3), which is strengthened to read

$$2\langle Av, v \rangle + 2(F(v), v) + \|\Sigma(v)\|_R \leq \alpha - \beta \|v\|_V^2, \ \forall \ v \in V, \tag{7.33}$$

for some constants α and $\beta > 0$. Then there exists an unique invariant measure μ for the solution process.

Proof.　We shall apply Theorem 5.5 to show the existence of an invariant measure. To verify the Feller property, suppose u_t^g and u_t^h of (7.22) with initial conditions $\xi = g$ and h in H, respectively. Then it follows from the energy equation and the monotonicity condition (D.4) that the difference $u_t^g - u_t^h$ satisfies

$$E\|u_t^g - u_t^h\|^2 = \|g - h\|^2 + E\int_0^t \{2\langle A(u_s^g - u_s^h), u_s^g - u_s^h\rangle$$

$$+2(F(u_s^g) - F(u_s^h), u_s^g - u_s^h) + \|\Sigma(u_s^g) - \Sigma(u_s^h)\|_R^2\}ds$$

$$\leq \|g - h\|^2 + \lambda \int_0^t E\|u_s^g - u_s^h\|^2 ds,$$

for some constant $\lambda > 0$. By means of Gronwall's inequality, we get

$$E\|u_t^g - u_t^h\|^2 \leq e^{\lambda T}\|g - h\|^2, \text{ for } 0 \leq t \leq T. \tag{7.34}$$

Since any $\Phi \in \mathbf{C}_b(H)$ can be approximated pointwise by a sequence of functions in $\mathbf{C}_b^1(H)$, to verify the Feller property, it suffices to take a bounded Lipschitz-continuous function Φ. Then, for $s, t \in [0, T]$ and $g, h \in H$,

$$\|(P_t\Phi)(g) - (P_s\Phi)(h)\| \leq \|[(P_t - P_s)\Phi](g)\| + \|(P_t\Phi)(g) - (P_t\Phi)(h)\|$$

$$\leq E\|\Phi(u_t^g) - \Phi(u_s^g)\| + E\|\Phi(u_t^g) - \Phi(u_t^h)\|$$

$$\leq 2k\{E\|u_t^g - u_s^g\|^2)^{1/2},$$

where k is the Lipschitz constant. Making use of the fact that u_t^g is mean-square continuous and (7.34), the Feller property follows.

Now we will show that the condition (7.31) in Theorem 5.5 is satisfied. By the energy equation and condition (7.33), we can obtain

$$E\|u_t^h\|^2 = \|h\|^2 + E\int_0^t \{2\langle Au_s^h, u_s^h\rangle + 2(F(u_s^h), u_s^h) + \|\Sigma(u_s^h)\|_R^2\}ds$$

$$\leq \|h\|^2 + \alpha t - \beta E\int_0^t \|u_s^h\|_1^2 ds,$$

which implies

$$E\int_0^t \|u_s^h\|_V^2 ds \leq \frac{1}{\beta}(\alpha t + \|h\|^2). \tag{7.35}$$

Since u_t^h is a V-valued process, its probability measure μ_t^h has support in V so that $P(h, t, A_N) = P\{\|u_t^h\|_V > N\}$. By invoking the Chebyshev inequality and (7.35), we deduce that

$$\frac{1}{T}\int_0^T P(h, t, A_N)dt = \frac{1}{T}\int_0^T P\{\|u_t^h\|_V > N\}dt$$

$$\leq \frac{1}{N^2 T}\int_0^T E\|u_t^h\|_V^2 dt \leq \frac{1}{\beta N^2 T}(\alpha T + \|h\|^2),$$

which converges to zero as $N \to \infty$ uniformly in $T \geq 1$. Therefore condition (7.31) in Theorem 5.5 is met and $\mu_t^h \rightharpoonup \mu$ for any $h \in H$. Moreover, by Lemma 5.2, the invariant measure μ is unique. $\square$

Remarks: The results in Theorems 5.5 and 5.6 were adopted from the paper [15]. In fact it can shown that the invariant measure μ has support in V. This and more general existence and uniqueness results can be found in the aforementioned paper.

(**Example 5.1**) Consider the reaction-diffusion equation in $D \subset \mathbf{R}^d$:

$$
\begin{aligned}
\frac{\partial u}{\partial t} &= \sum_{j,k=1}^d \frac{\partial}{\partial x_j} [a_{jk}(x) \frac{\partial u}{\partial x_k}] - c(x)u + f(u,x) \\
&\quad + \sigma(x)\dot{W}(x,t), \quad x \in D,\, t \in (0,T), \\
u|_{\partial D} &= 0, \quad u(x,0) = h(x),
\end{aligned}
\tag{7.36}
$$

where the coefficients $a_{jk}(x), c(x), \sigma(x)$ and $f(u,x)$ are bounded continuous on D. Suppose that

(1) There exist positive constants $\alpha_1, \alpha_2, \alpha_3$ such that

$$
\alpha_1 |\xi|^2 \leq \sum_{j,k=1}^d a_{jk}(x)\xi_j \xi_k \leq \alpha_2 |\xi|^2, \quad \forall\, \xi \in \mathbf{R}^d
$$

and $c(x) \geq \alpha_3$, for all $x \in D$.

(2) There exist $C_1, C_2 > 0$ such that

$$
\begin{aligned}
|f(r,x)| &\leq C_1(1+|r|^2), \\
|f(r,x) - f(s,x)| &\leq C_2|r-s|, \quad \forall\, x \in D,\, r,s \in \mathbf{R},
\end{aligned}
$$

and $rf(r,x) \leq 0, \forall x \in D$.

(3) The Wiener random field $W(x,t)$ has a bounded continuous correlation function $r(x,y)$ such that $\int_D r(x,x)\sigma^2(x)dx = \alpha < \infty$.

As before, let $H = L^2(D)$ and $V = H_0^1$, here, with the norm $\|v\|_V^2 = \|v\|^2 + \|\nabla v\|^2$. Set $Au = \sum_{j,k=1}^d \frac{\partial}{\partial x_j} [a_{jk} \frac{\partial u}{\partial x_k}] - cu$, $F(u) = f(u,\cdot)$ and $\Sigma = \sigma(\cdot)$.
Then, under the above assumptions, it is easy to check all conditions for Theorem 5.6 are satisfied. In particular, to verify the condition (7.33), we

make use of the assumptions (1)–(3) to obtain

$$2\langle Av, v\rangle + 2(F(v), v) + \|\Sigma(v)\|_R$$

$$= -2\int_D \{\sum_{j,k=1}^{d} a_{jk}(x)\frac{\partial v(x)}{\partial x_j}\frac{\partial v(x)}{\partial x_k} + c(x)u^2(x)\}dx$$

$$+2\int_D v(x)f(v, x)dx + \int_D r(x, x)\sigma^2(x)dx$$

$$\leq -2\int_D \{\alpha_1|\nabla v(x)|^2 + \alpha_3|v(x)|^2\}dx + \alpha$$

$$\leq \alpha - \beta\|v\|_V^2,$$

with $\beta = 2(\alpha_1 \wedge \alpha_3)$. Therefore, by Theorem 5.6, there exists a unique invariant measure for this problem.

(**Example 5.2**) Referring to (Example 7.2) in Chapter Six, consider the stochastic reaction-diffusion equation in domain $D \subset \mathbf{R}^3$ with a cubic nonlinearity:

$$\frac{\partial u}{\partial t} = (\kappa\Delta - \alpha)u - \gamma u^3 + \sum_{k=1}^{3} \frac{\partial u}{\partial x_k}\frac{\partial}{\partial t}W_k(x, t), \tag{7.37}$$

$$u|_{\partial D} = 0, \quad u(x, 0) = h(x),$$

where κ, α, γ are given positive constants and $W_k(x, t), k = i, \cdots, d$, are independent Wiener random fields with continuous covariance functions $r_k(x, y)$ such that

$$\sup_{x,y\in D} |r_k(x, y)| \leq r_0, \quad k = 1, 2, 3. \tag{7.38}$$

Let the spaces H and V be defined as in (Example 5.1). It was shown that the equation (7.37) has a unique strong solution $u \in L^2(\Omega; \mathbf{C}([0, T]; H)) \cap L^2(\Omega_T; H_0^1)$. Similar to (Example 4.1), we can obtain the following estimate:

$$2\langle Av, v\rangle + 2(F(v), v) + \|\Sigma(v)\|_R$$
$$\leq -(2\kappa - r_0)\|\nabla v\|^2 - 2\alpha\|v\|^2 - 2\gamma|v|_4^4,$$

which implies the condition (7.33) of Theorem 5.6 provided that $r_0 < 2\kappa$. In this case the equation (7.37) has a unique invariant measure.

7.6 Small Random Perturbation Problems

Consider the equation (7.22) with a small noise:

$$du_t^\epsilon = Au_t^\epsilon dt + F(u_t^\epsilon)dt + \epsilon\Sigma(u_t^\epsilon)dW_t, \quad t > 0, \tag{7.39}$$
$$u_0^\epsilon = h,$$

where u_t^ϵ shows the dependence of the solution on a small parameter $\epsilon \in (0,1)$, and $h \in H$. The equation (7.39) is regarded as a random perturbation of the deterministic equation:

$$
\begin{aligned}
du_t &= Au_t dt + F(u_t)dt, \quad t > 0, \\
u_0 &= h
\end{aligned}
\tag{7.40}
$$

We wish to show that, under suitable conditions, the seemingly obvious fact: $u_t^\epsilon \to u_t$ on $[0,T]$ in some probabilistic sense. To this end, we recall conditions B and conditions D in Theorem 6-7.5 in Section 6.7, which are assumed to hold for the coefficients of equation (7.39).

Theorem 6.1   Assume the conditions for Theorem 6-7.5 hold and, in addition, there is $b > 0$ such that

$$
\|\Sigma(v)\|_R^2 \leq b(1 + \|v\|_V^2), \quad \forall v \in V.
\tag{7.41}
$$

Then the perturbed solution u_t^ϵ of equation (7.39) converges uniformly on [0,T] to the solution u_t of equation (7.40) in mean-square. In fact there exists a constant $C_T > 0$ such that

$$
E \sup_{0 \leq t \leq T} \|u_t^\epsilon - u_t\|^2 \leq C_T \epsilon (1 + \|h\|^2), \quad \forall \, \epsilon \in (0,1).
\tag{7.42}
$$

Proof.   By invoking Theorem 6-7.5, equation (7.39) has a unique strong solution satisfying the energy equation

$$
\begin{aligned}
\|u_t^\epsilon\|^2 = \|h\|^2 &+ 2\int_0^t \langle Au_s^\epsilon, u_s^\epsilon\rangle ds + 2\int_0^t (F(u_s^\epsilon), u_s^\epsilon)ds \\
&+ \epsilon^2 \int_0^t \|\Sigma(u_s^\epsilon)\|_R^2 ds + 2\epsilon \int_0^t (u_s^\epsilon, \Sigma(u_s^\epsilon)dW_s).
\end{aligned}
\tag{7.43}
$$

Making use of similar estimates as in the proof of the existence theorem, we can deduce from (7.43) that there is $C_1(T) > 0$ such that

$$
E \int_0^T \|u_t^\epsilon\|_V^2 dt \leq C_1(T),
\tag{7.44}
$$

where $C_1(T)$ is independent of ϵ. Let $v^\epsilon = u_t^\epsilon - u_t$. Then v^ϵ satisfies

$$
\begin{aligned}
dv_t^\epsilon &= Av_t^\epsilon dt + [F(u_t^\epsilon) - F(u_t)]dt + \epsilon\Sigma(u_t^\epsilon)dW_t, \quad t \geq 0, \\
v_0^\epsilon &= 0.
\end{aligned}
\tag{7.45}
$$

By applying the Itô formula, we get

$$
\begin{aligned}
\|v_t^\epsilon\|^2 = 2\int_0^t \langle Av_s^\epsilon, v_s^\epsilon\rangle ds &+ 2\int_0^t (F(u_s^\epsilon) - F(u_t), v_s^\epsilon)ds \\
&+ \epsilon^2 \int_0^t \|\Sigma(u_s^\epsilon)\|_R^2 ds + 2\epsilon \int_0^t (v_s^\epsilon, \Sigma(u_s^\epsilon)dW_s).
\end{aligned}
\tag{7.46}
$$

By making use of condition (D.4) (assuming $\delta \geq 0$), (7.41) and a submartingale inequality, the above yields

$$E \sup_{0 \leq t \leq T} \|v_t^\epsilon\|^2 \leq 2\delta E \int_0^T \|v_t^\epsilon\|^2 dt + \epsilon^2 E \int_0^T \|\Sigma(u_t^\epsilon)\|_R^2 dt$$

$$+ 6\epsilon \{ E \int_0^T (\Sigma(u_t^\epsilon) R \Sigma^*(u_t^\epsilon) v_t^\epsilon, v_t^\epsilon) dt \}^{1/2}$$

$$\leq 2\delta E \int_0^T \|v_t^\epsilon\|^2 dt + b (\epsilon^2 + 18\epsilon) E \int_0^T (1 + \|u_t^\epsilon\|_V^2) dt$$

$$+ \frac{1}{2} E \sup_{0 \leq t \leq T} \|v_t^\epsilon\|^2,$$

or, in view of (7.44),

$$E \sup_{0 \leq t \leq T} \|v_t^\epsilon\|^2 \leq 4\delta \int_0^T E \sup_{0 \leq s \leq t} \|v_s^\epsilon\|^2 dt$$

$$+ 4b\epsilon(18 + \epsilon)(C_1(T) + T).$$

It follows from Gronwall's inequality, there exists $C_T > 0$ such that, for any $\epsilon \in (0, 1)$,

$$E \sup_{0 \leq t \leq T} \|v_t^\epsilon\|^2 \leq C_T \epsilon,$$

with $C_T = 80b(C_1 + T)e^{4\delta T}$. □

Remark: In comparison with (7.42), if we take the expectation of (7.48), we can obtain the estimate:

$$\sup_{0 \leq t \leq T} E\|u_t^\epsilon - u_t\|^2 \leq \tilde{C}_T \epsilon^2,$$

for some $\tilde{C}_T > 0$.

Let μ^ϵ denote the probability measure for u^ϵ in $\mathbf{C}([0, T]; H)$. By Theorem 6.1, since u^ϵ converges to u in mean-square, μ^ϵ converges weakly to $\mu^0 = \delta_u$, the Dirac measure concentrated at u. We are interested in the rate of convergence, such as an estimate for the probability

$$\mu^\epsilon(B_\delta^c) = P\{ \sup_{0 \leq t \leq T} \|u_t^\epsilon - u_t\| > \delta \}, \tag{7.47}$$

for small $\epsilon, \delta > 0$, where $B_\delta^c = H \setminus B_\delta$ and

$$B_\delta = \{v \in H : \sup_{0 \leq t \leq T} \|v_t - u_t\| \leq \delta \}.$$

Before taking up this question, we need a useful exponential estimate which will be presented as a lemma.

Lemma 6.2   Let M_t be a continuous H-valued martingale in $[0, T]$ with the local characteristic operator Q_t such that

$$\sup_{0 \leq t \leq T} Tr Q_t \leq N < \infty, \quad a.s., \tag{7.48}$$

for some $N > 0$. Then the following estimate holds

$$P\{\sup_{0 \leq s \leq t} \|M_s\| \geq r\} \leq 3 \exp\{-\frac{r^2}{4Nt}\}, \tag{7.49}$$

for any $t \in (0, T]$ and $r > 0$.

Proof.   Introduce a functional Φ_λ on H as follows:

$$\Phi_\lambda(v) = (1 + \lambda\|v\|^2)^{1/2}, \quad v \in H, \tag{7.50}$$

which depends on a positive parameter λ. Then the first two derivatives of Φ_λ are given by $\Phi_\lambda'(v) = \lambda\Phi_\lambda^{-1}(v)v$ and $\Phi_\lambda''(v) = \lambda\Phi_\lambda^{-1}(v)I - \lambda^2\Phi_\lambda^{-3}(v)(v \otimes v)$, where I is the identity operator on H and $\otimes$ denotes the tensor product. Apply the Itô formula to $\Phi_\lambda(M_t)$ to get

$$
\begin{aligned}
\Phi_\lambda(M_t) &= 1 + \lambda \int_0^t \Phi_\lambda^{-1}(M_s)(M_s, dM_s) \\
&\quad + \frac{1}{2} \int_0^t \{\lambda\Phi_\lambda^{-1}(M_s)TrQ_s - \lambda^2\Phi_\lambda^{-3}(M_s)(Q_sM_s, M_s)\}ds \\
&\leq 1 + \xi_t^\lambda + \frac{1}{2} \int_0^t \{\lambda\Phi_\lambda^{-1}(M_s)TrQ_s \\
&\quad + \lambda^2[\Phi_\lambda^{-2}(M_s) - \Phi_\lambda^{-3}(M_s)](Q_sM_s, M_s)\}ds,
\end{aligned}
\tag{7.51}
$$

where we set

$$\xi_t^\lambda = \lambda \int_0^t \Phi_\lambda^{-1}(M_s)(M_s, dM_s) - \frac{1}{2}\lambda^2 \int_0^t \Phi_\lambda^{-2}(M_s)(Q_sM_s, M_s)ds. \tag{7.52}$$

It is easy to check that $\Phi_\lambda^{-\alpha}(v) \leq 1$ and

$$[\Phi_\lambda^{-2}(v) - \Phi_\lambda^{-3}(v)](Q_sv, v) \leq \frac{1}{\lambda}TrQ_s,$$

for any $\alpha \geq 0, \lambda > 0$ and $v \in V$, so that (7.51) yields

$$
\begin{aligned}
\Phi_\lambda(M_t) &\leq 1 + \xi_t^\lambda + \lambda \int_0^t TrQ_s ds \\
&\leq (1 + \lambda NT) + \xi_t^\lambda,
\end{aligned}
\tag{7.53}
$$

for $0 < t \leq T$ by condition (7.48). Define

$$Z_t^\lambda = \exp\{\xi_t^\lambda\}, \quad t \in [0, T], \tag{7.54}$$

which is known to be an exponential martingale [75] so that

$$EZ_t^\lambda = EZ_0^\lambda = 1. \tag{7.55}$$

Now, by (7.50) and (7.53),

$$
\begin{aligned}
P\{ \sup_{0 \le s \le t} \|M_s\| \ge r \} &= P\{ \sup_{0 \le s \le t} \Phi_\lambda(M_t) \ge (1 + \lambda r^2)^{1/2} \} \\
&\le P\{ \sup_{0 \le s \le t} Z_s^\lambda \ge \exp[(1 + \lambda r^2)^{1/2} - (1 + \lambda N t)] \}.
\end{aligned}
\tag{7.56}
$$

By applying the Chebyshev inequality and Doob's martingale inequality, the above gives rise to

$$P\{ \sup_{0 \le s \le t} \|M_s\| \} \ge r \} \le \exp[-(1 + \lambda r^2)^{1/2} + (1 + \lambda N t)]\},$$

which holds for any $\lambda > 0$. In particular, by choosing

$$\lambda = (\frac{r}{2Nt})^2 - \frac{1}{r^2} > 0,$$

the inequality (7.56) yields the estimate (7.49) when $r^2 > 2Nt$. Since, for $r^2 \le 2NT$, (7.49) is trivially true, the desired estimate holds for any $r > 0$. $\square$

Remarks: The exponential inequality was first obtained in [13] where $M_t = \int_0^t \Sigma_s dW_s$ is assumed to be a stochastic integral in H. The extension to a continuous martingale is obvious. Under some suitable conditions, we shall apply this lemma to the randomly perturbed stochastic evolution equation to obtain results for the exponential rate of convergence and the a.s. convergence.

Theorem 6.3 Assume the conditions for Theorem 6-7.5 hold, provided that $\delta \le 0$ in condition (D.4) so that

$$\langle A(u - v), u - v \rangle + (F(u) - F(v), u - v) \le 0, \ \forall \ u, v \in V, \tag{7.57}$$

and there is $\kappa > 0$ such that

$$\|\Sigma(v)\|_R^2 \le \gamma^2, \ \forall \ v \in V. \tag{7.58}$$

Then the solution measure μ^ϵ for (7.39) converges weakly to $\mu = \delta_u$ at an exponential rate. More precisely, for any $r > 0$ and $t \in (0, T]$, we have

$$P\{ \sup_{0 \le s \le t} \|u_s^\epsilon - u_s\| \ge r \} \le 3 \exp\{-\frac{r^2}{4\gamma^2 \epsilon^2 t}\}. \tag{7.59}$$

Furthermore the perturbed solution u_t^ϵ converges a.s. to u_t uniformly in $[0, T]$ as $\epsilon \to 0$.

Proof.   The proof is quite similar to that of Theorem 6.1 and Lemma 6.2. Let $v_t^\epsilon = u_t^\epsilon - u_t$ and Φ_λ is given by (7.50). Again, by applying the Itô formula, we obtain

$$
\begin{aligned}
\Phi_\lambda(v_t^\epsilon) = 1 + \lambda \int_0^t \Phi_\lambda^{-1}(v_t^\epsilon)[\langle Av_s^\epsilon, v_s^\epsilon \rangle + (F(u_s^\epsilon) - F(u_s), v_s^\epsilon)]ds \\
+ \lambda \epsilon \int_0^t \Phi_\lambda^{-1}(v_s^\epsilon)(v_s^\epsilon, \Sigma(u_s^\epsilon)dW_s) + \frac{\epsilon^2}{2} \int_0^t \{\lambda \Phi_\lambda^{-1}(v_s^\epsilon)\|\Sigma(u_s^\epsilon)\|_R^2 \\
- \lambda^2 \Phi_\lambda^{-3}(v_s^\epsilon)(\Sigma(u_s^\epsilon)R\Sigma^*(u_s^\epsilon)v_s^\epsilon, v_s^\epsilon)\}ds.
\end{aligned}
\tag{7.60}
$$

By (7.57) and (7.58), we can deduce from (7.60) that

$$
\begin{aligned}
\Phi_\lambda(v_t^\epsilon) \leq\; & 1 + \eta_t^\lambda + \frac{\epsilon^2}{2} \int_0^t \{\lambda \Phi_\lambda^{-1}(v_s^\epsilon)\|\Sigma(u_s^\epsilon)\|_R^2 \\
& + \lambda^2[\Phi_\lambda^{-2}(v_s^\epsilon) - \Phi_\lambda^{-3}(v_s^\epsilon)]\,(\Sigma(u_s^\epsilon)R\Sigma^*(u_s^\epsilon)v_s^\epsilon, v_s^\epsilon)\}ds \\
\leq\; & (1 + \lambda \gamma^2 \epsilon^2 T) + \eta_t^\lambda,
\end{aligned}
\tag{7.61}
$$

where

$$
\begin{aligned}
\eta_t^\lambda =\; & \lambda \epsilon \int_0^t \Phi_\lambda^{-1}(v_s^\epsilon)(v_s^\epsilon, \Sigma(u_s^\epsilon)dW_s) \\
& - \frac{1}{2}\lambda^2 \epsilon^2 \int_0^t \Phi_\lambda^{-2}(v_s^\epsilon)\,(\Sigma(u_s^\epsilon)R\Sigma^*(u_s^\epsilon)v_s^\epsilon, v_s^\epsilon)\}ds.
\end{aligned}
\tag{7.62}
$$

Similar to the estimate (7.56), we can get

$$
P\{ \sup_{0 \leq s \leq t} \|u_s^\epsilon - u_s\| \geq r\} \leq \exp[-(1 + \lambda r^2)^{1/2} + (1 + \lambda \gamma^2 \epsilon^2 T)]\},
$$

which yields the desired result (7.59) after setting $\lambda = (r/2\gamma \epsilon T)^2 - (1/r^2)$.

To show the a.s. convergence, we can choose $\epsilon = \epsilon_n = n^{-1}$ and $\delta = \delta_n = n^{-2}$ in (7.59). It follows from the Borel-Cantalli lemma that $u_t^{\epsilon_n} \to u_t$ a.s. in H uniformly over [0,T].   □

Remarks: The exponential estimates presented in this section are known as a large deviations estimate. A precise estimate of the rate of convergence is given by the rate function in the large deviations theory, which will be discussed briefly in the next section.

7.7   Large Deviations Problems

For ease of discussion, we will only consider the special case of equation (7.39) when the noise is additive. Then it becomes

$$
\begin{aligned}
du_t^\epsilon &= Au_t^\epsilon dt + F(u_t^\epsilon)dt + \epsilon dW_t, \quad t \geq 0, \\
u_0^\epsilon &= h,
\end{aligned}
\tag{7.63}
$$

where W_t is a R-Wiener process in H.

Let $\nu = \mathcal{L}\{W\}$ denote the R-Wiener measure in $\mathbf{C}([0, T]; H)$ and let $\nu^\epsilon = \mathcal{L}\{W^\epsilon\}$ with $W^\epsilon = \epsilon W$. Given $\phi \in \mathbf{C}_b([0, T] \times H)$ and B being a Borel set in H, we are interested in an asymptotic evaluation of the integral

$$J^\epsilon(B) = \int_B \exp\{-\phi(v)/\epsilon^2\}\nu^\epsilon(dv) \qquad (7.64)$$

as $\epsilon \to 0$. To gain some intuitive idea, suppose that W is replaced by a sequence $\{\xi_k\}$ of independent, identically distributed Gaussian random variables with mean zero and variance σ^2. Then the integral (7.64) would have taken the form

$$J_n^\epsilon(B) = C_n(\epsilon) \int_B \exp\{-\frac{I(x)}{\epsilon^2}\}dx, \qquad (7.65)$$

where $C_n(\epsilon) = (1/2\pi\epsilon^2)^{n/2}$, $x = (x_1, \cdots, x_n)$, B is a Borel set in $\mathbf{R}^n$ and

$$I(x) = \phi(x) + \frac{1}{2\sigma}|x|^2.$$

Then, since L^p-norm tends to L^∞-norm as $p \to \infty$, we obtain

$$\lim_{\epsilon\downarrow 0} \epsilon^2 \log J_n^\epsilon(B) = \lim_{p\to\infty} \log\{[\int_B \{\exp[-I(x)]dx\}^p\}^{1/p} = -\inf_{x\in B} I(x),$$

or $\gamma(B) = \inf_{x\in B} I(x)$ is the exponential rate of convergence for the integral (7.65). Therefore it is reasonable to call I a *rate function*. Now let W_t be a standard Brownian motion in one dimension, and $\{t_0 = 0, t_1, \cdots, t_n = T\}$ is a partition of $[0, T]$. Then the joint probability density for $(W_{t_1}, \cdots, W_{t_n})$ is given by

$$p(x(t_1), \cdots, x(t_n)) = C_n \exp\{-\sum_{k=1}^n \frac{|x(t_1) - x(t_n)|^2}{2\sigma^2(t_k - t_{k-1})}\},$$

where $C_n = \prod_{j=1}^n [2\pi\sigma^2(t_j - t_{j-1})]^{-1/2}$. In spite of the fact that a Brownian path is nowhere differentiable, heuristically, as the mesh size of the partition goes to zero, the exponent of the density tends to $-(1/2\sigma^2)\int_0^T |\dot{x}(t)|^2 dt$, with $\dot{x} = \frac{d}{dt}x$. Therefore, for a one-dimensional Brownian motion, it seems plausible to guess that the rate function is given by

$$I(x) = \phi(x) + \frac{1}{2\sigma^2} \int_0^T |\dot{x}(t)|^2 dt,$$

if the above integral makes sense. This kind of formal argument can be generalized to the case of a R-Wiener process in H. However, to make it precise, one needs the *large deviations theory* [23].

Let I be an extended real-valued functional on $\mathbf{C}([0, T] \times H)$. Then I is said to be a *rate function* if

(1) $I : \mathbf{C}([0,T] \times H) \to [0,\infty]$ is lower semicontinuous.

(2) For any $r > 0$, the set $\{u \in \mathbf{C}([0,T] \times H) : I(u) \leq r\}$ is compact.

A family of probability measures $\{P^\epsilon\}$ on $\mathbf{C}([0,T];H)$ is said to obey the *large deviations principle* if there exists a rate function I such that

(1) For each closed set F,

$$\limsup_{\epsilon \downarrow 0} \epsilon^2 \log P^\epsilon(F) \leq - \inf_{v \in F} I(v).$$

(2) For any open set G,

$$\liminf_{\epsilon \downarrow 0} \epsilon^2 \log P^\epsilon(G) \geq - \inf_{v \in G} I(v).$$

After the digression, let us return to the Wiener measures $\{\nu^\epsilon\}$. We denote by $\mathcal{R}(H)$ the range of $R^{1/2}$, which is a Hilbert-Schmidt operator. Introduce the rate function as follows:

$$I(v) = \frac{1}{2} \int_0^T \|R^{-1/2}\dot{v}_t\|^2 dt, \tag{7.66}$$

if $v_0 = 0, \dot{v}_t = \frac{d}{dt}v_t \in \mathcal{R}(H)$ such that $R^{-1/2}\dot{v}_t \in L^2((0,T);H)$, and set $I(v) = \infty$, otherwise. Then it can be shown that [19]

Lemma 7.1   The family $\{\nu^\epsilon\}$ of Wiener measures on $\mathbf{C}([0,T];H)$ obeys the large deviations principle with the rate function given by (7.66).   □

Now consider the family $\{\mu^\epsilon\}$ of solution measures for the equation (7.63). A useful tool in dealing with such a problem is Varadhan's contraction principle [80] which says that

Lemma 7.2   Let $\mathcal{S}$ map $\mathbf{C}([0,T];H)$ into itself continuously and let $Q^\epsilon = P^\epsilon \circ \mathcal{S}^{-1}$. If $\{P^\epsilon\}$ obeys the large deviations principle with rate function I, so is the family $\{Q^\epsilon\}$ with rate function $J(v)$ given by

$$J(v) = \{I(u) : \mathcal{S}(u) = v\}. \qquad\qquad \square \qquad (7.67)$$

Instead of (7.63), consider the deterministic equation:

$$u_t = h + \int_0^t [Au_s + F(u_s)]ds + v_t, \tag{7.68}$$

where $h \in H$ and $v \in \mathbf{C}([0,T];H)$ with $\frac{d}{dt}v \in L^2((0,T);V')$. Then, by conditions on A and F, it is known that the integral equation (7.68) has a unique solution $u \in \mathbf{C}([0,T]; H)$. Moreover, given $h \in H$, the solution u depends continuously on v. We write

$$u_t = \mathcal{S}_t(v)$$

so that the solution operator $\mathcal{S} : \mathbf{C}([0,T]; H) \to \mathbf{C}([0,T]; H)$ is continuous. Also, from (7.68), we have

$$v_t = \mathcal{S}_t^{-1}(u) = u_t - h - \int_0^t [Au_s + F(u_s)]ds. \tag{7.69}$$

In view of Lemma 7.1 and (7.66)–(7.69), we can apply Lemma 7.2 to conclude that

Theorem 7.3 The family $\{\mu^\epsilon\}$ of solution measures on $\mathbf{C}([0,T]; H)$ for the equation (7.63) obeys the large deviations principle with the rate function given by

$$J(u) = \frac{1}{2} \int_0^T \| R^{-1/2}[\dot{u}_t - Au_t - F(u_t)] \|^2 dt, \tag{7.70}$$

if $u \in \mathbf{C}([0,T]; H)$ with $u_0 = h$ and $[\dot{u}_t - Au_t - F(u_t)] \in \mathcal{R}(H)$ such that the above integral converges, and set $J(u) = \infty$, otherwise. $\qquad\square$

As a simple application, consider the convergence rate $\gamma(B_\delta^c)$ for $\mu^\epsilon(B_\delta^c)$ given by (7.47). The exact rate can be determined by the equation (7.70) as follows:

$$\gamma(B_\delta^c) = \frac{1}{2} \inf_{u \in B_\delta^c} \int_0^T \| R^{-1/2}[\dot{u}_t - Au_t - F(u_t)] \|^2 dt.$$

Though the exact rate γ is difficult to compute, the above constrained minimization problem may be solved approximately.

Remarks: Theorem 7.1 is one of the large deviations problems for stochastic partial differential equations perturbed by a Gaussian noise. When $\Sigma(v)$ is nonconstant, we have a multiplicative noise. Even though the basic idea is similar, technically, such problems are more complicated (see [10] and [74], among others). The problem of exit time distributions is treated in [19]. A thorough treatment of the large deviations theory for Itô's equations in finite dimensions is given in the books [23] and [28].

Chapter 8

Further Applications

8.1 Introduction

In the previous two chapters, suggested by the parabolic and hyperbolic Itô equations, we studied the existence, uniqueness and asymptotic behavior of solutions to some stochastic evolution equations in a Hilbert space. Even though we have given a number of examples to demonstrate some applications of the general results obtained therein, they are relatively simple. In this chapter we will present several more examples to show a wider range of applicability of the general results. A major source of applications for stochastic partial differential equations comes from the statistical theory of turbulence in fluid dynamics and the related problems. All of our examples to be presented are given in this area, even though there are other applications, such as in biology and systems science (see, e.g., [6], [22], [26], [73]).

In this chapter we shall describe five applied problems. In Section 7.2 the stochastic Burgers' equation, as a simple model in turbulence, is studied by means of a Hopf transformation. It is shown that the random solution can be obtained explicitly by the stochastic Feynman-Kac formula. Then the mild solution of the Schrödinger equation with a random potential is analyzed in Section 7.3. Its connection to the problem of wave propagation in random media is elucidated. The model problem given in Section 7.4 is concerned with the vibration of a nonlinear elastic beam excited by the turbulent wind. It will be shown that the existence of a unique strong solution with a H^2-regularity holds in any finite time interval. In the following section we consider the linear stability of the Cahn-Hillard equation with damping under a random perturbation. This equation arises from some dynamical phase transition problem. We obtain the conditions on the physical parameters so that the equilibrium solution is a.s. exponentially stable. The last section deals with the model equations for turbulence, the randomly forced Navier-Stokes equations. As in the deterministic case, the existence and uniqueness questions in three dimensions remain open. We shall sketch a proof of the existence theorem in two dimensions and show that there exists an invariant measure for this problem.

8.2 Stochastic Burgers and Related Equations

The Burgers equation was first introduced to mimic the Navier-Stokes equations in hydrodynamics. As a simple model of turbulence, a randomly forced Burgers equation in one dimension was proposed [30]. It can be related to following nonlinear parabolic Itô equation:

$$\frac{\partial v}{\partial t} = \frac{\nu}{2}\Delta v + \frac{1}{2}|\nabla v|^2 + \dot{V}(x,t), \quad x \in \mathbf{R}^d, t \in (0,T),$$

$$v(x,0) = h(x),$$

(8.1)

where $\dot{V}(x,t) = b(x,t) + \sigma(x,t)\dot{W}(x,t)$, or

$$V(x,t) = \int_0^t b(x,s)ds + \int_0^t \sigma(x,s)W(x,ds)$$

(8.2)

is a $\mathbf{C}^m$-semimartingale. For $d = 2$, this equation was used to model the dynamic growth of a liquid interface under random perturbations [30]. In (8.1), $v(x,t)$ denotes the surface height with respect to a reference plane, ν the surface tension, ∇v the gradient of v, and $h(x)$ is the initial height. Assume that the solution of (8.1) is smooth. Let $\mathbf{u} = -\nabla v$. It follows from (8.1) that $\mathbf{u}$ satisfies

$$\frac{\partial \mathbf{u}}{\partial t} = \frac{\nu}{2}\Delta\mathbf{u} - (\mathbf{u}\cdot\nabla)\mathbf{u} - \nabla\dot{V}(x,t), \quad x \in \mathbf{R}^d, t \in (0,T),$$

$$\mathbf{u}(x,0) = -\nabla h(x),$$

(8.3)

which is known as a d-dimensional Burgers' equation perturbed by a conservative random force for the velocity field $\mathbf{u}(x,t)$ in $\mathbf{R}^d$. This equation was analyzed as a stochastic evolution equation in a Hilbert space [20]. Here we shall consider its solution as a continuous random field and adopt the method of probabilistic representation discussed in Chapter Three. To this end we first connect the equation (8.1) to the linear parabolic equation with a multiplicative white noise in the Stratonovich sense:

$$\frac{\partial \varphi}{\partial t} = \frac{\nu}{2}\Delta\varphi + \frac{1}{\nu}\varphi \circ \dot{V}(x,t), \quad x \in \mathbf{R}^d, t \in (0,T),$$

$$\varphi(x,0) = g(x).$$

(8.4)

By applying the Itô formula, it can be shown that the solutions of the equations (8.1) and (8.4) are related by

$$v(x,t) = \nu \log \varphi(x,t), \qquad v(x,0) = h(x) = \nu \log g(x).$$

(8.5)

It follows that the solution of the Burgers' equation can be expressed as

$$\mathbf{u}(x,t) = -\nu \frac{\nabla \varphi(x,t)}{\varphi(x,t)}, \qquad \mathbf{u}(x,0) = -\nabla h(x). \qquad (8.6)$$

For simplicity, we set $\nu = 1$. By Theorem 4-4.2, suppose that $g \in \mathbf{C}_b^{m,\delta}$, $W(x,t)$ is a $\mathbf{C}^m$-Wiener random field, $b(\cdot,t)$ and $\sigma(\cdot,t)$ are continuous adapted $\mathbf{C}^{m,\delta}$-processes such that they satisfy the conditions for Theorem 4-4.2 for $m \geq 3$. Then the equation (8.4) has a unique $\mathbf{C}^m$-solution given by the Feynman-Kac formula:

$$\varphi(x,t) = E_z\{g[x+z(t)] \exp \int_0^t V[x+z(t) - z(s), ds]\}, \qquad (8.7)$$

where E_z denotes the partial expectation with respect to the Brownian motion $z(t)$. Since $g(x) = e^{h(x)} > 0$, by Theorem 4-5.1, the solution $\varphi(x,t)$ of (8.4) is strictly positive so that the function v given by (8.5) is well defined. Moreover $v(\cdot,t)$ is a continuous $\mathbf{C}^m$-process for $m \geq 3$, and it satisfies the equation (8.1). This, in turn, implies that $\mathbf{u}(\cdot,t) \in \mathbf{C}^{m-1}$ is the solution of the Burgers equation (8.3). In view of (8.6) and (8.7), the solution $\mathbf{u}$ of (8.3) has the probabilistic representation

$$\mathbf{u}(x,t) = -\nu \frac{E_z\{\nabla g[x+z(t)] + g[x+z(t)]\nabla \Phi(x,t)\} \exp \Phi(x,t)}{E_z\{g[x+z(t)] \exp \Phi(x,t)\}},$$

where $\Phi(x,t) = \int_0^t V[x+z(t) - z(s), ds]$.

8.3 Random Schrödinger Equation

In quantum mechanics, the Schrödinger equation plays a prominent role. Here we consider such an equation with a time-dependent random potential:

$$\frac{\partial u}{\partial t} = i\{\alpha \Delta u + \beta \dot{V}(x,t)u\}, \quad x \in \mathbf{R}^d, \, t \in (0,T),$$
$$u(x,0) = h(x), \qquad (8.8)$$

where $i = \sqrt{-1}$; α, β are real parameters, h is the initial state and $\dot{V}$ is a random potential. In particular we assume that

$$\dot{V}(x,t) = b(x) + \dot{W}(x,t), \qquad (8.9)$$

where $b(x)$ is a bounded function on $\mathbf{R}^d$ and $W(x,t)$ is a Wiener random field with covariance function $r(x,y)$. To treat the equation (8.8) as a stochastic evolution in a Hilbert space, we let H be the space of complex-valued functions in $L^2(\mathbf{R}^d)$ with inner product $(f,g) = \int_{\mathbf{R}^d} f(x)\overline{g(x)}dx$, where the over-bar denotes the complex conjugate. Then we rewrite (8.8) as a stochastic equation in H^{-1}:

$$du_t = Au_t dt + \beta u_t dW_t,$$
$$u_0 = h,$$

(8.10)

where we set $A = i(\alpha\Delta + \beta b)$. Suppose that

(1) The functions $h \in H$ and $b \in L^p(\mathbf{R}^d)$ for $p \geq (2 \vee d/2)$.

(2) The covariance function $r(x,y)$, $x,y \in \mathbf{R}^d$, is bounded and continuous.

Under condition (1), it is known that $A : H^2 \to H$ generates a group of unitary operators $\{G_t\}$ on H (p. 225, [70]). Therefore the equation (8.10) can be written as

$$u_t = G_t h + \beta \int_0^t G_{t-s} u_s dW_s.$$

(8.11)

As mentioned before, even though Theorem 6-6.5 was proved in a real Hilbert space, the result is also valid for the complex case. Here, for $g \in H$, $F_t(g) \equiv 0$ and $\Sigma_t(g) = \beta g$ is linear with

$$\|\Sigma_t(g)\|_R^2 = \beta^2 \|R^{1/2}g\|^2 = \beta^2 \int_{\mathbf{R}^d} r(x,x)|g(x)|^2 dx.$$

In view of condition (2), we have

$$\|\Sigma_s(g)\|_R^2 \leq C\|g\|^2, \quad \|\Sigma_s(g) - \Sigma_s(h)\|_R^2 \leq C\|g-h\|^2,$$

for any $g, h \in H$, where $C = \beta^2 \sup_x r(x,x)$. Therefore the conditions for the above-mentioned theorem are satisfied and we can conclude that the equation (8.8) has a unique mild solution $u(\cdot, t)$ which is a continuous H-valued process such that

$$E\{\sup_{0 \leq t \leq T} \|u(\cdot, t)\|^p\} \leq K(p, T)(1 + \|f\|^p),$$

for some constant $K > 0$ and any $p \geq 2$. The stochastic Schrödinger equation also arises from a certain problem in wave propagation through random media. For electromagnetic wave propagation through the atmosphere, by a quasi-static approximation, the time-harmonic wave function $v(y)$, $y \in \mathbf{R}^3$, is assumed to satisfy the random Helmholtz equation [44]:

$$\triangle v + k^2 \eta(y)v = 0,$$

(8.12)

where k is the wave number and $\eta(y)$ is a random field, known as the refractive index. On physical grounds, the wave function must satisfy a radiation condition. Let $y = (x_1, x_2, t)$, where t denotes the third space variable. For an optical beam wave transmission through a random slap: $\{y \in \mathbf{R}^3 : 0 \leq t \leq T\}$, by a so-called forward-scattering approximation [45], assume $v(x,t) = u(x,t)e^{ikt}$ with a large k and an initial aperture field $v(x,0) = h(x)$. When this product solution is substituted into equation (8.12) and the term $\frac{\partial^2}{\partial t^2}v$ is neglected in comparison with $2ik\frac{\partial}{\partial t}v$, it yields the random Schrödinger equation:

$$\frac{\partial u}{\partial t} = \frac{i}{2k}\{\Delta u + k^2\eta(x,t)u\}, \quad x \in \mathbf{R}^2, t \in (0,T),$$

$$u(x,0) = h(x).$$

(8.13)

In the "Markovian" approximation, let $E\eta(x,t) = b(x)$ and $\eta(x,t) - b(x) = \dot{W}(x,t)$. Then the equation (8.13) is a special case of the stochastic Schrödinger equation for $d = 2$ with $\alpha = 1/2k$ and $\beta = k/2$. This approximation has been used extensively in engineering applications.

8.4 Nonlinear Stochastic Beam Equations

Semilinear stochastic beam equations arise as mathematical models to describe the nonlinear vibration of an elastic panel excited by, say, an aerodynamic force. In the presence of air turbulence, a simplified model is governed by the following nonlinear stochastic integro-differential equation subject to the given initial-boundary conditions [14]:

$$\frac{\partial^2}{\partial t^2}u(x,t) = [\alpha + \beta \int_0^l |\frac{\partial}{\partial y}u(y,t)|^2 dy]\frac{\partial^2}{\partial x^2}u(x,t) - \gamma\frac{\partial^4}{\partial x^4}u(x,t)$$

$$+q(u, \partial_t u, \partial_x u, x, t) + \sigma(u, \partial_t u, \partial_x u, x, t)\dot{W}(x,t),$$

(8.14)

$$u(0,t) = u(l,t) = 0, \quad \frac{\partial}{\partial x}u(0,t) = \frac{\partial}{\partial x}u(l,0) = 0,$$

$$u(x,0) = g(x), \quad \frac{\partial}{\partial t}u(x,0) = h(x),$$

for $0 \le t < T$, $0 \le x \le l$, where α, β, γ are positive constants, and g, h are given functions. The Wiener random field $W(x,t)$ is continuous for $0 \le x \le l$, with covariance function $r(x,y)$ which is bounded and continuous for $x, y \in [0,l]$. The functions $q(\mu, y, z, x, t)$ and $\sigma(\mu, y, z, x, t)$ are continuous for $x \in (0,l), t \in [0,T]$ and $\mu, y, z \in \mathbf{R}$, such that they satisfy $q(0,0,0,x,t) = \sigma(0,0,0,x,t) = 0$, and the linear growth and the Lipschitz conditions:

$$|q(\mu, y, z, x, t)|^2 + |\sigma(\mu, y, z, x, t)|^2 \le C(1 + \mu^2 + y^2 + z^2),$$

(8.15)

and

$$|q(\mu, y, z, x, t) - q(\mu', y', z', x, t)|^2 + |\sigma(\mu, y, z, x, t) - p(\mu', y', z', x, t)|^2$$

$$\le K(|\mu - \mu'|^2 + |y - y'|^2|z - z'|^2),$$

(8.16)

for any $t \in [0,T]$, $x \in [0,l]$, $\mu, y, z, \mu', y', z' \in \mathbf{R}$ and for some positive constants C and K.

To regard (8.14) as a second-order stochastic evolution equation, introduce the Hilbert spaces $H = L^2(0,l)$, $H^m = H^m(0,l)$ and $V = H_0^2 = H^2$ satisfying the homogeneous boundary conditions. Define the following operators

$$Au = \alpha \frac{\partial^2 u}{\partial x^2} - \gamma \frac{\partial^4 u}{\partial x^4},$$

$$B(u) = \beta \left[\int_0^l |\frac{\partial}{\partial y} u(y,t)|^2 dy \right] \frac{\partial^2 u}{\partial x^2}, \tag{8.17}$$

and

$$\Gamma_t(u,v) = q(u,v,\partial_x u, \cdot, t),$$

$$\Sigma_t(u,v) = \sigma(u,v,\partial_x u, \cdot, t). \tag{8.18}$$

Then $A : V \to V' = H^{-2}$ is a continuous and coercive linear operator, because of the fact that

$$\langle A\phi, \phi \rangle \le -\gamma \| \frac{\partial^2 \phi}{\partial x^2} \|^2 - \alpha \| \frac{\partial \phi}{\partial x} \|^2, \quad \phi \in V, \tag{8.19}$$

and $\inf_{\phi \in H_0^2} \{ \| \frac{\partial^2 \phi}{\partial x^2} \|^2 / \| \phi \|^2 \} = \lambda > 0$.

Let $H^{1,4}$ denote the L^4-Sobolev space of first order, namely, the set of L^4-functions ϕ on $(0,l)$ with generalized derivatives $\frac{d}{dx}\phi \in L^4(0,l)$. Then we have $B : H_0^2 \cap H^{1,4} \to H$. Since the imbedding $H_0^2 \hookrightarrow H^{1,4}$ is continuous, the nonlinear operator $B : V \to H$ is locally bounded. To show it is locally Lipschitz continuous, we consider

$$\|B(\phi) - B(\varphi)\| = \beta \| \left[\int_0^l (|\frac{\partial}{\partial y}\phi(y,t)|^2 - |\frac{\partial}{\partial y}\varphi(y,t)|^2) dy \right] \frac{\partial^2 \phi}{\partial x^2}$$

$$+ \left[\int_0^l |\frac{\partial}{\partial y}\varphi(y,t)|^2 dy \right] (\frac{\partial^2 \phi}{\partial x^2} - \frac{\partial^2 \varphi}{\partial x^2}) \|$$

$$\le \beta \| \frac{\partial^2 \phi}{\partial x^2} \| \int_0^l (|\frac{\partial}{\partial y}\phi(y,t)|^2 - |\frac{\partial}{\partial y}\varphi(y,t)|^2)| \, dy$$

$$+ \beta \left[\int_0^l |\frac{\partial}{\partial y}\varphi(y,t)|^2 dy \right] \| \frac{\partial^2 \phi}{\partial x^2} - \frac{\partial^2 \varphi}{\partial x^2} \|,$$

so that

$$\|B(\phi) - B(\varphi)\| \le \beta \{ \| \frac{\partial^2 \phi}{\partial x^2} \| (\| \frac{\partial \phi}{\partial x} \| + \| \frac{\partial \varphi}{\partial x} \|) \| \frac{\partial \phi}{\partial x} - \frac{\partial \varphi}{\partial x} \|$$

$$+ \| \frac{\partial \varphi}{\partial x} \|^2 \| \frac{\partial^2 \phi}{\partial x^2} - \frac{\partial^2 \varphi}{\partial x^2} \| \},$$

for any $\phi, \varphi \in V$. It follows that there exists a constant $C > 0$ such that

$$\|B(\phi) - B(\varphi)\| \le C(\|\phi\|_2^2 + \|\varphi\|_2^2)\|\phi - \varphi\|_2, \tag{8.20}$$

which shows that B is locally Lipschitz continuous. By conditions (8.15) and (8.16), we can show that $\Gamma_t : V \times H \to H$ and $\Sigma_t : V \times H \to \mathcal{L}(H)$ such that, for any $t \in [0, T], \phi, \phi' \in V$ and $\varphi, \varphi' \in H$,

$$\|\Gamma_t(\phi, \varphi) - \Gamma_t(\phi', \varphi')\|^2 \le C_1(\|\phi - \varphi\|_2^2 + \|\phi' - \varphi'\|^2), \tag{8.21}$$

and

$$\|\Sigma_t(\phi, \varphi) - \Sigma_t(\phi', \varphi')\|_R^2 \le C_2(\|\phi - \varphi\|_2^2 + \|\phi' - \varphi'\|^2), \tag{8.22}$$

where C_1, C_2 are some positive constants.

Now rewrite the equation (8.14) as a first-order system:

$$\begin{aligned}
du_t &= v_t dt, \\
dv_t &= [A_t u_t + F_t(u_t, v_t)]dt + \Sigma_t(u_t, v_t)dW_t, \\
u_0 &= g, \quad v_0 = h,
\end{aligned} \tag{8.23}$$

where we let

$$F_t(u, v) = B(u) + \Gamma_t(u, v). \tag{8.24}$$

For any $\phi. \in \mathbf{C}([0, T]; V) \cap \mathbf{C}^1([0, T]; H)$, by noticing (8.21), (8.22) and (8.24), we can deduce that

$$\begin{aligned}
\int_0^t (F_s(\phi_s, \dot{\phi}_s), \dot{\phi}_s)ds &= \int_0^t (B(\phi_s), \dot{\phi}_s)ds + \int_0^t (\Gamma_s(\phi_s, \dot{\phi}_s), \dot{\phi}_s)ds \\
&\le -\frac{\beta}{4}(\|\partial_x \phi_t\|^4 - \|\partial_x \phi_0\|^4) + \frac{1}{2}\int_0^t \{\|(\Gamma_s(\phi_s, \dot{\phi}_s)\|^2 + \|\dot{\phi}_s\|^2\}ds \\
&\le \frac{\beta}{4}\|\partial_x \phi_0\|^4 + \frac{1}{2}\int_0^t \{C_1\|\phi_s\|_2^2 + (1 + C_1)\|\dot{\phi}_s\|^2\}ds,
\end{aligned}$$

and

$$\int_0^t \|\Sigma_s(\phi_s, \dot{\phi}_s)\|_R^2 ds \le C_2 \int_0^t (\|\phi_s\|_2^2 + \|\dot{\phi}_s\|^2)ds,$$

where $\dot{\phi}_t = \frac{d}{dt}\phi_t$ and $\partial_x \phi = \frac{\partial}{\partial x}\phi$. It follows that there exists a constant $\kappa > 0$ such that

$$\int_0^t \{(F_s(\phi_s, \dot{\phi}_s), \dot{\phi}_s) + \|\Sigma_s(\phi_s, \dot{\phi}_s)\|_R^2\}ds \le \kappa\{1 + \mathbf{e}(\phi_s, \dot{\phi}_s)\}ds, \tag{8.25}$$

where $\mathbf{e}(\phi, \dot{\phi}) = \|\phi\|_2^2 + \|\dot{\phi}\|^2$ is the energy function.

In view of (8.19), (8.20), (8.21), (8.22) and (8.25), the conditions for Theorem 6-8.5 are fulfilled, except for the monotonicity condition, which can be verified easily. Therefore, for $g \in H_0^2, h \in H$, the equation (8.14) has a unique solution $u \in L^2(\Omega; \mathbf{C}([0, T]; H_0^2))$ with $\partial_t u \in L^2(\Omega; \mathbf{C}([0, T]; H))$. In fact, from the energy equation, we can deduce that $u \in L^4(\Omega \times (0, T); H^1)$. It is also possible to study the asymptotic properties, such as the ultimate boundedness and stability, of solutions to this equation by the relevant theorems in Chapter Six.

8.5   Stochastic Stability of Cahn-Hilliard Equation

The Cahn-Hilliard equation was proposed as a mathematical model to describe the dynamics of pattern formation in phase transition (p. 147, [78]). In a bounded domain $D \subset \mathbf{R}^d$ with a smooth boundary ∂D, this equation, with damping, and its initial-boundary conditions read

$$\frac{\partial}{\partial t} v(x,t) + \nu \Delta^2 v + \alpha \Delta v + \beta v \; - \gamma \Delta v^3 = 0,$$

$$\frac{\partial v}{\partial n}|_{\partial D} = \frac{\partial \Delta v}{\partial n}|_{\partial D} = 0, \tag{8.26}$$

$$v(x,0) = g(x),$$

where Δ and $\Delta^2 = \Delta\Delta$ denote the Laplacian and the bi-harmonic operators, respectively; α, β, γ and ν are positive parameters; $\frac{\partial}{\partial n} v$ denotes the normal derivative of v to the boundary and g is a given function. Let $H = L^2(D)$ and let

$$V = H_0^2 = \{\phi \in H^2(D) : \frac{\partial \phi}{\partial n}|_{\partial D} = 0\}.$$

Then it is known (p. 151, [78]) that, in the variational formulation, for $g \in H$, the problem (8.26) has a unique strong solution $v \in \mathbf{C}([0,T]; H) \cap L^2((0,T); V)$ for any $T > 0$. Suppose that this problem has an equilibrium solution $v = \varphi(x)$ in V satisfying the equation

$$\nu \Delta^2 \varphi + \alpha \Delta \varphi + \beta \varphi - \gamma \Delta \varphi^3 = 0, \tag{8.27}$$

subject to the homogeneous boundary conditions as in (8.26). We are interested in the linear stability of this pattern φ perturbed by a state-dependent white noise. To this end, let $v = \varphi + u$, where u is a small disturbance. Then, in view of (8.26) and (8.27), the linearized equation for u satisfies the following system:

$$\frac{\partial}{\partial t} u(x,t) + \nu \Delta^2 u + \alpha \Delta u + \beta u - 3\gamma \Delta(\varphi^2 u) = \sigma u \dot{W}(x,t),$$

$$\frac{\partial u}{\partial n}|_{\partial D} = \frac{\partial \Delta u}{\partial n}|_{\partial D} = 0, \tag{8.28}$$

$$u(x,0) = h(x),$$

where the constant $\sigma > 0$, $h \in H$ and $W(x,t)$ is Wiener random field with a bounded continuous covariance function $r(x,y)$, for $x, y \in D$. Define

$$Au = -\{\nu \Delta^2 u + \alpha \Delta u + \beta u\} + 3\gamma \Delta(\varphi^2 u), \tag{8.29}$$

and $\Sigma(u) = \sigma u$. Then we consider (8.28) as a stochastic evolution equation in V' as follows

$$du_t = Au_t dt + \Sigma(u_t)dW_t,$$
$$u_0 = h. \tag{8.30}$$

The linear operator $A : V \to V' = H^{-2}$ is continuous. In fact we will show that, under suitable conditions, it is coercive. For $g \in V$, using integration by parts, we can obtain

$$\langle Ag, g \rangle = -\{\nu \langle \triangle^2 g, g \rangle + \alpha \langle \triangle g, g \rangle + \beta(g, g)\} + 3\gamma(\triangle(\varphi^2 g), g)$$
$$\leq -\nu\|\triangle g\|^2 + \alpha\|\triangle g\|\,\|g\| - \beta\|g\|^2 + 3\gamma\varphi_0^2\|\triangle g\|\,\|g\|,$$

where we set $\varphi_0 = \sup_{x \in D} |\varphi(x)|$. It follows that, by recalling the inequality: $|ab| \leq \epsilon|a|^2 + (1/4\epsilon)|b|^2$ for any $\epsilon > 0$,

$$2\langle Ag, g \rangle + \|\Sigma(g)\|_R^2 \leq -2(\nu - \alpha\epsilon_1 - 3\gamma\varphi_0^2\epsilon_2)\|\triangle g\|^2$$
$$-2(\beta - \alpha/4\epsilon_2 - 3\gamma\varphi_0^2/4\epsilon_2 - \sigma^2 r_0/2)\|g\|^2, \tag{8.31}$$

for any $\epsilon_1, \epsilon_2 > 0$, and $r_0 = \sup_{x \in D} |r(x,x)|$. Making use of the fact that the V-norm $\|g\|_2$ is equivalent to the norm $\{\|\triangle g\|^2 + \|g\|^2\}^{1/2}$ (p. 150, [78]), the inequality (8.31) implies the coercivity condition (D.3) for Theorem 6-7.5 provided that

$$(\nu - \alpha\epsilon_1 - 3\gamma\varphi_0^2\epsilon_2) > 0. \tag{8.32}$$

Moreover, since the equation (8.28) is linear, the monotonicity condition (D.4) is also satisfied. It is possible to choose ϵ_1, ϵ_2 properly such that the condition (8.32) is met. Then, for $h \in H$, Theorem 6-7.5 ensures that the equation (8.28) has a unique solution $u \in L^2(\Omega; C([0,T]; H)) \cap L^2(\Omega \times (0,T); H_0^2)$. Furthermore, assume that

$$(\beta - \alpha/4\epsilon_1 - 3\gamma\varphi_0^2/4\epsilon_2 - \sigma^2 r_0/2) > 0. \tag{8.33}$$

By appealing to Theorem 7-4.3, it is easy to show that the equilibrium solution $v = \varphi$ of (8.26), or the null solution of (8.28), is a.s. exponentially stable. We claim that $\Phi(u) = \|u\|^2$ is a Lyapunov functional for the problem (8.30) under conditions (8.32) and (8.33). This is indeed the case since, by noticing (8.31),

$$\mathcal{L}\Phi(u) = 2\langle Au, u \rangle + \|\Sigma(u)\|_R^2 \leq -\lambda\Phi(u), \tag{8.34}$$

for any $u \in V$ and for some $\lambda > 0$. For instance, choose $\epsilon_1 = (\nu/4\alpha)$ and $\epsilon_2 = \nu/6\gamma\varphi_0^2$. Then we can deduce that

$$\mathcal{L}\Phi(u) \leq -\frac{1}{4}\nu\|\triangle u\|^2 - \lambda\|u\|^2,$$

with $\lambda = (\beta - \alpha^2/\nu - 9\gamma^2\varphi_0^4/2\nu - \sigma^2 r_0/2)$. Therefore, if

$$\beta > \alpha^2/\nu + 9\gamma^2\varphi_0^4/2\nu + \sigma^2 r_0/2,$$

then the conditions (8.32), (8.33) and (8.34) are satisfied and the equilibrium solution φ to (8.26) is a.s. exponentially stable.

8.6 Invariant Measures for Stochastic Navier-Stokes Equations

We will consider randomly perturbed Navier-Stokes equations in two space dimensions, a turbulence model for an incompressible fluid [83]. In particular one is interested in the question on the existence of an invariant measure. The velocity $u = (u_1; u_2)$ and the pressure p of the turbulent flow in domain $D \in \mathbf{R}^2$ are governed by the stochastic system:

$$\frac{\partial}{\partial t} u_i(x,t) + \sum_{j=1}^{2} u_i \frac{\partial u_i}{\partial x_j} = -\frac{1}{\rho} \frac{\partial p}{\partial x_i} + \nu \sum_{j=1}^{2} \frac{\partial^2 u_i}{\partial x_j^2} + \dot{W}_i(x,t),$$

$$\sum_{j=1}^{2} \frac{\partial u_j}{\partial x_j} = 0, \quad x \in D, \ t > 0, \tag{8.35}$$

$$u_i|_{\partial D} = 0, \quad u_i(x,0) = g_i(x), \quad i = 1, 2,$$

where ν is the kinematic viscosity, ρ is the fluid density, the initial data $g_i's$ are given functions, and W_1, W_2 are Wiener random fields with covariance functions $r_{ij}, i, j = 1, 2$. In a vectorial notation, the above equations take the form:

$$\frac{\partial}{\partial t} u(x,t) + (u \cdot \nabla)u = -\frac{1}{\rho}\nabla p + \nu \triangle u + \dot{W}(x,t),$$

$$\nabla \cdot u = 0, \quad x \in D, \ t > 0, \tag{8.36}$$

$$u|_{\partial D} = 0, \quad u(x,0) = g(x),$$

where $W = (W_1; W_2)$ and $g = (g_1; g_2)$.

To set up the above system as a stochastic evolution equation, it is customary to introduce the following function spaces [83]. Let $\mathbb{L}^p = L^p(D) \times L^p(D) = [L^p(D)]^2$,

$$\mathbb{S} = \{v \in [\mathbf{C}_0^\infty(D)]^2 : \nabla \cdot v = 0\},$$
$$\mathbb{H} = \text{the closure of } \mathbb{S} \text{ in } \mathbb{L}^2,$$
$$\mathbb{V} = \text{the closure of } \mathbb{S} \text{ in } [H_0^1]^2,$$
$$\mathbb{H}^\perp = \{v = \nabla p, \text{ for some } p \in H^1(D)\},$$

and $\mathbb{V}'$ denotes the dual space of $\mathbb{V}$. Again the norms in $\mathbb{H}$ and $\mathbb{V}$ will be denoted by $\|\cdot\|$ and $\|\cdot\|_1$, respectively. It is known that the space $\mathbb{L}^2$ has the direct sum decomposition:

$$\mathbb{L}^2 = \mathbb{H} \oplus \mathbb{H}^\perp.$$

Let $\Pi : \mathbb{L}^2 \to \mathbb{H}$ be an orthogonal projection. For $v \in \mathbb{S}$, define the operators A and $F(\cdot)$ by $Av = \nu\Pi\triangle v$ and $F(v) = -\Pi(v \cdot \nabla)v$. Then they can be extended to be continuous operators from $\mathbb{V}$ to $\mathbb{V}'$. After applying the projector Π to equation (8.27), we obtain the stochastic evolution equation in $\mathbb{V}'$ of the form:

$$du_t = [Au_t + F(u_t)]dt + dV_t,$$
$$u_0 = g,$$
(8.37)

where $V_t = \Pi W_t$, and $W_t = W(\cdot, t)$ is assumed to be a $\mathbb{H}$-valued Wiener process with covariance operator Q, and $g \in \mathbb{H}$.

The existence and uniqueness of solution to the above equation has been studied by several authors (see, e.g., [83]). Here we sketch a proof given in [62] by a method of monotone truncation. First define the bilinear operator B from $\mathbb{V} \times \mathbb{V} \to \mathbb{V}'$ as follows

$$\langle B(u,v), w \rangle = \sum_{i,j=1}^{2} \int_D u_i \frac{\partial v_j}{\partial x_i} w_j dx, \quad u, v, w \in \mathbb{V},$$
(8.38)

and set $B(u) = B(u, u)$. Then we have

$$\langle B(u,v), w \rangle = -\langle B(u,w), v \rangle,$$
(8.39)

which implies $\langle B(u,v), v \rangle = 0$. First it is easy to check that $A : V \to V'$ is coercive. Notice that $F(u) = -B(u)$. Making use of (8.39) and some Sobolev inequalities, the following estimates can be obtained [83]:

$$|\langle F(u), w \rangle| = |\langle B(u,u), w \rangle| \le C_1 \|u\|_1^2 \|w\|_1,$$

and

$$\begin{aligned}
|\langle F(u) - F(v), w \rangle| &\le |\langle B(u,w) - B(v,w), v \rangle| \\
&\le |\langle B(u-v, w), u \rangle| + |\langle B(v,w), u-v \rangle| \\
&\le C_2(\|u\|_1 + \|v\|_1)\|u - v\|_1 \|w\|_1.
\end{aligned}$$

Hence the operator $F : \mathbb{V} \to \mathbb{V}'$ is locally bounded and Lipschitz continuous.

For uniqueness, let $u, v \in L^2(\Omega; \mathbf{C}([0,T]; \mathbb{H})) \cap L^2(\Omega \times (0,T); \mathbb{V})$ be two solutions of (8.37), as in the deterministic case, the following estimate holds a.s. for some $\gamma > 0$,

$$\|u_t - v_t\|^2 \le \|u_0 - v_0\|^2 \exp\{\gamma \int_0^t |u_s|_4^4 ds\}.$$

Here $|u_s|_4$ denotes the $\mathbb{L}^4(D)$−norm of u_s, which exists because the imbedding $\mathbb{V} \hookrightarrow \mathbb{L}^4$ is continuous for $d = 2$. It follows that $u_t = v_t$ a.s. and the solution is unique.

To show the existence of a solution, we mollify the operator F by using a $\mathbb{L}^4$-truncation. Let $\mathbf{B}_n = \{u \in \mathbb{L}^4(D) : |u|_4 \le n\}$ be a ball in $\mathbb{L}^4$ centered at the origin with radius n. It can be shown [62] that the operator $A + F(\cdot)$ is monotone in $\mathbf{B}_n$ due to the following estimate:

$$\langle A(u - v), u - v \rangle + \langle F(u) - F(v), u - v \rangle \leq \frac{16n^2}{\nu^3} \|u - v\|^2 - \frac{\nu}{2} \|u - v\|_1^2.$$

Now introduce a mollified operator F_n defined by

$$F_n(u) = \left(\frac{n}{n \vee |u|_4} \right)^4 F(u),$$

which is well defined in $\mathbb{V}$. In fact it can be shown that $A + F_n(\cdot)$ is a monotone operator from $\mathbb{V}$ into $\mathbb{V}'$. Let us approximate the equation (8.37) by the following one:

$$du^n(\cdot, t) = [Au^n(\cdot, t) + F_n(u^n(\cdot, t))]dt + dV_t,$$
$$u_0 = g, \tag{8.40}$$

where the covariance operator Q of V_t is assumed to satisfy the trace condition

$$Tr\, Q \leq q_0, \tag{8.41}$$

for some constant $q_0 > 0$. In view of the coercivity and the monotonicity conditions for A and F_n indicated above, we can apply Theorem 6-7.5 to the problem (8.40) and claim that it has a unique strong solution $u^n \in L^2(\Omega; \mathbf{C}([0, T]; \mathbb{H})) \cap L^2(\Omega \times (0, T); \mathbb{V})$, for any $T > 0$. Furthermore, noting the property (8.39), the following energy equation holds

$$\|u_t^n\|^2 = \|u_0\|^2 + 2 \int_0^t \langle Au_s^n, u_s^n \rangle ds + 2 \int_0^t (u_s^n, \Sigma dW_s) + t\, Tr\, Q. \tag{8.42}$$

From this equation, similar to a coercive stochastic evolution equation, we can deduce that

$$E\{ \sup_{0 \leq t \leq T} \|u_t^n\|^2 + 2\nu \int_0^t \|u_s^n\|_1^2 ds \} \leq C_T\, Tr\, Q,$$

for some constant $C_T > 0$. Therefore the sequence $\{u^n\}$ is bounded in $X_T = L^2(\Omega; \mathbf{C}([0, T]; \mathbb{H})) \cap L^2(\Omega \times (0, T); \mathbb{V})$, and there exists a subsequence, still denoted by $\{u^n\}$, that converges weakly to u in X_T. As $n \to \infty$, we have $Au^n \to Au$ and $F_n(u^n) \to F_0$ weakly in $L^2(\Omega \times (0, T); \mathbb{V}')$. By taking the limit in (8.40), we obtain

$$du_t = [Au_t + F_0(t)]dt + dV_t,$$
$$u_0 = g. \tag{8.43}$$

It remains to show that $F_0(t) = F(u_t)$ for each $t \in (0, T)$ a.s. The idea of the proof is similar to that of Theorem 6-7.5. A complete proof can be found in [62]. Therefore we can conclude that the problem (8.35) for stochastic

Navier-Stokes equations has a unique strong solution $u \in X_T$, which is known to be a Markov process in H with the Feller property. By passing to a limit as $n \to \infty$, the energy equation (8.42) yields

$$\|u_t\|^2 = \|u_0\|^2 + 2 \int_0^t \langle Au_s, u_s \rangle ds + 2 \int_0^t (u_s, dV_s) + t\, Tr\, Q. \qquad (8.44)$$

Finally, by following the approach in [15], we will show that there exists an invariant measure for the problem (8.35). Since A is coercive and it satisfies

$$\langle Av, v \rangle \leq -\lambda \|v\|_1^2, \quad v \in \mathbb{V},$$

for some $\lambda > 0$, we can deduce from the energy equation (8.44) that

$$E\|u_T\|^2 + 2\lambda\nu \int_0^T \|u_t\|_1^2 dt \leq \|g\|^2 + T\, Tr\, Q.$$

It follows that, for any $u_0 = g \in \mathbb{H}$ and $T_0 > 0$, there exists a constant $M > 0$ such that

$$\frac{1}{T} \int_0^T E\|u_t^g\|_1^2 dt \leq M, \quad \text{for any } T > T_0, \qquad (8.45)$$

where u_t^g denotes the solution u_t with $u_0 = g$. By means of the Chebyshev inequality and (8.45), we obtain

$$\lim_{R \to \infty} \sup_{T > T_0} \{\frac{1}{T} \int_0^T P(\|u_t^g\|_1 > R) dt\}$$
$$\leq \lim_{R \to \infty} \sup_{T > T_0} \{\frac{1}{R^2 T} \int_0^T E\|u_t^g\|_1^2 dt\} = 0. \qquad (8.46)$$

Therefore we can apply Theorem 7-5.5 to conclude that there exists an invariant measure μ on the Borel field of $\mathbb{H}$. Moreover it can be shown that the μ is supported in $\mathbb{V}$ and, if g is a $\mathbb{H}$-valued random variable with the invariant distribution μ, the solution u_t will be a stationary Markov process. The uniqueness question is not answered here but was treated in [25] by a different approach.

Chapter 9

Diffusion Equations in Infinite Dimensions

9.1 Introduction

In finite dimensions, it is well known that the solution of an Itô equation is a diffusion process and the conditional expectation of a smooth function of the solution process satisfies a diffusion equation in $\mathbf{R}^d$, known as the Kolmogorov equation. Therefore it is quite natural to explore such a relationship for the stochastic partial differential equations. For example, consider the randomly perturbed heat equation (3.26)in a bounded domain. By means of the eigenfunctions expansion, the solution is given by $u(\cdot, t) = \sum_{k=1}^{\infty} u_t^k e_k$, where u_t^k, $k = 1, 2, \cdots$, satisfy the infinite system of Itô equations (3.31) in $[0, T]$:

$$du_t^k = -\lambda_k u_t^k dt + \sigma_k dw_t^k, \quad u_0^k = h_k, \quad k = 1, 2, \ldots$$

Let $u^n(\cdot, t) = \sum_{k=1}^{n} u_t^k e_k$ be the n-term approximation of u. Then $u^n(\cdot, t)$ is an Ornstein-Uhlenbeck process, a time-homogeneous diffusion process in $\mathbf{R}^n$. Let $(y_1, \cdots, y_n) \in \mathbf{R}^n$. Then, for a smooth function Φ on $\mathbf{R}^n$, the expectation function $\Psi(y_1, \cdots, y_n, t) = E\left\{\Phi(u_t^1, \cdots, u_t^n) | u_0^1 = y_1, \cdots, u_0^n = y_n\right\}$ satisfies the Kolmogorov equation:

$$\frac{\partial \Psi}{\partial t} = \mathcal{A}_n \Psi, \quad \Psi(y_1, \cdots, y_n, 0) = \Phi(y_1, \cdots, y_n), \tag{9.1}$$

where $\mathcal{A}_n$ is the infinitesimal generator of the diffusion process $u^n(\cdot, t)$ given by

$$\mathcal{A}_n = \frac{1}{2} \sum_{k=1}^{n} \sigma_k^2 \frac{\partial^2}{\partial y_k^2} - \sum_{k=1}^{n} \lambda_k y_k \frac{\partial}{\partial y_k}.$$

Formally, as $n \to \infty$, we get

$$\mathcal{A} = \frac{1}{2} \sum_{k=1}^{\infty} \sigma_k^2 \frac{\partial^2}{\partial y_k^2} - \sum_{k=1}^{\infty} \lambda_k y_k \frac{\partial}{\partial y_k}. \tag{9.2}$$

The corresponding Kolmogorov reads

$$\frac{\partial \Psi}{\partial t} = \mathcal{A}\Psi, \quad \Psi(y_1, y_2, \cdots, 0) = \Phi(y_1, y_2 \cdots). \tag{9.3}$$

237

However the meaning of this infinite-dimensional differential operator $\mathcal{A}$ and the associated Kolmogorov equation is not clear and needs to be clarified.

The early work on the connection between a diffusion process in a Hilbert space and the infinite-dimensional parabolic and elliptic equations was done by Baklan [3] and Daleskii [18]. More refined and in-depth studies of such problems in an abstract Wiener space was initiated by Gross [33] and carried on by Piech [70] and Kuo [52]. Later, in the development of the Malliavin calculus, further progress had been made in the area of analysis in Wiener spaces [85], in particular, the Wiener-Sobolev spaces. More recently, this subject has been studied systematically in the book [21].

For the analysis in Wiener spaces, when $\lambda_k = \sigma_k = 1, \forall k$, $\mathcal{A}$ will play the role of the infinite-dimensional Laplacian (or Laplace-Beltrami operator, and $(-\mathcal{A})$ is known as a number operator. In contrast, there are two technical problems in dealing with the diffusion equation associated with the stochastic partial differential equations. First the differential operator $\mathcal{A}$ has singular coefficients. This can be seen from the right-hand side of equation (9.2) in which the eigenvalues $\lambda_k \to \infty$, while $\sigma_k \to 0$ as $k \to \infty$. Secondly it is well known that, in infinite dimension, there exists no analog of Lebesque measure, a consequence of the basic fact that a closed unit ball in an infinite-dimensional space is non-compact. For a class of problems to be considered later, by introducing a Gauss-Sobolev space based on an invariant measure for the stochastic partial differential equation of interest, the two technical problems raised above have been resolved satisfactorily [11].

This chapter is organized as follows. In Section 9.2, we show that the solutions of stochastic evolution equations are Markov diffusion processes. The properties of the associated semigroups and the Kolmogorov equations are studied in a classical setting. To facilitate a L^2- theory, we introduce the Gauss-Sobolev spaces in Section 9.3 based on a reference Gaussian measure μ. For a linear stochastic evolution equation, the associated Ornstein-Uhlenbeck semigroup in $L^2(\mu)$ is analyzed in Section 9.4 by means of Hermite polynomial functionals. Then, by making use of this semigroup, the mild solution of a Kolmogorov equation is considered in Section 9.5. Also the strong solutions of the parabolic and elliptic equations are studied in a Gauss-Sobolev space. The final section is concerned with the Hopf equations for some simple stochastic evolution equations. Similar to a Kolmogorov equation, they are functional differential equations which govern the characteristic functionals for the solution process.

9.2 Diffusion Processes and Kolmogorov Equations

Consider the autonomous stochastic evolution equation as in Chapter Seven:

$$\begin{aligned} du_t &= Au_t dt + F(u_t)dt + \Sigma(u_t)dW_t, \quad t \geq 0, \\ u_0 &= \xi, \end{aligned} \tag{9.4}$$

where $A : V \to V'$, $F : H \to H$, $\Sigma : V \to \mathcal{L}_R^2$, W_t is a R-Wiener process in K, and ξ is $\mathcal{F}_0$-measurable random variable in H. Suppose that the coefficients are regular so that it has a unique strong solution. For ease of reference, we reiterate a set of sufficient conditions given in Chapter Six as the following conditions G:

(G.1) $A : V \to V'$ is a closed linear operator with domain $\mathcal{D}(A)$ dense in V, and there exist constant $\alpha > 0$ such that $|\langle Av, v \rangle| \leq \alpha \|v\|_V^2$, $\forall\, v \in V$.

(G.2) There exist positive constants $C_1, C_2 > 0$ such that

$$\|F(v)\|^2 + \|\Sigma(v)\|_R^2 \leq C_1(1 + \|v\|_V^2),$$

$$\|F(v) - F(v')\|^2 + \frac{1}{2}\|\Sigma(v) - \Sigma(v')\|_R^2 \leq C_2\|v - v'\|_V^2, \; \forall\, v, v' \in V.$$

(G.3) There exist constants $\beta > 0, \gamma$ such that

$$\langle Av, v \rangle + (F(v), v) + \frac{1}{2}\|\Sigma(v)\|_R^2 \leq -\beta\|v\|_V^2 + \gamma\|v\|^2, \; \forall\, v \in V.$$

(G.4) There exists constant $\delta \geq 0$ such that

$$\langle A(v - v'), v - v' \rangle + (F(v) - F(v'), v - v') + \frac{1}{2}\|\Sigma(v) - \Sigma(v')\|_R^2$$
$$\leq \delta\|v - v'\|^2, \; \forall\, v, v' \in V.$$

(G.5) W_t is a R-Wiener process in K with $TrR < \infty$.

Theorem 2.1 Under conditions $(G.1) - (G.5)$, the (strong) solution u_t of equation (9.4) is a time-homogeneous Markov process and it depends continuously on the initial data in mean-square. Moreover, for $g, h \in H$, let u_t^g and u_t^h denote the solutions with $\xi = g$ and $\xi = h$, respectively. Then there exists $b > 0$ such that

$$E\|u_t^g - u_s^h\|^2 \leq b(|t - s| + \|g - h\|^2), \tag{9.5}$$

for any $s, t \in [0, \infty)$.

Proof. The conditions $(G.1) - (G.5)$ are sufficient for Theorem 6-7.4 to hold. Hence the problem (9.4) has a unique strong solution $u \in L^2(\Omega; \mathbf{C}([0,T]; H)) \cap L^2(\Omega_T; V)$ such that the energy equation holds:

$$\|u_t\|^2 = \|h\|^2 + 2\int_0^t \langle Au_s, u_s \rangle ds + 2 \int_0^t (F(u_s), u_s) ds \qquad (9.6)$$
$$+ 2 \int_0^t (u_s, \Sigma(u_s) dW_s) + \int_0^t \|\Sigma(u_s)\|_R^2 \, ds.$$

Similar to an Itô's equation in $\mathbf{R}^n$, the fact that u_t is a homogeneous Markov diffusion process is easy to show. We will verify only the inequality (9.5). Clearly we have, for $s < t$,

$$E \|u_t^g - u_s^h\|^2 \le 2\{E \|u_t^g - u_s^g\|^2 + E \|u_s^g - u_s^h\|^2\}. \qquad (9.7)$$

It follows from (9.4), (9.6) and condition (G.3) that

$$E \|u_t^g - u_s^g\|^2 = 2E \int_s^t \{\langle Au_r^g, u_r^g \rangle + (F(u_r^g), u_r^g) + \frac{1}{2}\|\Sigma(u_r^g)\|_R^2\} \, dr \qquad (9.8)$$
$$\le 2|\gamma| \int_s^t E \|u_s^g\|^2 \, ds.$$

In the proof of Theorem 6-7.4, we found that

$$E \sup_{0 \le t \le T} \|u_t^g\|^2 \le K_1,$$

for some constant $K_1 > 0$, so that (9.8) yields

$$E \|u_t^g - u_s^g\|^2 \le 2|\gamma| K_1 |t - s|. \qquad (9.9)$$

Similarly, by (9.4), (9.6) and condition (G.4), we have

$$E \|u_s^g - u_s^h\|^2 = \|g - h\|^2 + 2E \int_0^s \{\langle A(u_r^g - u_r^h), u_r^g - u_r^h \rangle$$
$$+ (F(u_r^g) - F(u_r^h), u_r^g - u_r^h) + \frac{1}{2}\|\Sigma(u_r^g) - \Sigma(u_r^h)\|_R^2\} \, dr$$
$$\le \|g - h\|^2 + 2\delta \int_0^s E \|u_r^g - u_r^h\|^2 \, dr,$$

which, by the Gronwall inequality, implies that

$$E \|u_s^g - u_s^h\|^2 \le K_2\|g - h\|^2, \qquad (9.10)$$

for some $K_2 > 0$. Now, in view of (9.7), (9.9) and (9.11), the desired inequality (9.5) follows. $\qquad \square$

As in Section 7.5, let $P(h, t, B)$, $t \ge 0$, $B \in \mathcal{B}(H)$, be the transition probability function for the time-homogeneous Markov process u_t, and let P_t denote the corresponding transition operator. With the aid of this theorem, we can show that

Lemma 2.2 The transition probability $P(h, t, \cdot)$ for the solution process has the Feller property and the associated transition operator $P_t : X = \mathbf{C}_b(H) \to \mathbf{C}_b(H)$ satisfies the following strongly continuous semigroup properties:

(1) $P_0 = I$, (I denotes the identity operator on X).

(2) $P_{t+s} = P_t P_s$, for every $s, t \geq 0$. (Semigroup property).

(3) $\lim_{t \downarrow 0} P_t \Phi = \Phi$, for any $\Phi \in X$. (Strong continuity). $\qquad\square$

The semigroup $\{P_t, t \geq 0\}$ given above is known as a *Markov semigroup*. Recall the definition of a (strong) Itô functional introduced in Section 7.2 in connection with a generalized Itô formula. The sets of time-dependent and -independent Itô functionals on H are denoted by $\mathbb{C}(H \times [0, T])$ and $\mathbb{C}(H)$, respectively. Let $\mathbb{C}_b^2(H)$ denote the subset of Itô functionals Φ in $\mathbb{C}(H)$ such that the Fréchet derivatives $\Phi' : V \to V$ and $\Phi'' : V \to \mathcal{L}(H)$ are bounded. Similarly, $\mathbb{C}_b^{2,1}(H)$ denotes the subset of Itô functionals Ψ in $\mathbb{C}(H \times [0, T])$ such that $\Psi(\cdot, t) \in \mathbb{C}_b^2(H)$ and $\frac{\partial}{\partial t}\Psi(\cdot, t)$ is bounded in $V \times [0, T]$.

Rename the second-order differential operator $\mathcal{L}$ in (7.9) as $\mathcal{A}$ given by

$$\mathcal{A}\Phi = \frac{1}{2}Tr\left[\Sigma(v)R\Sigma^\star(v)\Phi''(v)\right] + \langle Av + F(v), \Phi'(v)\rangle, \qquad (9.11)$$

for $\Phi \in \mathbb{C}_b^2(H)$ and $v \in V$.

Now consider the expectation functional

$$\Psi(v, t) = E\,\Phi(u_t^v), \quad v \in V, \qquad (9.12)$$

where u_t^v denotes the solution of equation (9.4) with $u_0^v = v$. To derive the Kolmogorov equation for Ψ, we apply the Itô formula for a strong solution to get, for $t \in (0, T)$,

$$\Phi(u_t^v) = \Phi(v) + \int_0^t \mathcal{A}\Phi(u_s^v)ds + \int_0^t (\Phi'(u_s^v), \Sigma(u_s^v)dW_s). \qquad (9.13)$$

If we can show that

$$E\,\mathcal{A}\Phi(u_t^v) = \mathcal{A}\Psi(t, v), \ \forall\, v \in V, \qquad (9.14)$$

then, by taking the expectation of (9.13), one would easily arrive at a diffusion equation for Ψ. To show (9.14), we have to impose additional smoothness assumptions on the coefficients as follows.

(H.1) $A : V \to V'$ is a self-adjoint linear operator with domain $\mathcal{D}(A)$ dense in V, and there exist constants $\alpha, \beta > 0$ and γ such that $|\langle Au, v \rangle| \leq \alpha \|u\|_V \|v\|_V$, and

$$\langle Av, v \rangle \leq \gamma \|v\|^2 - \beta \|v\|_V^2, \ \forall u, v \in V.$$

(H.2) $F : H \to V$ and $\Sigma : H \to \mathcal{L}(K, V)$ such that

$$\|F(v)\|_V^2 + \|\Sigma(v)\|_{V,R}^2 \leq \delta (1 + \|v\|_V^2), \ \forall v \in V,$$

for some constant $\delta > 0$, where $\|\Sigma(v)\|_{V,R}^2 = Tr [\Sigma(v) R\Sigma^*(v)]_V$ and $Tr [\cdot]_V$ denotes the trace in V.

(H.3) There is $\kappa > 0$ such that

$$\|F(u) - F(v)\|_V^2 + \frac{1}{2} \|\Sigma(u) - \Sigma(v)\|_{V,R}^2 \leq \kappa \|u - v\|_V^2, \ \forall u, v \in V.$$

Then the corresponding solution is also more smooth and the following lemma holds.

Lemma 2.3 Let conditions G and (H.1)–(H.3) be satisfied. Then, if $u_0 = v \in V$, the solution u_t^v of the problem (9.4) belongs to $L^2(\Omega; \mathbf{C}([0, T]; V)) \cap L^2(\Omega \times [0, T]; \mathcal{D}(A))$.

Proof. We will sketch the proof. Let $B = (-A)^{1/2}$ be the square-root operator. Then it is known that $B : V \to H$ is a positive, self-adjoint and closed linear operator [70]. Let $P_n : H \to V_n$ and $P_n' : V' \to V_n$ be the projection operators introduced in the proof of Theorem 6-7.5. Then $B_n = BP_n' : V' \to V$ is bounded with $B_n Av = AB_n v$ for $v \in V$. Notice $P_n' h = P_n h$ for $h \in H$. Apply B_n to equation (9.4) to get

$$u_t^n = v_n + \int_0^t [Au_s^n + B_n F(u_s)] \, ds + \int_0^t B_n \Sigma(u_s) dW_s, \tag{9.15}$$

where we set $u_t^n = B_n u_t$ and $v_n = B_n v$. Now consider the linear equation:

$$\mu_t = Bv + \int_0^t [A\mu_s + BF(u_s)] \, ds + \int_0^t B\Sigma(u_s) dW_s. \tag{9.16}$$

By conditions G and H, we can apply Theorem 6-7.4 to assert that the above equation has a unique strong solution $\mu \in L^2(\Omega; \mathbf{C}([0, T]; H)) \cap L^2(\Omega \times [0, T]; V)$. From (9.15) and (9.16) it follows that $\nu_t^n = (\mu_t - u_t^n)$ satisfies

$$\nu_t^n = (I - P_n)Bv + \int_0^t [A\nu_s^n + (I - P_n)BF(u_s)] \, ds$$
$$+ \int_0^t (I - P_n)B\Sigma(u_s) dW_s. \tag{9.17}$$

By making use of the energy equation for (9.17), it can be shown that

$$E \left\{ \sup_{0 \le t \le T} \|\nu_t - u_t^n\|^2 + \int_0^T \|\nu_t - u_t^n\|_V^2 dt \right\} \to 0,$$

so that $u_t^n \to Bu_t = \mu_t$ in mean-square, uniformly in t, as $n \to \infty$. On the other hand, since B is a closed linear operator, we have $u_t^n = B_n u_t \to Bu_t$ so that $\mu = Bu$. In view of the fact that V-norm $\|\cdot\|_V$ is equivalent to B-norm $\|B \cdot \|$, the conclusion of the theorem follows. $\qquad\square$

With the aid of the above lemma, we can show that the expectation functional Ψ satisfies a diffusion equation in V.

Theorem 2.4 Let the conditions in Lemma 2.3 hold true. Suppose that $\Phi \in \mathbb{C}_b^2(H)$ is an Itô functional on H satisfying the following properties: There exist positive constants b and c such that

$$\|\Phi'(v)\|_V + \|\Phi''(v)\|_{\mathcal{L}} \le b, \qquad \text{and} \tag{9.18}$$

$$\|\Phi'(u) - \Phi'(v)\|_V + \|\Phi''(u) - \Phi''(v)\|_{\mathcal{L}} \le c\|u - v\|_V, \tag{9.19}$$

for all $u, v \in V$. Then $\mathcal{A}$ defined by (9.11) is the infinitesimal generator of the Markov semigroup P_t and the expectation functional Ψ defined by (9.12) satisfies the Kolmogorov equation:

$$\begin{aligned} \frac{\partial}{\partial t}\Psi(v,t) &= \mathcal{A}\Psi(v,t), \quad v \in V, \\ \Psi(v,0) &= \Phi(v), \quad v \in H. \end{aligned} \tag{9.20}$$

Proof. By taking the expectation of the equation (9.13), we obtain

$$\Psi(v,t) = \Phi(v) + \int_0^t E\,\mathcal{A}\Phi(u_s^v)ds. \tag{9.21}$$

For $u, v \in V$, by making use of conditions (H.1), (9.18) and (9.19), we can get

$$\begin{aligned} |\langle Au, \Phi'(u)\rangle - \langle Av, \Phi'(v)\rangle| &\le |\langle A(u-v), \Phi'(u)\rangle| + |\langle Av, \Phi'(u) - \Phi'(u)\rangle| \\ &\le \alpha(\|u-v\|_V\|\Phi'(u)\|_V + \|v\|_V\|\Phi'(u) - \Phi'(v)\|_V) \\ &\le \alpha(b + c\|v\|_V)\|u - v\|_V, \end{aligned}$$

or

$$|\langle Au, \Phi'(u)\rangle - \langle Av, \Phi'(v)\rangle| \le b_1(1 + \|v\|_V)\|u - v\|_V, \tag{9.22}$$

with $b_1 = \alpha(b \vee c) > 0$. Similarly we can show that

$$|(F(u), \Phi'(u)) - (F(v), \Phi'(v))| \le b_2(1 + \|v\|_V)\|u - v\|_V, \tag{9.23}$$

for some $b_2 > 0$. Let

$$Q(v) = \Sigma(v)R\Sigma^\star(v), \tag{9.24}$$

and consider

$$
\begin{aligned}
&|Tr\,[Q(u)\Phi''(u)] - Tr\,[Q(v)\Phi''(v)]\,| \\
&\le |Tr\,[(Q(u) - Q(v))\Phi''(v)]| + |Tr\,[Q(u)(\Phi''(u) - \Phi''(v))]\,| \\
&\le \|\Phi''(v)\|_{\mathcal{L}}\|\Sigma(u) - \Sigma(v)\|_R^2 + \|\Sigma(u)\|_R^2\,\|\Phi''(u) - \Phi''(v)\|_{\mathcal{L}}.
\end{aligned}
$$

By invoking conditions (H.2), (9.18) and (9.19), one can deduce from the above inequality that

$$|Tr\,[Q(u)\Phi''(u)] - Tr\,[Q(v)\Phi''(v)]\,| \le b_3\,(1 + \|v\|_V)\,\|u - v\|_V, \tag{9.25}$$

for some constant $b_3 > 0$. Now rewrite (9.21) as

$$\frac{1}{t}[\Psi(v,t) - \Phi(v)] = \mathcal{A}\Phi(v) + \frac{1}{t}\int_0^t \Theta(v,s)ds, \tag{9.26}$$

where $\Theta(v,s) = E\,\Phi(u_s^v) - \Phi(v)$. By taking equations (9.22)–(9.26) into account, we can get a constant $C_1 > 0$ such that

$$|\Theta(v,s)| \le C_1 E\,(1 + \|u_s^v\|_V)\,\|u_s^v - v\|_V$$

$$\le 2C_1(1 + E\,\|u_s^v\|_V^2)^{1/2}\,(E\,\|u_s^v - v\|_V^2)^{1/2}.$$

By the regularity properties of u_t^v as depicted in Theorem 2.1 and Lemma 2.3, we can show that $\sup_{0 \le s \le T} E\,\|u_s^v\|_V^2 \le C_2$ and $E\,\|u_s^v - v\|_V^2 \le C_3 s$, for some $C_2, C_3 > 0$. It follows that

$$\Big|\int_0^t \Theta(v,s)ds\Big| \le C_4 t^{3/2},$$

for some $C_4 > 0$ and

$$\lim_{t \downarrow 0} \frac{1}{t}\int_0^t \Theta(v,s)ds = 0.$$

Therefore we can deduce from (9.26) that

$$\lim_{t \downarrow 0} \frac{1}{t}[\Psi(v,t) - \Phi(v)] = \lim_{t \downarrow 0} \frac{1}{t}[P_t\Phi(v) - \Phi(v)] = \mathcal{A}\Phi(v), \quad v \in V. \tag{9.27}$$

This shows that $\mathcal{A}$ is the infinitesimal generator of the Markov semigroup.

To derive the Kolmogorov equation (9.20), we claim that, under the conditions for the theorem, condition (H.4) in particular, $\Psi \in \mathbb{C}^{2,1}$ (see Theorem 9.4, [19]). By the semigroup property given in Lemma 2.2, we have $\Psi(\cdot, t + \tau) = P_\tau\Psi(\cdot, t)$, for $t, \tau \in (0, T)$. Thus, by (9.27), we obtain

$$\frac{\partial}{\partial t}\Psi(v,t) = \lim_{\tau \downarrow 0} \frac{1}{\tau}[P_\tau\Psi(v,t) - \Psi(v,t)] = \mathcal{A}\Psi(v,t), \quad v \in V.$$

By the mean-square continuity of the solution, the fact that $\lim_{t \downarrow 0} \Psi(v, t) = \Phi(v)$ is easy to verify. $\qquad \qquad \qquad \qquad \qquad \qquad \qquad \qquad \qquad \qquad \square$

Remarks:

(1) In finite-dimension, for $H = \mathbf{R}^n$, the infinitesimal generator $\mathcal{A}$ can be defined everywhere in H. In contrast, here in infinite-dimension, $\mathcal{A}$ is only defined in a subset V of H. As to be seen, even though V is dense in H, it may be a set of measure zero with respect to a reference measure on $(\Omega, \mathcal{B}(H))$.

(2) For partial differential equations in $\mathbf{R}^n$, the technique of integration by parts, such as the Green's formula, plays an important role in analysis. Without the standard Lebesgue measure in infinite-dimension, to develop a L_p-theory, it is paramount to choose a reference measure which is most suitable for this purpose. As we will show, in some special cases, an invariant measure μ for the associated stochastic evolution equation seems to be a natural choice.

(3) For the Markov semigroup P_t on $\mathbf{C}_b(H)$, clearly the set $\mathbf{C}_b^2(H)$ is in the domain $\mathcal{D}(\mathcal{A})$ of its generator $\mathcal{A}$. Equipped with a proper reference measure μ, a semigroup theory can be developed in a $L^2(\mu)$-Sobolev space.

9.3 Gauss-Sobolev Spaces

In the study of parabolic partial differential equations in $\mathbf{R}^d$, the Lebesgue's L^p-spaces play a prominent role. To develop a L^p-theory for infinite-dimensional parabolic equations, the Kolmogorov equation in particular, an obvious substitute for the Lebesgue measure is a Gaussian measure. For $p = 2$, such spaces will be called *Gauss-Sobolev spaces*, which will be introduced in what follows.

Let H denote a real separable Hilbert space with inner product $(\cdot, \cdot)$ and norm $\|\cdot\|$ as before. Let $\mu \in \mathcal{N}(0, \Lambda)$ be a Gaussian measure on $(H, \mathcal{B}(H))$ with mean zero and nuclear covariance operator Λ. Since $\Lambda : H \to H$ is a positive-definite, self-adjoint operator with $Tr\, \Lambda < \infty$, its square-root operator $\Lambda^{1/2}$ is a positive self-adjoint Hilbert-Schmidt operator on H. Introduce the inner product

$$(g, h)_0 = (\Lambda^{-1/2} g, \Lambda^{-1/2} h), \quad \text{for} \quad g, h \in \Lambda^{1/2} H. \tag{9.28}$$

Let $H_0 \in H$ denote the Hilbert subspace which is the completion of $\Lambda^{1/2} H$ with respect to the norm $\|g\|_0 = (g, g)_0^{1/2}$. Then H_0 is dense in H and the inclusion map $i : H_0 \hookrightarrow H$ is compact. The triple (i, H_0, H) forms an abstract Wiener space [53]. In this setting, one will be able to make use of several results in the Wiener functional analysis.

Let $\mathbb{H} = L^2(H, \mu)$ denote the Hilbert space of Borel measurable functionals on the probability space with inner product $[\cdot, \cdot]$ defined by

$$[\Phi, \Theta] = \int_H \Phi(v)\Theta(v)\mu(dv), \quad \text{for} \quad \Phi, \Theta \in \mathbb{H}, \tag{9.29}$$

and norm $\|\Phi\| = [\Phi, \Phi]^{1/2}$.

In $\mathbb{H}$ we choose a complete orthonormal system $\{\varphi_k\}$ such that $\varphi_k \in H_0$, $k = 1, 2, \cdots$. Introduce a class of simple functionals on H as follows. A functional $\Phi : H \to \mathbf{R}$ is said to be a *smooth simple functional*, or a *cylinder functional* if there exists a $\mathbf{C}^\infty$-function ϕ on $\mathbf{R}^n$ and n-continuous linear functionals $\ell_1, \cdots, \ell_N$ on H such that

$$\Phi(h) = \phi(h_1, \cdots, h_n), \tag{9.30}$$

where $h_i = \ell_i(h), i = 1, \cdots, n$. The set of all such functionals will be denoted by $\mathcal{S}(\mathbb{H})$. Obviously, if ϕ is a polynomial, then the corresponding polynomial functional Φ belongs to $\mathcal{S}(\mathbb{H})$. For the linear functionals, one may take $\ell_i(h) = (h, \varphi_i)$. However, as to be seen, this is not a proper one for our purpose .

Now we are going to introduce the Hermite polynomials in $\mathbf{R}$. Let p_k denote the Hermite polynomial of degree k defined by

$$p_k(x) = (-1)^k (k!)^{-1/2} e^{x^2/2} \frac{d^k}{dx^k} e^{-x^2/2}, \quad k = 1, 2, \cdots, \tag{9.31}$$

with $p_0 \equiv 1$. It is well known that $\{p_k\}$ is complete orthonormal basis for $L^2(\mathbf{R}, (1/\sqrt{2\pi})e^{-x^2/2}dx)$. Denote $\mathbf{n} = (n_1, n_2, \cdots, n_k, \cdots)$, where $n_k \in \mathbb{Z}^+$, the set of non-negative integers, and let $\mathbf{Z} = \{\mathbf{n} = (n_1, n_2, \cdots, n_k, \cdots) : n_k \in \mathbb{Z}^+, n = |\mathbf{n}| = \sum_{k=1}^\infty n_k < \infty\}$, so that, in $\mathbf{n}$, $n_k = 0$ except for a finite number of $n_k's$. Define Hermite polynomial functionals, or simply Hermite functionals, on H as

$$\mathcal{H}_\mathbf{n}(h) = \prod_{k=1}^\infty p_{n_k}[\ell_k(h)], \tag{9.32}$$

for $n = 0, 1, 2, \cdots$, and $r = 1, 2, \cdots$. Clearly the Hermite polynomials $\mathcal{H}_\mathbf{n}$ are smooth simple functionals. As in one dimension, they span the space $\mathbb{H}$.

Lemma 3.1 Let $\ell_k(h) = (h, \Lambda^{-1/2}\varphi_k), k = 1, 2, \cdots$. Then the set $\{\mathcal{H}_\mathbf{n}\}$ of all Hermite polynomials on H forms a complete orthonormal system for $\mathbb{H} = L^2(H, \mu)$. Hence the set of all such functionals are dense in $\mathbb{H}$. Moreover we have the direct sum decomposition:

$$\mathbb{H} = \mathcal{K}_0 \oplus \mathcal{K}_1 \oplus \cdots \oplus \mathcal{K}_j \oplus \cdots = \oplus_{j=0}^\infty \mathcal{K}_j,$$

where $\mathcal{K}_j$ is the subspace of $\mathbb{H}$ spanned by $\{\mathcal{H}_\mathbf{n} : |\mathbf{n}| = j\}$.

Proof. As pointed out earlier, we may regard μ as a Wiener measure in the Wiener space (i, H_0, H). Notice that, if we let $\tilde{\varphi}_k = \Lambda^{1/2} \varphi_k, k = 1, 2, \cdots$, then

$$(\tilde{\varphi}_j, \tilde{\varphi}_k)_0 = (\varphi_j, \varphi_k) = \delta_{jk}$$

so that $\{\tilde{\varphi}_k\}$ is a complete orthonormal system for H_0. Also we can check that

$$[\ell_j, \ell_k] = (\tilde{\varphi}_j, \tilde{\varphi}_k)_0, \quad j, k = 1, 2, \cdots.$$

Therefore, by appealing to the Wiener-Itô decomposition, we can follow a proof in the paper [71] or in (Prop.1.2 [85]) to reach the desired conclusion. $\square$

Let Φ be a smooth simple functional given by (9.30). Then the Fréchet derivatives, $D\Phi = \Phi'$ and $D^2\Phi = \Phi''$ can be computed as follows:

$$
\begin{aligned}
(D\Phi(h), v) &= \sum_{k=1}^{n} [\partial_k \phi(h_1, \cdots, h_n)] \ell_k(v), \\
(D^2\Phi(h)u, v) &= \sum_{j,k=1}^{n} [\partial_j \partial_k \phi(h_1, \cdots, h_n)] \ell_j(u) \ell_k(v),
\end{aligned}
\tag{9.33}
$$

for any $u, v \in H$, where $\partial_k \phi = \frac{\partial}{\partial h_k} \phi$. Similarly, for $m > 2$, $D^m\Phi(h)$ is a m-linear form on $(H)^k = (H \times \overset{(m)}{\cdots} \times H)$ with inner product $(\cdot, \cdot)_m$. We have $[D^m\Phi(h)](v_1, \cdots, v_n) = (D^m\Phi(h), v_1 \otimes, \cdots, \otimes v_n)_m$, for $h, v_1, \cdots, v_n \in H$.

Now, for $v \in H$, let $T_v : H \to H$ be a translation operator so that

$$T_v h = v + h, \quad \text{for} \quad h \in H.$$

Denote by $\mu_v = \mu \circ T_v^{-1}$, the measure induced by T_v. Then the following is known as the Cameron-Martin formula [81].

Lemma 3.2 If $v \in H_0$, the measures μ and μ_v are equivalent ($\mu \sim \mu_v$), and the Radon-Nikodym derivative is given by

$$\frac{d\mu_v}{d\mu}(\xi) = \exp\{(\Lambda^{-1/2}v, \Lambda^{-1/2}\xi) - \frac{1}{2}\|\Lambda^{-1/2}v\|^2\}, \text{ for } \xi \in H, \mu - a.s., \tag{9.34}$$

where $\Lambda^{-1/2}$ denotes a pseudo-inverse.

Proof. Let ρ_k denote the eigenvalues of Λ with eigenfunction φ_k, for $k = 1, 2, \cdots$. For $v \in H_0$, it can be expressed as

$$v = \sum_{k=1}^{\infty} v_k \tilde{\varphi}_k,$$

where $\tilde{\varphi}_k = \Lambda^{1/2}\varphi_k$ and $v_k = (v, \tilde{\varphi}_k)_0$ is such that

$$\|v\|_0^2 = \|\Lambda^{-1/2}v\|^2 = \sum_{k=1}^{\infty} v_k^2 < \infty.$$

Let v^n be the n-term approximation of v:

$$v^n = \sum_{k=1}^{n} v_k \tilde{\varphi}_k$$

and define

$$h^n = \Lambda^{-1}v^n = \sum_{k=1}^{n} \rho_k^{-1/2} v_k \varphi_k.$$

Let us define a sequence of measures ν_n by

$$\nu_n(d\xi) = \exp\{(\xi, h^n) - \frac{1}{2}(\Lambda h^n, h^n)\}\,\mu(d\xi).$$

Since $\xi \in \mathcal{N}(0, \Lambda)$ is a centered Gaussian random variable in H, for positive integers m, n, we have

$$\begin{aligned}
E\,|(\xi, h^m) - (\xi, h^n)|^2 &= E\,|(\xi, h^m - h^n)|^2 \\
&= (\Lambda(h^m - h^n), h^m - h^n) = \|\Lambda^{-1/2}(v_m - v_n)\| \\
&= \|v_m - v_n\|_0^2 \to 0,
\end{aligned}$$

as $m, n \to \infty$. Therefore (ξ, h^n) converges in $L^2(\mu)$ to a random variable $\varphi(\xi)$, which will be written as $(\Lambda^{-1/2}v, \Lambda^{-1/2}\xi)$. Also it can be shown that $(\Lambda h^n, h^n) \to (\Lambda^{-1/2}v, \Lambda^{-1/2}v) = \|\Lambda^{-1/2}v\|^2$. Since all random variables involved are Gaussian, we can deduce that

$$\lim_{n\to\infty} \exp\{(\Lambda^{-1/2}h, \Lambda^{-1/2}\xi) - \frac{1}{2}(\Lambda h^n, h^n)\}$$

$$= \exp\{(\Lambda^{-1/2}v, \Lambda^{-1/2}\xi) - \frac{1}{2}\|\Lambda^{-1/2}v\|^2\}$$

in $L^1(H, \mu)$. It follows that ν_n converges weakly to ν. To show $\nu = \mu_v$, it is equivalent to verifying they have a common characteristic functional. We know that the characteristic functional for μ_v is given by

$$\Psi(\eta) = \exp\{i(v, \eta) - \frac{1}{2}(\Lambda\eta, \eta)\}, \quad \eta \in H.$$

Let $\Psi_n(\eta)$ be the characteristic functional for ν_n, which can be computed as follows

$$\begin{aligned}
\Psi_n(\eta) &= \int_H \exp\{i(\eta, \xi) + (\xi, h^n) - \frac{1}{2}(\Lambda h^n, h^n)\}\,\mu(d\xi) \\
&= \exp\{i(\Lambda h^n, \eta) - \frac{1}{2}(\Lambda\eta, \eta)\}
\end{aligned}$$

which converges to the characteristic functional $\Psi(\eta)$ as $n \to \infty$. $\square$

To develop a L^2-theory for parabolic or elliptic equations in infinite dimensions, we need an analog of Green's formula in finite dimensions. Therefore it is necessary to perform an integration by parts with respect to the Gaussian measure μ. With the aid of the above lemma, we are able to verify the following integration-by-parts formulas.

Theorem 3.3 Let $\Phi \in \mathcal{S}(\mathbb{H})$ be a smooth simple functional and let $\mu \in \mathcal{N}(0, \Lambda)$ be a Gaussian measure in H. Then, for any $g, h \in H$, the following formulas hold:

$$\int_H (v, h)\Phi(v)\mu(dv) = \int_H (D\Phi(v), \Lambda h)\mu(dv), \tag{9.35}$$

$$\int_H (v, g)(v, h)\Phi(v)\mu(dv) = \int_H \Phi(v)(\Lambda g, h)\mu(dv) \\ + \int_H (D^2\Phi(v)\Lambda g, \Lambda h)\mu(dv). \tag{9.36}$$

Proof. We will only sketch a proof of (9.35). The formula (9.36) can be shown in a similar fashion.

For $u \in H_0$, we have, by Lemma 3.2,

$$\int_H \Phi(v + u)\mu(dv) = \int_H \Phi(v)\mu_u(dv) \\ = \int_H \Phi(v)Z_u(v)\mu(dv), \tag{9.37}$$

where we let $Z_u(v) = \dfrac{d\mu_u}{d\mu}(v)$. Therefore, letting $g = \Lambda h$ and making use of (9.34) and (9.37), we can verify (9.35) by the following computations:

$$\int_H (D\Phi(v), \Lambda h)\mu(dv) = \int_H (D\Phi(v), g)\mu(dv)$$
$$= \int_H \frac{\partial}{\partial \epsilon}\Phi(v + \epsilon g)|_{\epsilon=0}\mu(dv) = \frac{\partial}{\partial \epsilon}\int_H \Phi(v + \epsilon g)\mu(dv)|_{\epsilon=0}$$
$$= \int_H \Phi(v)\frac{\partial}{\partial \epsilon}Z_{\epsilon g}(v)|_{\epsilon=0}\mu(dv) = \int_H (\Lambda^{-1/2}v, \Lambda^{-1/2}g)\Phi(v)\mu(dv)$$
$$= \int_H (v, h)\Phi(v)\mu(dv),$$

where, since Φ is smooth, the interchanges of integration and differentiation with respect to ϵ can be justified by the dominated convergence theorem. $\square$

Remark: In view of the proof, the integration-by-parts formulas (9.35) and (9.36) can actually be shown to hold for $\Phi \in \mathbf{C}_b^2(H)$ as well.

As a generalization of the Sobolev space $H^k(D)$ with $D \subset \mathbf{R}^d$, we now introduce the Gauss-Sobolev spaces. In the previous chapters, we see that $H^k(D)$ can be defined either by the L^2-integrability of the functions and its derivatives up to order k, or else by the boundedness of its Fourier series or integral via the spectral decomposition. This is also true for Wiener-Sobolev spaces, due to a theorem of Meyer [65]. Here, in the spirit of Chapter Three, we will introduce the $L^2(\mu)$-Sobolev spaces based on a Fourier series representation.

By Lemma 3.1, for $\Phi \in \mathbb{H}$, it can be represented as

$$\Phi(v) = \sum_{\mathbf{n} \in \mathbf{Z}} \phi_{\mathbf{n}} \mathcal{H}_{\mathbf{n}}(v), \tag{9.38}$$

where $\phi_{\mathbf{n}} = [\Phi, \mathcal{H}_{\mathbf{n}}]$ and $\sum_{\mathbf{n}} |\phi_{\mathbf{n}}|^2 < \infty$. Let $\alpha_{\mathbf{n}} = \alpha_{n_1 \cdots n_k, \cdots}$ be a sequence of positive numbers with $\alpha_{\mathbf{n}} > 0$, such that $\alpha_{\mathbf{n}} \to \infty$ as $n \to \infty$. Define

$$\begin{aligned}
\|\Phi\|_{k,\alpha} &= \{\sum_{\mathbf{n} \in \mathbf{Z}} (1 + \alpha_{\mathbf{n}})^k |\phi_{\mathbf{n}}|^2\}^{1/2}, \quad k = 1, 2, \cdots, \\
\|\Phi\|_{0,\alpha} &= \|\Phi\| = \{\sum_{\mathbf{n} \in \mathbf{Z}} |\phi_{\mathbf{n}}|^2\}^{1/2},
\end{aligned} \tag{9.39}$$

which is $L^2(\mu)$-norm of Φ. For the given sequence $\alpha = \{\alpha_{\mathbf{n}}\}$, let $\mathbb{H}_{k,\alpha}$ denote the completion of $\mathbb{S}(H)$ with respect to the norm $\|\cdot\|_{k,\alpha}$. Then $\mathbb{H}_{k,\alpha}$ is called a *Gauss-Sobolev space* of order k with parameter α. The dual space of $\|\cdot\|_{k,\alpha}$ is denoted by $\|\cdot\|_{-k,\alpha}$.

Remarks:

(1). In later applications, the sequence $\{\alpha_{\mathbf{n}}\}$ will be fixed and we shall simply denote $\mathbb{H}_{k,\alpha}$ by $\mathbb{H}^k$.

(2). If we choose the sequence $\{\alpha_{\mathbf{n}}\}$ to be a sequence of non-negative integers, then $\mathbb{H}^k$'s coincide with the $L^2(\mu)$-Sobolev spaces with the Wiener measure μ.

9.4 Ornstein-Uhlenbeck Semigroup

For simplicity, we shall be concerned mainly with the Markov semigroup associated with the linear stochastic equation:

$$\begin{aligned}
du_t &= Au_t dt + dW_t, \quad t > 0, \\
u_0 &= h \in H.
\end{aligned} \tag{9.40}$$

Here we assume that $A : V \to V'$ is a self-adjoint, strictly negative operator satisfying conditions (B.1)–(B.3) in Section 6.7, and W_t is a R-Wiener process in H. Then we know that the solution is given by

$$u_t^h = G_t h + \int_0^t G_{t-s} \, dW_s, \tag{9.41}$$

where G_t is the associated Green's operator, and the distribution of the solution u_t^h is a Gaussian measure μ_t^h in H with mean $G_t h$ and covariance operator

$$\Gamma_t = \int_0^t G_s R G_s ds.$$

By Theorem 7-5.4, there exists a unique invariant measure μ for the equation (9.40), which is a centered Gaussian measure in H with covariance operator Λ given by

$$\Lambda = \int_0^\infty G_t R G_t \, dt. \tag{9.42}$$

By Theorem 2.1, the solution of (9.40) is a time-homogeneous Markov process with transition operator P_t. For $\Phi \in \mathbb{H}$, we have

$$(P_t \Phi)(h) = \int_H \Phi(v) \mu_t^h (dv) = E \, \Phi(u_t^h). \tag{9.43}$$

Suppose that $(-A)$ and R have the same set of eigenfunctions $\{e_k\}$ with positive eigenvalues $\{\lambda_k\}$ and $\{\rho_k\}$, respectively. Then R commutes with G_t and the covariance operator given by (9.42) can be evaluated to give

$$\Lambda v = R \int_0^\infty G_{2t} v \, dt = \frac{1}{2} R(-A)^{-1} v, \quad v \in H.$$

Notice that, if $ARv = RAv$ for $v \in \mathcal{D}(A)$, then A and R have a common set of eigenfunctions. Therefore the following lemma holds.

Lemma 4.1 Suppose that A and R satisfy the following:

(1) $A : V \to V'$ is self-adjoint and there is $\beta > 0$ such that

$$\langle Av, v \rangle \le -\beta \|v\|_V^2, \quad \forall v \in V.$$

(2) A commutes with R in the domain of A.

Then the equation (9.40) has a unique invariant measure μ, which is Gaussian measure on H with zero mean and covariance operator $\Lambda = \frac{1}{2} R(-A)^{-1} = \frac{1}{2}(-A)^{-1} R$. $\qquad \square$

Remarks: In what follows, this invariant measure μ will be used for the Gauss-Sobolev spaces. Even though the assumption on the commutativity of

A and R is not essential, it simplifies the subsequent analysis. Notice that, by condition (1), the eigenvalues of $(-A)$ are strictly positive, $\lambda_k \geq \delta$, for some $\delta > 0$.

Let $\Phi \in \mathbb{S}(H)$ be a smooth simple functional. By (9.30), it takes the form

$$\Phi(h) = \phi(\ell_1(h), \cdots, \ell_n(h)),$$

where $\ell_k(h) = (h, \Lambda^{-1/2} e_k)$. Define a differential operator $\mathcal{A}_0$ on $\mathbb{S}(H)$ by

$$\mathcal{A}_0\Phi(v) = \frac{1}{2} Tr\,[RD^2\Phi(v)] + \langle Av, D\Phi(v) \rangle, \ v \in H, \tag{9.44}$$

which is well defined, since $D\Phi \in \mathcal{D}(A)$ and $\langle Av, D\Phi(v) \rangle = (v, A\,D\Phi(v))$.

Lemma 4.2 Let P_t be the transition operator as defined by (9.42). Then the following properties hold:

(1) $P_t : \mathbb{S}(H) \to \mathbb{S}(H)$ for $t \geq 0$.

(2) $\{P_t, t \geq 0\}$ is a strongly continuous semigroup on $\mathbb{S}(H)$ so that, for any $v \in \mathbb{S}(H)$, we have $P_0 = I$, $P_{t+s}v = P_t P_s v$, for all $t, s \geq 0$, and $\lim_{t \downarrow 0} P_t v = v$.

(3) $\mathcal{A}_0$ is the infinitesimal generator of P_t so that

$$\lim_{t \downarrow 0} \frac{1}{t}(P_t - I)v = \mathcal{A}_0 v.$$

Proof. For $\Phi \in \mathbb{S}(H)$, we have

$$\begin{aligned} (P_t\Phi)(h) &= E\,\phi[\ell_1(u_t^h), \cdots, \ell_n(u_t^h)] \\ &= E\,\phi[\ell_1(G_t h) + \ell_1(v_t), \cdots, \ell_n(G_t h) + \ell_n(v_t)], \end{aligned} \tag{9.45}$$

where $v_t = \int_0^t G_{t-s}\,dW_s \in \mathcal{N}(0, \Lambda_t)$ with $\Lambda_t = (1/2)A^{-1}R(I - G_{2t})$. Notice that $\xi_i, i = 1, \cdots, n$, are jointly Gaussian random variables with mean zero and covariance matrix $\sigma_t = [\sigma_t^{ij}]_{n \times n}$. By a translation $\xi_i \to \xi_i - \ell_i(h_t)$ and Lemma 3.2, we can rewrite (9.45) as

$$(P_t\Phi)(h) = \int_{\mathbf{R}^n} \phi(x)\exp\{-(\sigma_t^{-1}x, \ell_i(h_t)) - \frac{1}{2}|\sigma_t^{-1}\ell(h_t)|^2\}\mu_t^n(dx), \tag{9.46}$$

where $h_t = G_t h$; x and $\ell(h) = (\ell_1(h), \cdots, \ell_n(h)) \in \mathbf{R}^n$ for $h \in H$, and μ_t^n is the joint Gaussian distribution for $(\xi_1, \cdots, \xi_n)$. Since $\phi \in \mathbf{C}^\infty(\mathbf{R}^n)$ is of a polynomial growth, we can deduce from (9.46) that $(P_t\Phi) \in \mathbb{S}(H)$.

The semigroup properties (2) follow from the Markov property of the solution u_t, as shown in Lemma 2.2. To verify (3), let us first assume that

$\Phi \in \mathbb{S}(H)$ is bounded. By applying the Itô formula and then taking expectation, we obtain

$$E\left\{\Phi(u_t^v) - \Phi(v)\right\} = \int_0^t E\,\mathcal{A}_0\Phi(u_s^v)\,ds,$$

so that, by the boundedness, continuity and Lebesgue's theorem, we can show that

$$\lim_{t\downarrow 0}\frac{1}{t}(P_t - I)\Phi(v) = \lim_{t\downarrow 0}\frac{1}{t}E\left\{\Phi(u_t^v) - \Phi(v)\right\}$$

$$= \lim_{t\downarrow 0}\frac{1}{t}\int_0^t E\,\mathcal{A}_0\Phi(u_s^v)\,ds = \mathcal{A}_0\Phi(v).$$

For any $\Phi \in \mathbb{S}(H)$, we can still proceed as before with the aid of a localizing stop time τ_N and replace t by $(t \wedge \tau_N)$ in the Itô formula. Then we let $N \to \infty$.
□

Lemma 4.3 Let $\mathcal{H}_{\mathbf{n}}(h)$ be a Hermite polynomial functional given by (9.32). Then the following hold:

$$\mathcal{A}_0\mathcal{H}_{\mathbf{n}}(h) = -\lambda_{\mathbf{n}}\,\mathcal{H}_{\mathbf{n}}(h), \tag{9.47}$$

and

$$P_t\mathcal{H}_{\mathbf{n}}(h) = \exp\{-\lambda_{\mathbf{n}}t\}\,\mathcal{H}_n^r(h), \tag{9.48}$$

for any $\mathbf{n} = (n_1, \cdots, n_k, \cdots) \in \mathbf{Z}$ and $h \in H$, where

$$\lambda_{\mathbf{n}} = \sum_{k=1}^{\infty} n_k\lambda_k.$$

Proof. First let $n = n_k$ so that

$$\mathcal{H}_n(h) = p_{n_k}(\xi_k), \quad \xi_k = \ell_k(h) = (\Lambda^{-1/2}e_k, h).$$

Then $D\mathcal{H}_n(h) = p'_{n_k}(\xi_k)(\Lambda^{-1/2}e_k)$, $D^2\mathcal{H}_n(h) = p''_{n_k}(\xi_k)(\Lambda^{-1/2}e_k \otimes \Lambda^{-1/2}e_k)$. Hence, by Lemma 4.1 and the fact that A and R have the same eigenfunctions e_k with eigenvalues $-\lambda_k$ and ρ_k respectively, we have

$$\langle Ah, D\mathcal{H}_{\mathbf{n}}(h)\rangle = (h, AD\mathcal{H}_{\mathbf{n}}(h)) = -\lambda_k\xi_k p'_{n_k}(\xi_k), \tag{9.49}$$

$$Tr\,[RD^2\mathcal{H}_{\mathbf{n}}(h)] = \rho_k(\Lambda^{-1/2}e_k, e_k)p''_{n_k}(\xi_k)$$
$$= 2\lambda_i p''_{n_k}(\xi_k), \tag{9.50}$$

which imply that $\mathcal{H}_{\mathbf{n}} \in \mathbb{S}(H)$ and

$$\mathcal{A}_0\mathcal{H}_{\mathbf{n}}(h) = \lambda_k[p''_{n_k}(\xi_k) - \xi_k p'_{n_k}(\xi_k)]$$
$$= -n_k\lambda_k p_{n_k}(\xi_k), \tag{9.51}$$

where use was made of the identity for a Hermite polynomial:

$$p_k''(x) - xp_k'(x) = -kp_k(x).$$

In general, consider

$$\mathcal{H}_{\mathbf{n}}(h) = \prod_{k=1}^{\infty} p_{n_k}(\xi_k).$$

By means of the differentiation formulas (9.33), we can compute

$$\begin{aligned}
D\mathcal{H}_{\mathbf{n}}(h) &= \sum_{k=1}^{\infty} \partial_k \mathcal{H}_{\mathbf{n}}(h)(\Lambda^{-1/2}e_k) \\
&= \mathcal{H}_{\mathbf{n}}(h) \sum_{k=1}^{\infty} \frac{p_{n_k}'(\xi_k)}{p_{n_k}(\xi_k)}(\Lambda^{-1/2}e_k),
\end{aligned} \tag{9.52}$$

$$\begin{aligned}
D^2\mathcal{H}_{\mathbf{n}}(h) &= \sum_{i,j=1}^{\infty} \partial_i \partial_j \mathcal{H}_n^r(h)(\Lambda^{-1/2}e_i) \otimes (\Lambda^{-1/2}e_j) \\
&= \mathcal{H}_{\mathbf{n}}(h) \sum_{i=1}^{\infty} \frac{p_{n_i}''(\xi_i)}{p_{n_i}(\xi_i)}(\Lambda^{-1/2}e_i) \otimes (\Lambda^{-1/2}e_i) \\
&\quad + \{\text{terms with} \quad i \neq j\}.
\end{aligned} \tag{9.53}$$

We can deduce from (9.50) to (9.53) that

$$\langle Ah, D\mathcal{H}_{\mathbf{n}}(h) \rangle = -\mathcal{H}_n^r(h) \sum_{k=1}^{\infty} \lambda_k \xi_k \frac{p_{n_k}'(\xi_k)}{p_{n_k}(\xi_k)},$$

and

$$Tr\,[RD^2\mathcal{H}_{\mathbf{n}}(h)] = 2\mathcal{H}_{\mathbf{n}}(h) \sum_{k=1}^{\infty} \lambda_k \frac{p_{n_k}''(\xi_k)}{p_{n_k}(\xi_k)},$$

from which we obtain (9.47),

$$\mathcal{A}_0 \mathcal{H}_{\mathbf{n}}(h) = -\left(\sum_{k=1}^{\infty} n_k \lambda_k\right) \mathcal{H}_{\mathbf{n}}(h).$$

To show (9.48), since $\mathcal{H}_{\mathbf{n}} \in \mathbb{S}(H)$, we can apply Lemma 4.2 to get

$$\begin{aligned}
\frac{d}{dt} P_t \mathcal{H}_{\mathbf{n}} &= \mathcal{A}_0 P_t \mathcal{H}_{\mathbf{n}} \\
&= P_t \mathcal{A}_0 \mathcal{H}_{\mathbf{n}} = -\lambda_{\mathbf{n}} P_t \mathcal{H}_{\mathbf{n}},
\end{aligned}$$

with $P_0 \mathcal{H}_{\mathbf{n}} = \mathcal{H}_{\mathbf{n}}$. It follows that

$$P_t \mathcal{H}_{\mathbf{n}} = e^{-\lambda_{\mathbf{n}}t} \mathcal{H}_{\mathbf{n}}$$

as to be shown. $\square$

Lemma 4.4 Assume the conditions for Lemma 4.3 hold. Then, for any $\Phi, \Psi \in \mathbb{S}(H)$, the following Green's formula holds:

$$\int_H (\mathcal{A}_0 \Phi) \Psi d\mu = \int_H \Phi (\mathcal{A}_0 \Psi) d\mu = -\frac{1}{2} \int_H (RD\Phi, D\Psi) d\mu. \qquad (9.54)$$

Proof. For $\Phi, \Psi \in \mathbb{S}(H)$, we have

$$\begin{aligned} Tr\,[RD^2\Phi(h)] &= \sum_{k=1}^n (D^2\Phi(h)e_k, Re_k(h)) \\ &= \sum_{k=1}^n \rho_k (D^2\Phi(h)e_k, e_k), \end{aligned} \qquad (9.55)$$

for some $n \geq 1$. Now it is clear that

$$(D^2\Phi(h)e_k, e_k) = (D(D\Phi(h)), e_k), e_k),$$

and

$$\Psi(D^2\Phi(h)e_k, e_k) = (D[\Psi(D\Phi(h)), e_k)], e_k) - (D\Phi(h), e_k)(D\Psi(h), e_k).$$

Therefore, by the integration-by-parts formula given in Theorem 3.3, we can get

$$\begin{aligned} & \int_H \Psi(h)(D^2\Phi(h)e_k, e_k)\mu(dh) + \int_H (D\Phi(h), e_k)(D\Psi(h), e_k)\mu(dh) \\ =& \int_H (D[\Psi(D\Phi(h), e_k)], e_k)\mu(dh) \\ =& \int_H \Psi(h)(\Lambda^{-1}e_k, h)(D\Phi(h), e_k)\mu(dh), \end{aligned} \qquad (9.56)$$

where Λ^{-1} is a pseudo-inverse. It follows from (9.55) and (9.56) that

$$\begin{aligned} \int_H \Psi\, Tr\,[RD^2\Phi]d\mu &= \sum_{k=1}^\infty \rho_k \int_H \Psi\,(D^2\Phi e_k, e_k)d\mu \\ &= \sum_{k=1}^\infty \rho_k \{\int_H \Psi(\Lambda^{-1}e_k, h)(D\Phi, e_k)d\mu - \int_H (D\Phi, e_k)(D\Psi, e_k)d\mu\} \\ &= -2\int_H \Psi \sum_{k=1}^\infty \langle Ah, e_k\rangle(D\Phi, e_k)d\mu - \int_H \sum_{k=1}^\infty \rho_k(D\Phi, e_k)(D\Psi, e_k)d\mu, \\ &= -2\int_H \Psi\langle Ah, D\Phi\rangle d\mu - \int_H (RD\Phi, D\Psi)d\mu, \end{aligned}$$

which yields the Green's formula (9.54). $\qquad \square$

Now recall the definition of the Gauss-Sobolev space $\mathbb{H}_{k,\alpha}$ defined in the last section. Let $\alpha_{\mathbf{n}} = \lambda_{\mathbf{n}}$ be fixed from now on and denote the corresponding $\mathbb{H}_{\alpha,k}$ simply by $\mathbb{H}_k$ with norm $\|\| \cdot \|\|_k$ defined by

$$\|\|\Phi\|\|_k = \{\sum_{\mathbf{n}}(1+\lambda_{\mathbf{n}})^k |\phi_{\mathbf{n}}|^2\}^{1/2}, \quad k = 0, 1, 2, \cdots, \tag{9.57}$$

with $\|\|\Phi\|\|_0 = \|\|\Phi\|\|$.

Theorem 4.5   Let the conditions on A and R in Lemma 4.1 hold. Then $P_t : \mathbb{H} \to \mathbb{H}$, for $t \geq 0$, is a contraction semigroup with the infinitesimal generator $\tilde{A}$. The domain of $\tilde{A}$ contains $\mathbb{H}_2$ and we have $\tilde{A} = \mathcal{A}_0$ in $\mathbb{S}(H)$.

Proof. The semigroup properties are easy to verify. To show P_t is a contraction map, Let $\Phi \in \mathbb{H}$. Since

$$(P_t\Phi)(v) = E\,\Phi(u_t^v) = \int_H \Phi(v+h)\mu_t(dh),$$

where $\mu_t \in \mathcal{N}(0, \Lambda)$, we have

$$\begin{aligned}
\|P_t\Phi\|^2 &= \int_H [P_t\Phi(v)]^2 \mu(dv) \\
&= \int_H [\int_H \Phi(h)P(t,v,dh)]^2 \mu(dv) \\
&\leq \int_H \int_H \Phi^2(h)P(t,v,dh)\mu(dv) \\
&= \|\|\Phi\|\|^2,
\end{aligned}$$

where use was made of the Cauchy-Schwarz inequality and a property of the invariant measure μ. Hence $P_t : \mathbb{H} \to \mathbb{H}$ is a contraction mapping.

Let $\tilde{A}$ denote the infinitesimal generator of P_t with domain $\mathcal{D}(\tilde{A})$ dense in $\mathbb{H}$ [70]. Clearly, by Lemma 4.2, $\tilde{A}\Phi = \mathcal{A}_0\Phi$, for $\Phi \in \mathbb{S}(H)$. For any $\Phi \in \mathbb{H}_2$, by writing Φ in terms of Hermite polynomials, it can be shown that $\tilde{A}\Phi \in \mathbb{H}$ so that $\mathbb{H}_2 \subset \mathcal{D}(\tilde{A})$. $\qquad\square$

Theorem 4.6   Let the conditions for Theorem 4.5 hold true. The differential operator $\mathcal{A}_0$ defined by (9.44) in $\mathbb{S}(H)$ can be extended to be a self-adjoint linear operator $\mathcal{A}$ in $\mathbb{H}$ with domain $\mathbb{H}_2$.

Proof. Let $\mathbb{V}_N$ be the closed linear span of $\{\mathcal{H}_{\mathbf{n}}, |\mathbf{n}| \leq N\}$, and let $\mathcal{P}_N : \mathbb{H} \to \mathbb{V}_N$ be a projection operator defined by

$$\mathcal{P}_N\Phi = \sum_{|\mathbf{n}|\leq N} \phi_{\mathbf{n}}\mathcal{H}_{\mathbf{n}}, \quad \Phi \in \mathbb{H},$$

with $\phi_\mathbf{n} = [\Phi, \mathcal{H}_\mathbf{n}]$. Let $\mathcal{A}_N = \mathcal{A}_0 \mathcal{P}_N$. Then, for $\Phi \in \mathbb{H}_2$,

$$\mathcal{A}_N \Phi = - \sum_{|\mathbf{n}| \leq N} \lambda_\mathbf{n} \phi_\mathbf{n} \mathcal{H}_\mathbf{n}.$$

Thus, for any positive integers $N < N'$,

$$\|\mathcal{A}_{N'}\Phi - \mathcal{A}_N \Phi\|^2 = \sum_{N \leq |\mathbf{n}| \leq N'} \lambda_\mathbf{n}^2 |\phi_\mathbf{n}|^2$$
$$\leq \|\mathcal{P}_{N'}\Phi - \mathcal{P}_N \Phi\|_2^2 \to 0,$$

as $N, N' \to \infty$. Hence $\{\mathcal{A}_N \Phi\}$ is a Cauchy sequence in $\mathbb{H}$ and denote its limit by $\mathcal{A}\Phi$, that is,

$$\lim_{N \to \infty} \mathcal{A}_N \Phi = \mathcal{A}\Phi, \quad \Phi \in \mathbb{H}_2.$$

In the meantime, by passing through the sequences $\mathcal{P}_N \Phi$ and $\mathcal{P}_N \Psi$, one can show by invoking the integral identity (9.54) that, for $\Phi, \Psi \in \mathbb{H}_2$,

$$[\mathcal{A}\Phi, \Psi] = [\Phi, \mathcal{A}\Psi]$$
$$= -\frac{1}{2} \int_H (RD\Phi, D\Psi)d\mu,$$

so that $[\mathcal{A}\Phi, \Phi] \leq 0$. Therefore $\mathcal{A}$ is a symmetric, semi-bounded operator in $\mathbb{H}$ with a dense domain $\mathbb{H}_2$. By appealing to Friedrich's extension theorem (p. 317, [86]), it admits a self-adjoint extension, still to be denoted by $\mathcal{A}$, with domain $\mathcal{D}(\mathcal{A}) = \mathbb{H}_2$. $\qquad\square$

Remarks: Since both $\tilde{A}$ and $\mathcal{A}$ are extensions of $\mathcal{A}_0$ to a domain containing $\mathbb{H}_2$, they must coincide there. Also it is noted that the above result is a generalization for the case of the Wiener measure. If $K = H, A = R = I$, and μ is the Wiener measure in H, then $\lambda_\mathbf{n}$ is a non-negative integer, and $(-\mathcal{A})$ is known as the *number operator* [70].

9.5 Parabolic Equations and Related Elliptic Problems

In view of the Green's formula (9.54), one naturally associates the differential operator $\mathcal{A}$ with a bilinear (Dirichlet) form as in finite dimensions. To this end, let $\mathbb{H}_k$ be the Gauss-Sobolev space of order k with norm $\|\cdot\|_k$ and denote its dual space by $\mathbb{H}_{-k}$ with norm $\|\cdot\|_{-k}$. Thus we have

$$\mathbb{H}_k \subset \mathbb{H} \subset \mathbb{H}_{-k}, \quad k \geq 0.$$

The duality between $\mathbb{H}_k$ and $\mathbb{H}_{-k}$ has the following pairing

$$\langle\langle \Psi, \Phi \rangle\rangle_k, \quad \Phi \in \mathbb{H}_k, \quad \Psi \in \mathbb{H}_{-k}.$$

As before we set $\mathbb{H}_0 = \mathbb{H}$, $\|\cdot\|_0 = \|\cdot\|$, and $\langle\langle\cdot,\cdot\rangle\rangle_1 = \langle\langle\cdot,\cdot\rangle\rangle$, $\langle\langle\cdot,\cdot\rangle\rangle_0 = ((\cdot,\cdot))$.

Now consider a nonlinear perturbation of equation (9.40):

$$du_t = Au_t dt + F(u_t) dt + dW_t, \quad t > 0,$$
$$u_0 = h \in H, \tag{9.58}$$

where $F : H \to H$ is Lipschitz continuous and of linear growth. Then the associated Kolmogorov equation takes the form:

$$\frac{\partial}{\partial t}\Psi(v,t) = \mathcal{A}\Psi(v,t) + (F(v), D\Psi(v,t)), \quad \text{a.e.}\, v \in V,$$
$$\Psi(v,0) = \Phi(v), \quad v \in H, \tag{9.59}$$

where, as defined in Theorem 4.5, $\mathcal{A} : \mathbb{H}_2 \to \mathbb{H}$ is given by

$$\mathcal{A}\Phi = \frac{1}{2}Tr\,[RD^2\Phi(v)] + \langle Av, D\Phi(v)\rangle, \quad \text{a.e.}\, v \in H. \tag{9.60}$$

For a regular F to be specified later, the additional term $(F(v), D\Psi(v,t))$ can be defined μ-a.e. $v \in H$. It appears that the above differential operator is a special case of $\mathcal{A}$ defined by equation (9.11). However there is noticeable difference between them in that the latter was defined only for $v \in V$, while the operator (9.60) is defined μ-a.e. v in H. Moreover the classical solution of the Kolmogorov equation (9.59) requires a $\mathbf{C}_b^2(H)$-datum Φ [19]. Here we may allow Φ to belong to $\mathbb{H}$, but the solution will be given in a generalized sense. In particular we will consider the mild and the strong solutions of equation (9.59) as in finite dimensions.

Let $\lambda > 0$ be a parameter. By changing Ψ_t to $e^{\lambda t}\Psi_t$ in (9.59), the new Ψ_t will satisfy the equation:

$$\frac{\partial}{\partial t}\Psi(v,t) = \mathcal{A}_\lambda\Psi(v,t) + (F(v), D\Psi(v,t)), \quad \text{a.e.}\, v \in V,$$
$$\Psi(v,0) = \Phi(v), \quad v \in H, \tag{9.61}$$

where $\mathcal{A}_\lambda = (\mathcal{A} - \lambda I)$ and I is the identity operator in $\mathbb{H}$. Clearly the problems (9.59) and (9.61) are equivalent, as far as the existence and uniqueness questions are concerned. It is more advantageous to consider the transformed problem (9.61).

We first consider the case of a mild solution of (9.61). By a semigroup property given in Theorem 4.5, we rewrite the equation (9.61) in the integral form:

$$\Psi(v,t) = e^{-\lambda t}(P_t\Phi)(v) + \int_0^t e^{-\lambda(t-s)}[P_{t-s}(F, D\Psi_s)](v)\,ds, \tag{9.62}$$

where we denote $\Phi = \Phi(\cdot)$ and $\Psi_s = \Psi(\cdot, s)$. The following lemma will be needed to prove the existence theorem.

Lemma 5.1 Let $\Psi \in L^2((0,T); \mathbb{H})$. Then, for any $\lambda > 0$, there exists $C_\lambda > 0$ such that

$$\||\int_0^t e^{-\lambda(t-s)} P_{t-s} \Psi_s \, ds\||^2 \leq C_\lambda \int_0^T \||\Psi_s\||_{-1}^2 ds. \tag{9.63}$$

Proof. By Lemma 3.1, we can expand $\Psi(s)$ into a series in Hermite polynomials:

$$\Psi_s = \sum_{\mathbf{n}} \psi_{\mathbf{n}}(s) \mathcal{H}_{\mathbf{n}}, \tag{9.64}$$

so that

$$\int_0^t e^{-\lambda(t-s)} P_{t-s} \Psi_s \, ds = \sum_{\mathbf{n}} \int_0^t e^{-(\lambda+\lambda_{\mathbf{n}})(t-s)} \psi_{\mathbf{n}}(s) \, ds \, \mathcal{H}_{\mathbf{n}}, \tag{9.65}$$

where $\psi_{\mathbf{n}}(s) = [\Psi_s, \mathcal{H}_{\mathbf{n}}]$. It follows from (9.65) that

$$\||\int_0^t e^{-\lambda(t-s)} P_{t-s} \Psi_s ds\||^2 = \sum_{\mathbf{n}} |\int_0^t e^{-(\lambda+\lambda_{\mathbf{n}})(t-s)} \psi_{\mathbf{n}}(s)|^2 \, ds$$

$$\leq \sum_{\mathbf{n}} \frac{1}{2(\lambda+\lambda_{\mathbf{n}})} \int_0^t |\psi_{\mathbf{n}}(s)|^2 \, ds$$

$$\leq \frac{1}{2(\lambda \wedge 1)} \sum_{\mathbf{n}} \int_0^T \frac{1}{(1+\lambda_{\mathbf{n}})} |\psi_{\mathbf{n}}(s)|^2 \, ds = C_\lambda \int_0^T \||\Psi_s\||_{-1}^2 ds,$$

where $C_\lambda = 1/2(\lambda \wedge 1)$. This verifies (9.63). $\qquad \square$

Theorem 5.2 Suppose that $F : H \to H$ satisfies the usual Lipschitz continuity and linear growth conditions such that, for any $\Phi \in \mathbb{H}$, $v \in V$,

$$\||(F(v), D\Phi(v))\||_{-1}^2 \leq C \||\Phi(v)\||^2, \quad v \in V, \tag{9.66}$$

for some $C > 0$. Then, for $\Phi \in \mathbb{H}$, the initial-value problem (9.59) has a unique mild solution $\Psi \in \mathbf{C}([0,T]; \mathbb{H})$.

Proof. Let $\mathbb{X}_T$ denote the Banach space $\mathbf{C}([0,T]; \mathbb{H})$ with the sup-norm

$$\||\Psi\||_T = \sup_{0 \leq t \leq T} \||\Psi_t\||. \tag{9.67}$$

Let $\mathbb{Q}$ be a linear operator in $\mathbb{X}_T$ defined by

$$\mathbb{Q}_t \Psi = e^{-\lambda t} P_t \Phi + \int_0^t e^{-\lambda(t-s)} P_{t-s}(F, D\Psi_s) ds, \tag{9.68}$$

for any $\Psi \in \mathbb{X}_T$. Then, by Theorem 4.5 and Lemma 5.1, we have

$$\||Q_t\Psi\||^2 \leq 2\{\|| P_t\Phi\||^2 + \||\int_0^t e^{-\lambda(t-s)}P_{t-s}(F,D\Psi_s)\,ds\||^2$$

$$\leq 2\{\||\Phi\||^2 + C_\lambda \int_0^t \||(F,D\Psi_s)\||_{-1}^2\,ds\}$$

$$\leq 2\||\Phi\||^2 + C_1 \int_0^t \||\Psi_s\||^2\,ds,$$

for some $C_1 > 0$. Hence

$$\||Q\Psi\||_T \leq C_2(1 + \||\Psi\||_T),$$

for some positive C_2 depending on Φ, λ and T. Therefore the map $Q : \mathbb{X}_T \to \mathbb{X}_T$ is well defined. To show the map is contraction for a small t, let $\Psi, \Psi' \in \mathbb{X}_T$. Then

$$\||Q_t\Psi - Q_t\Psi'\||^2 = \||\int_0^t e^{-\lambda(t-s)}P_{t-s}[(F,D\Psi_s) - (F,D\Psi_s')]\,ds\||^2$$

$$\leq C_\lambda \int_0^t \||[(F,D\Psi_s - D\Psi_s')\||_{-1}^2\,ds$$

$$\leq C_3 \int_0^t \||\Psi_s - \Psi_s'\||^2\,ds,$$

for some $C_3 > 0$. It follows that

$$\||Q\Psi - Q\Psi'\||_T \leq \sqrt{C_3 T}\||\Psi - \Psi'\||_T,$$

which shows that Q is a contraction map for a small T. Hence the Cauchy problem (9.61) has a unique mild solution. □

We now consider the strong solutions via a variational formulation. Let us introduce a bilinear form $\mathbf{a}$ on $\mathbb{H}_1 \times \mathbb{H}_1$ defined by

$$\mathbf{a}(\Phi_1, \Phi_2) = \frac{1}{2}\int_H (RD\Phi_1, D\Phi_2)\,d\mu, \quad \text{for } \Phi_1, \Phi_2 \in \mathbb{H}_2, \qquad (9.69)$$

which, by the Green's formula, has the following property:

$$\mathbf{a}(\Phi_1, \Phi_2) = -\langle\langle\mathcal{A}\Phi_1, \Phi_2\rangle\rangle = -\langle\langle\mathcal{A}\Phi_2, \Phi_1\rangle\rangle. \qquad (9.70)$$

In view of (9.69) and (9.70), by setting $\Phi_1 = \Phi_2 = \Phi$ and expanding Φ in terms of Hermite polynomials, it can be shown that

$$\mathbf{a}(\Phi, \Phi) = \frac{1}{2}\||R^{1/2}D\Phi\||^2 = \||\Phi\||_1^2 - \||\Phi\||^2. \qquad (9.71)$$

From (9.69) and (9.71), it follows that

$$|\mathbf{a}(\Phi_1, \Phi_2)| \leq \frac{1}{2} \|| R^{1/2} D\Phi_1 \|| \, \|| R^{1/2} D\Phi_2 \|| \tag{9.72}$$
$$\leq \|| \Phi_1 \||_1 \|| \Phi_2 \||_1.$$

The inequalities (9.71) and (9.72) show that the bilinear form is continuous on $\mathbb{H}_1 \times \mathbb{H}_1$ and it is $\mathbb{H}_1$-coercive. The following lemma, which is well known in the theory of partial differential equations (p. 41, [29]), is valid in a general Hilbert space setting.

Lemma 5.3 The Dirichlet form (9.69) defines a unique bounded linear operator $\hat{\mathcal{A}} : \mathbb{H}_1 \to \mathbb{H}_{-1}$ such that

$$\mathbf{a}(\Phi_1, \Phi_2) = -\langle\langle \hat{\mathcal{A}} \Phi_1, \Phi_2 \rangle\rangle, \quad \text{for} \quad \Phi_1, \Phi_2 \in \mathbb{H}_1, \tag{9.73}$$

and the restriction $\hat{\mathcal{A}}|_{\mathbb{H}_2} = \mathcal{A}$. □

For a sufficiently smooth F, define $\mathcal{B} : \mathbb{H}_2 \to \mathbb{H}$ as follows:

$$\mathcal{B}\Phi(v) = \mathcal{A}\Phi(v) + (F(v), D\Phi(v)), \quad \text{a.e. } v \in H. \tag{9.74}$$

To $\mathcal{B}$ we associate a bilinear form $\mathbf{b}$ defined by

$$\mathbf{b}(\Phi_1, \Phi_2) = \mathbf{a}(\Phi_1, \Phi_2) + \int_H (F(v), D\Phi_1(v)) \, \Phi_2(v) \, \mu(dv), \tag{9.75}$$

for all $\Phi_1, \Phi_2 \in \mathbb{H}_1$.

Then, similar to the above lemma, the following holds.

Lemma 5.4 Suppose that $F : H \to H_0$ is essentially bounded and there is $M > 0$ such that

$$\sup_{v \in H} \| R^{-1/2} F(v) \| \leq M, \quad \mu - \text{a.e.} \tag{9.76}$$

Then the bilinear form $\mathbf{b}$ on $\mathbb{H}_1 \times \mathbb{H}_1$ is continuous and $\mathbb{H}_1$-coercive. Moreover it defines uniquely a bounded linear operator $\hat{\mathcal{B}} : \mathbb{H}_1 \to \mathbb{H}$ so that

$$\mathbf{b}(\Phi_1, \Phi_2) = -\langle\langle \hat{\mathcal{B}} \Phi_1, \Phi_2 \rangle\rangle, \quad \text{for} \quad \Phi_1, \Phi_2 \in \mathbb{H}_1, \tag{9.77}$$

and the restriction $\hat{\mathcal{B}}|_{\mathbb{H}_2} = \mathcal{B}$.

Proof. Similar to Lemma 5.3, it suffices to show that the form $\mathbf{b}$ is continuous and coercive. To this end, consider the integral term in (9.75) and notice the condition (9.76) to get

$$\int_H (F, D\Phi_1)\Phi_2 \, d\mu \leq \int_H \| R^{-1/2} F \| \, \| R^{1/2} D\Phi_1 \| \, |\Phi_2| \, d\mu \tag{9.78}$$
$$\leq M \|| \Phi_1 \||_1 \|| \Phi_2 \|| \leq M(\varepsilon \|| \Phi \||_1^2 + \frac{1}{4\varepsilon} \|| \Phi \||^2),$$

for any $\varepsilon > 0$. The above two upper bounds together with (9.72), (9.73) and (9.75) show that the continuity and the coercivity of $\mathbf{b}$ by choosing ε so small that $M\varepsilon < \frac{1}{2}$. $\qquad\square$

First let us consider the associated elliptic problem. Introduce the elliptic operator

$$\mathcal{B}_\lambda = \mathcal{B} - \lambda I,$$

where $\lambda > 0$ and I is the identity operator in $\mathbb{H}$. For $Q \in \mathbb{H}$, consider the solution of the equation:

$$-\mathcal{B}_\lambda U = Q. \tag{9.79}$$

We associate $\mathcal{B}_\lambda$ to the bilinear form on $\mathbb{H}_1 \times \mathbb{H}_1$:

$$\mathbf{b}_\lambda(\Phi_1, \Phi_2) = \mathbf{b}(\Phi_1, \Phi_2) + \lambda[\Phi_1, \Phi_2]. \tag{9.80}$$

Then a function $U \in \mathbb{H}_1$ is said to be a strong solution to the elliptic equation (9.79) if the function U satisfies

$$-\langle\langle \mathcal{B}_\lambda U, \Phi \rangle\rangle = [Q, \Phi], \quad \forall \, \Phi \in \mathbb{H}_1. \tag{9.81}$$

The following existence theorem can be easily proved.

Theorem 5.5 Let the conditions for Lemma 5.4 be satisfied. Then, for $\Phi \in \mathbb{H}$, there exists $\lambda_0 > 0$ such that the elliptic equation (9.81) has a unique strong solution $U \in \mathbb{H}_1$.

Proof.   Consider the equation:

$$\mathbf{b}_\lambda(U, \Phi) = [Q, \Phi] \quad \forall \, \Phi \in \mathbb{H}_1. \tag{9.82}$$

By Lemma 5.4, it is easily seen that $\mathbf{b}_\lambda : \mathbb{H}_1 \times \mathbb{H}_1 \to \mathbf{R}$ is continuous. To show the coercivity, we invoke (9.71), (9.78) and (9.80) to get

$$\mathbf{b}_\lambda(\Phi, \Phi) \geq \|\Phi\|_1^2 + (\lambda - 1)\|\Phi\|^2 - \int_H |(F(v), D\Phi_1(v))\Phi_2(v)| \, \mu(dv)$$
$$\geq \|\Phi\|_1^2 + (\lambda - 1 - M/4\varepsilon)\|\Phi\|^2 - M\varepsilon\|\Phi\|_1^2$$
$$\geq \alpha\|\Phi\|_1^2,$$

where we chose ε small enough so that $\alpha = 1 - M\epsilon > 0$ and λ so large that $\lambda \geq \lambda_0 = 1 + M/4\varepsilon$. Hence the form is also $\mathbb{H}_1$-coercive. Now, in view of (9.81) and (9.82), we can apply the Lax-Milgram theorem (p. 92, [86]) to assert that there exists a continuous injective linear operator, denoted by $(-\mathcal{B}_\lambda)$ from $\mathbb{H}_1$ into $\mathbb{H}_{-1}$ and a unique $U \in \mathbb{H}_1$ such that

$$\mathbf{b}_\lambda(U, \Phi) = -\langle\langle \mathcal{B}_\lambda U, \Phi \rangle\rangle = [Q, \Phi] \quad \forall \, \Phi \in \mathbb{H}_1.$$

This proves that the elliptic equation (9.81) has a unique strong solution. $\square$

Next we consider the strong solution of the parabolic equation (9.61) with a nonhomogeneous term:

$$\frac{\partial}{\partial t}\Psi(v,t) = A_\lambda\Psi(v,t) + (F(v), D\Psi(v,t)) + \mathcal{G}_t, \quad \text{a.e. } v \in V, \tag{9.83}$$

$$\Psi(v,0) = \Phi(v), \quad v \in H,$$

where $\mathcal{G} \in L^2((0,T);\mathbb{H})$.

Let $\mathcal{X}$ denote the space of $\mathbb{H}$-valued functions Ψ on $(0,T)$, with $\Psi_t \in \mathbb{H}_1$ and $\frac{\partial}{\partial t}\Psi_t \in \mathbb{H}_{-1}$ a.e. t, such that

$$\|\Psi\|_{\mathcal{X}} = \{\int_0^T [\|\frac{\partial}{\partial t}\Psi_t\|_{-1}^2 + \|\Psi_t\|_1^2]dt\}^{1/2} < \infty.$$

Also introduce the space $\mathcal{Z} = L^2((0,T);\mathbb{H}_1) \times \mathbb{H}$ with norm:

$$\|(\Psi;\Phi)\|_{\mathcal{Z}} = \{\int_0^T \|\Psi_t\|_1^2 dt + \|\Phi\|^2\}^{1/2} < \infty,$$

for $\Psi \in L^2((0,T);\mathbb{H}_1)$, $\Phi \in \mathbb{H}$. Denote the space $\mathcal{Y} = \{(\Psi;\Phi) \in \mathcal{Z} : \Psi \in \mathcal{X}, \lim_{t\downarrow 0} \Psi_t = \Phi, \text{ and } \lim_{t\uparrow T} \Psi_t = 0 \text{ in } H\}$.

Let β_λ, β'_λ be two bilinear forms on $\mathcal{X} \times \mathcal{X}$ defined by

$$\beta_\lambda(\Psi,\Theta) = \int_0^T \{-\langle\langle\frac{\partial}{\partial t}\Theta_t, \Psi_t\rangle\rangle + \mathbf{b}_\lambda(\Psi_t, \Theta_t)\}dt, \tag{9.84}$$

and

$$\beta'_\lambda(\Psi,\Theta) = \int_0^T \{\langle\langle\frac{\partial}{\partial t}\Psi_t, \Theta_t\rangle\rangle + \mathbf{b}_\lambda(\Psi_t, \Theta_t)\}dt. \tag{9.85}$$

Lemma 5.6 Assume as in Lemma 5.4. Then, there exists $\lambda_0 > 0$ such that, for $\lambda > \lambda_0$ and for any $\Psi \in \mathcal{X}$, the following inequalities hold:

$$2\beta_\lambda(\Psi,\Psi) + \|\Psi_T\|^2 \geq \frac{1}{2}\int_0^T \|\Psi_t\|_1^2 dt + \|\Psi_0\|^2, \tag{9.86}$$

and

$$2\beta'_\lambda(\Psi,\Psi) + \|\Psi_0\|^2 \geq \frac{1}{2}\int_0^T \|\Psi_t\|_1^2 dt + \|\Psi_T\|^2. \tag{9.87}$$

Proof. The above inequalities follow easily from the equality:

$$2\int_0^T \langle\langle\frac{\partial}{\partial t}\Psi_t, \Psi_t\rangle\rangle dt = \|\Psi_T\|^2 - \|\Psi_0\|^2,$$

and the previous estimate on $\mathbf{b}_\lambda$. $\qquad\qquad\qquad\qquad\qquad\qquad\qquad$ □

With the aid of the above results, we can show the existence of a strong solution. Here, for $\Phi \in \mathbb{H}$ and $\mathcal{G} \in L^2((0,T); \mathbb{H})$, we say that Ψ is a strong solution if $\Psi = U \in \mathcal{X}$ and it satisfies the variational equation:

$$\beta_\lambda(U, \Theta) = \int_0^T [\mathcal{G}_t, \Theta_t]\, dt + [\Phi, \Theta_0], \quad \forall\, (\Theta; \Theta_0) \in \mathcal{Y}. \qquad (9.88)$$

Theorem 5.7 Let the conditions for Lemma 5.4 be satisfied, and let $\mathcal{G} \in L^2((0,T); \mathbb{H})$ such that

$$\int_0^T \||\mathcal{G}_t\||^2\, dt \le M, \qquad (9.89)$$

for some $M > 0$. Then, for $\Phi \in \mathbb{H}$, the parabolic equation (9.83) has a unique strong solution $U \in \mathbf{C}([0,T]; \mathbb{H}) \cap L^2((0,T); \mathbb{H}_1)$.

Proof. Based on the framework Gauss-Sobolev spaces developed so far, it is possible to give a proof similar to that for parabolic equations in finite dimensions. In particular we shall make use of an extended Lax-Milgram theorem due to J. L. Lions [58]. To proceed, we first regard β_λ as a bilinear form $\hat{\beta}_\lambda$ on $\mathcal{Z} \times \mathcal{Y}$ by setting

$$\hat{\beta}_\lambda(\hat{\Psi}, \hat{\Theta}) = \beta_\lambda(\Psi, \Theta), \quad \text{for } \hat{\Psi} = (\Psi; \Psi_0),\ \hat{\Theta} = (\Theta; \Theta_0).$$

It is easy to show that $\beta_\lambda : \mathcal{X} \times \mathcal{X} \to \mathbf{R}$ is continuous so that, for each fixed $\hat{\Theta} \in \mathcal{Y}$, the linear functional $\hat{\beta}_\lambda(\cdot, \hat{\Theta}) : \mathcal{Z} \to \mathbf{R}$ is continuous. The form $\hat{\beta}_\lambda$ is also coercive for $\lambda > \lambda_0$ in the sense that

$$\hat{\beta}_\lambda(\hat{\Psi}, \hat{\Psi}) \ge \frac{1}{4} \|\hat{\Psi}\|_{\mathcal{Z}}, \quad \forall\, \hat{\Psi} \in \mathcal{Y}.$$

This follows from the inequality (9.86) which, for $\hat{\Psi} \in \mathcal{Y}$, yields

$$\hat{\beta}_\lambda(\hat{\Psi}, \hat{\Psi}) \ge \frac{1}{4} \int_0^T \||\Psi_t\||_1^2 dt + \frac{1}{2} \||\Psi_0\||^2 \ge \frac{1}{4} \|\Psi\|_{\mathcal{Z}},$$

as claimed. Now, given $\mathcal{G} \in L^2((0,T); \mathbb{H})$ and $\Phi \in \mathbb{H}$, define the linear functional

$$\varphi(\hat{\Psi}) = \int_0^T [\Psi_t, \mathcal{G}_t]\, dt + [\Psi_0, \Phi].$$

Clearly $\varphi : \mathcal{Z} \to \mathbf{R}$ is continuous. This property together with the continuity and coercivity of $\hat{\beta}_\lambda$ imply that the Lax-Milgram-Lions theorem alluded to earlier can be applied to obtain an element $\hat{U} \in \mathcal{Z}$ such that

$$\hat{\beta}_\lambda(\hat{U}, \hat{\Theta}) = \varphi(\hat{\Theta}), \quad \forall\, \hat{\Theta} \in \mathcal{Y}. \qquad (9.90)$$

Since $\hat{\beta}_\lambda(\hat{\Psi},\hat{\Theta}) = \beta_\lambda(\Phi,\Theta)$, the equations (9.88) and (9.90) are equivalent. Therefore U is a strong solution of the equation (9.83).

For uniqueness, suppose U and $\tilde{U}$ are two solutions. Then the difference δU satisfies equation (9.83) with $\mathcal{G} = 0$ and $\Phi = 0$. Then, by an integration by parts, the associated variational equation can be written as

$$\beta'_\lambda(\delta U, \delta U) = 0,$$

which, in view of (9.87), implies that

$$\int_0^T |||\delta U_t|||_1^2 dt = 0.$$

Hence $U_t = \tilde{U}_t$ a.e. t, or the solution is unique. As in a Sobolev space setting, we have $U \in \mathbf{C}([0,T];\mathbb{H})$ due to the imbedding $\mathcal{X} \subset \mathbf{C}([0,T];\mathbb{H})$. $\square$

Remarks:

(1) As a corollary of the theorem, the Kolmogorov equation (9.59) has a unique solution $\Psi \in \mathbf{C}([0,T];\mathbb{H}) \cap L^2((0,T);\mathbb{H}_1)$ given by $\Psi_t = e^{\lambda t}U_t$ with $\mathcal{G} = 0$.

(2) In infinite dimensions, nonlinear parabolic equations can arise, for instance, from stochastic control of partial differential equations in the form of Hamilton-Jacobi-Bellman equations [21]. The analysis will be more complicated technically.

As an example, consider a stochastic reaction-diffusion equation in a bounded domain $D \subset \mathbf{R}^d$ as follows:

$$\frac{\partial u}{\partial t} = (\kappa \Delta - \alpha)u + \int_D k(x,y)f(u(y,t))dy + \dot{W}_t(x),$$

$$u|_{\partial D} = 0, \quad u(x,0) = v(x),$$

(9.91)

for $x \in D$, $t \in (0,T)$, where $k(x,y), v(x)$ are given functions, and $W_t(\cdot)$ is a Wiener process in $H = L^2(D)$ with a bounded covariance function $r(x,y)$. Let $A = (\kappa \Delta - \alpha)$, $F(v) = \int_D k(\cdot,y)f(v(y))dy$ and $V = H_0^1$. Suppose that

(1) $f : \mathbf{R} \to \mathbf{R}$ is bounded and uniformly Lipschitz continuous.

(2) Let K and R denote the integral operators with kernels $k(x,y)$ and $r(x,y)$, respectively. Assume that R commutes with Δ in H^2, and for some $C > 0$, we have

$$\|R^{-1/2}Kv\| \le C\|v\|, \quad v \in H.$$

Under the above assumptions, it can be shown that the conditions for Theorem 5.7 are satisfied. Let $\{\lambda_k\}$ and $\{\rho_k\}$ be the eigenvalues for $(-\Delta)$ and R, respectively, with the common eigenfunctions $\{e_k\}$. The invariant measure μ is a centered Gaussian measure with covariance function $r(x, y)$ given by

$$r(x, y) = \sum_{k=1}^{\infty} \frac{\rho_k}{2\lambda_k} e_k(x) e_k(y).$$

The associated Kolmogorov equation can be written down expressively as

$$
\begin{aligned}
\frac{\partial}{\partial t} \Psi(v, t) &= \frac{1}{2} \int_D \int_D r(x, y) \frac{\delta^2 \Psi(v, t)}{\delta v(x) \delta v(y)} \, dx dy \\
&+ \int_D [(\kappa \Delta - \alpha) v(x)] \frac{\delta \Psi(v, t)}{\delta v(x)} \, dx \\
&+ \int_D \int_D k(x, y) \frac{\delta \Psi(v, t)}{\delta v(x)} f(v(y)) \, dx dy, \quad \mu - \text{a.e. } v \in L^2(D),
\end{aligned}
\tag{9.92}
$$

$$\Psi(v, 0) = \Phi(v), \quad v \in L^2(D),$$

where $\dfrac{\delta \Phi(v)}{\delta v(x)}$ and $\dfrac{\delta^2 \Phi(v)}{\delta v(x) \delta v(y)}$ denote the variational (Volterra) derivatives of the first and second orders. By applying Theorem 5.7, we can conclude that the Kolmogorov equation (9.92) has a unique strong solution Ψ_t which is continuous in $L^2(\mathbb{H}, \mu)$ for $0 \leq t \leq T$ and it is square-integrable over $(0, T)$ in $\mathbb{H}_1$. Moreover the solution has the probabilistic representation: $\Psi(v, t) = E\{\Phi(u_t) | u_0 = v\}$.

Remark: An extensive treatment of the Kolmogorov equations and the parabolic equations in infinite dimensions can be found in the book [21], where many recent research papers on this subject are given in its bibliography.

9.6 Characteristic Functionals and Hopf Equations

Let us return to the autonomous stochastic evolution equation (9.4) which is recaptured below for convenience:

$$
\begin{aligned}
du_t &= A u_t dt + F(u_t) dt + \Sigma(u_t) dW_t, \quad t \geq 0, \\
u_0 &= \xi.
\end{aligned}
\tag{9.93}
$$

Assume the conditions for the existence and uniqueness of a strong solution in Theorem 6-7.5 are met. Given the initial distribution μ_0 for ξ, let μ_t denote the solution measure for u_t so that

$$\mu_t(S) = Prob.\{\omega : u_t(\omega) \in S\},$$

where S is a Borel subset of H. For every $t \in [0, T]$, the characteristic functional for the solution u_t is defined by

$$\Theta_t(\eta) = E\left\{e^{i(u_t,\eta)}\right\} = \int_H e^{i(v,\eta)}\mu_t(dv), \quad \eta \in H, \tag{9.94}$$

with

$$\Theta_0(\eta) = E\left\{e^{i(\xi,\eta)}\right\} = \int_H e^{i(v,\eta)}\mu_0(dv), \quad \eta \in H. \tag{9.95}$$

Here our goal is to study the time evolution of the solution measure μ_t by means of its characteristic functional Θ_t. In particular we are interested in the derivation of an evolution equation for Θ_t, if possible. Instead of working with the solution measure, the characteristic functional is much easier to deal with analytically. Moreover it may serve as the moment generating functional to compute moments of the solution. To do so, we need the following lemma, which can be proved as in finite dimension (see, e.g., [24]).

Lemma 6.1 Suppose that u_t has k finite moments such that

$$\sup_{0 \leq t \leq T} E\|u_t\|^k < \infty, \quad k = 1, 2, \cdots, m.$$

Then, for each $t \in [0, T]$, $\Theta_t(\cdot)$ is k-time continuously differentiable and the Fréchet derivatives $D^k\Theta_t$ are bounded in the sense that

$$\sup_{0 \leq t \leq T} \|D^k\Theta_t\|_{\mathcal{L}_2} < \infty, \quad k = 1, 2, \cdots, m,$$

where $\|\cdot\|_{\mathcal{L}_2}$ denotes the norm in $\mathcal{L}_2(H \times \overset{(k)}{\cdots} \times H; \mathbf{R})$, or the Hilbert-Schmidt norm of a k-linear form.

Conversely, if $\Theta_t(\eta)$ is k-time continuously differentiable at $\eta = 0$, then u_t has k finite moments and, for any $h_i \in H, i = 1, \cdots, k$,

$$E\left\{(u_t, h_1)\cdots(u_t, h_k)\right\} = (-i)^k D^k\Theta_t(0)(h_1 \otimes \cdots \otimes h_k). \quad \square \tag{9.96}$$

By Theorem 7-2.2, the Itô formula holds for a strong solution. By applying the formula to the exponential functional $\exp\{i(u_t, \eta)\}$ and then taking the expectation, we get, for $\eta \in V$,

$$\Theta_t(\eta) = E\left\{e^{i(\xi,\eta)} + \int_0^t e^{i(u_s,\eta)}[i\langle Au_s, \eta\rangle \right.$$
$$\left. +i(F(u_s), \eta) - \tfrac{1}{2}(Q_R(u_s)\eta, \eta)]ds\right\} \tag{9.97}$$
$$= \Theta_0(\eta) + E\int_0^t e^{i(u_s,\eta)}\{i\langle Au_s + F(u_s), \eta\rangle - \frac{1}{2}(Q_R(u_s)\eta, \eta)\}ds,$$

where $Q_R(v) = \Sigma(v)R\Sigma^\star(v)$.

Referring to (9.94) and (9.96), equation (9.97) can be rewritten as

$$\int_H e^{i\langle v,\eta\rangle}\mu_t(dv) = \int_H e^{i\langle v,\eta\rangle}\mu_0(dv) + \int_0^t\int_H e^{i\langle v,\eta\rangle}G(\eta,v)\mu_s(dv)ds, \quad (9.98)$$

for $t \in [0,T]$, $\eta \in V$, where

$$G(\eta,v) = i\langle Av + F(v),\eta\rangle - \frac{1}{2}(Q_R(v)\eta,\eta). \quad (9.99)$$

This equation is known sometimes as a generalized Hopf equation for equation (9.93). Given the initial distribution μ_0, the equation (9.98) can be regarded as an integral equation for the evolution of the solution measure μ_t. It can be shown that the equation has a unique solution. By a finite-dimensional projection in V_n and making use of some weak convergence results, the following theorem is proved in (pp. 378-381, [83]). Since the proof is somewhat long, it will not be reproduced here.

Theorem 6.2   Assume that the conditions for the existence Theorem 6-7.5 hold. Then, for a given initial probability measure μ_0 on H, there exists a unique probability measure μ_t on H such that it satisfies the equation (9.98) and the condition

$$\sup_{0\leq t\leq T}\int_H \|v\|^2\mu_t(dv) + \int_0^T\int_H \|v\|_V^2\mu_t(dv)dt < \infty. \qquad \square$$

It would be of great interest to obtain an equation for the characteristic functional Θ_t from equation (9.97). Unfortunately this is only possible when the coefficients $F(v)$ and $\Sigma(v)$ are of a polynomial type. For the purpose of illustration, we will consider only the linear equation with an additive noise:

$$\begin{aligned} du_t &= Au_tdt + \alpha u_tdt + dW_t, \quad t \geq 0, \\ u_0 &= \xi, \end{aligned} \quad (9.100)$$

where $A : V \to V'$ is self-adjoint, coercive linear operator as given before, $\alpha \in \mathbf{R}$, W_t is a H-valued Wiener process with trace-finite covariance operator R commuting with A, and the initial distribution $\mathcal{D}\{\xi\} = \mu_0$. In this case the equation (9.99) becomes

$$G(\eta,v) = i[\langle Av,\eta\rangle + \alpha(v,\eta)] - \frac{1}{2}(R\eta,\eta). \quad (9.101)$$

Now we need the following lemma.

Lemma 6.3   The following identities hold:

$$\langle\varphi, D\Theta_t(\eta)\rangle = i\int_H \langle\varphi,v\rangle e^{i\langle v,\eta\rangle}\mu_t(dv), \quad \forall\varphi \in V', \quad (9.102)$$

and

$$\langle D^2\Theta_t(\eta)g, h\rangle = -\int_H (g, v)(h, v)e^{i(v,\eta)}\mu_t(dv), \quad \forall g, h \in H, \qquad (9.103)$$

where $D\Phi(\eta)$ denotes the Fréchet derivative of Φ in $\eta \in V$.

Proof. Notice that, by the existence theorem for (9.93), μ_t is supported in V. Similar to the drift term in the Kolmogorov equation, the pairing $\langle\varphi, v\rangle$, though undefined for $v \in H$, may be defined μ_t-a.e. in H as a random variable.

In (9.102) and (9.103), by differentiating $\Theta_t(\eta)$ in the integral (9.94) and interchanging the differentiation and integration, these two identities can be easily verified. The formal approach can be justified by invoking Lemma 6.1, the Fubini theorem and the dominated convergence theorem. □

By means of this lemma, we can derive a differential equation for the characteristic functional as stated in the following theorem.

Theorem 6.4 For the linear equation (9.100), let the conditions for Lemma 5.4 hold. Then the characteristic functional Θ_t for the solution process u_t satisfies the Hopf equation:

$$\Theta_t(\eta) = \Phi(\eta) + \int_0^t \{\frac{1}{2} Tr[RD^2\Theta_s(\eta)] \qquad (9.104)$$
$$+\langle A\eta, D\Theta_s(\eta)\rangle + \alpha(\eta, D\Theta_s(\eta))\}ds,$$

for $t \in [0, T]$, $\eta \in V$, with $\Theta_0(\eta) = \Phi(\eta) = E\{e^{i(\xi,\eta)}\}$.

If $E\|\xi\|^2 < \infty$, then the equation has a unique strong solution, which satisfies the Hopf equation in the classical sense.

Proof. In view of (9.100), (9.101), by applying Lemma 6.3, the equation (9.97) leads to the Hopf equation (9.104).

Let $\tilde{W}_t$ be a Wiener process in H with covariance operator R and consider the stochastic equation;

$$\tilde{u}_t = \eta + \int_0^t (A + \alpha I)\tilde{u}_s ds + \tilde{W}_t. \qquad (9.105)$$

It is clear that the Hopf equation (9.104) is in fact the Komogorov equation for the solution $\tilde{u}_t$. It follows from Theorem 5.7 that there is a unique strong solution $\tilde{u}_t^\eta$. By the condition $E\|\xi\|^2 < \infty$, $\Phi \in \mathbf{C}_b^2(H)$, it follows that $\Theta_t(\eta) = E\Phi(\tilde{u}_t^\eta)$ is a strong solution, which, by Theorem 2.4, satisfies the Hopf equation in the strict sense. □

As a simple example, consider the linear parabolic Itô equation in $\mathbf{R}^d$:

$$\frac{\partial u}{\partial t} = (\Delta + \alpha)u + \dot{W}(x,t), \quad x \in \mathbf{R}^d, t \in (0, T), \qquad (9.106)$$
$$u(x, 0) = \xi(x),$$

where $W_t = W(\cdot, t)$ is a R-Wiener process in $H = L^2(\mathbf{R}^d)$, Δ commutes with R in H^2, and ξ is a random variable in H with moments of all orders. Therefore the characteristic functional $\Phi(\eta)$ is analytic at $\eta = 0$ and it has a Taylor series expansion:

$$\Phi(\eta) = 1 + \sum_{n=1}^{\infty} \frac{i^n}{n!} \Phi^{(n)}(\eta, \overset{(n)}{\cdots}, \eta), \qquad (9.107)$$

where

$$\Phi^{(n)} = D^n \Phi(0)$$

is a n-linear form on $(H)^n \overset{(n)}{=} H \times \cdots \times H$, and its kernel is denoted by $\phi^{(n)}(x^1, \cdots, x^{(n)})$, for $x^i \in \mathbf{R}^d$, $i = 1, \cdots, n$. $\phi^{(n)}$ is the n-point moment function of ξ

The Hopf equation (9.104) for equation (9.106) can be written as

$$\frac{\partial}{\partial t} \Theta_t(\eta) = \frac{1}{2} \int \int r(x, y) \frac{\delta^2 \Theta_t(\eta)}{\delta\eta(x)\delta\eta(y)} \, dx dy$$

$$+ \int [(\Delta + \alpha)\eta(x)] \frac{\delta\Theta_t(\eta)}{\delta\eta(x)} \, dx, \quad \mu - \text{a.e. } \eta \in H, \qquad (9.108)$$

$$\Theta_0(\eta) = \Phi(\eta), \quad \eta \in H,$$

Since ξ has all of its moments, so does the solution u_t. By a Taylor series expansion for $\Psi_t(\eta)$, we get

$$\Theta_t(\eta) = 1 + \sum_{n=1}^{\infty} \frac{i^n}{n!} \Theta_t^{(n)}(\eta, \overset{(n)}{\cdots}, \eta), \qquad (9.109)$$

where $\Theta_t^{(n)}$ is a n-linear form as yet to be determined. Let $\theta_t^{(n)}(x^1, \cdots, x^{(n)})$ denote its kernel, which is the moment function of order n for the solution u_t so that

$$\Theta_t^{(n)}(\eta, \cdots, \eta) = (\theta_t^{(n)}, \eta \otimes \overset{(n)}{\cdots} \otimes \eta).$$

By substituting the series (9.107) and (9.109) into the equation (9.108), one obtains the sequence of parabolic equations for the moment functions as follows:

$$\frac{\partial \theta_t^{(1)}}{\partial t} = (\Delta + \alpha)\theta_t^{(1)},$$

$$\theta_0^{(1)}(x) = \phi^{(1)}(x);$$

$$\frac{\partial \theta_t^{(2)}}{\partial t} = (\Delta_1 + \Delta_2 + 2\alpha)\theta_t^{(2)}(x^1, x^2) + r(x^1, x^2),$$

$$\theta_0^{(2)}(x^1, x^2) = \phi^{(2)}(x^1, x^2),$$

$$(9.110)$$

where Δ_i denotes the Laplacian in the variable x^i. For $n > 2$, the moment function θ_t^n satisfies

$$
\frac{\partial \theta_t^{(n)}}{\partial t} = \sum_{i=1}^{n} (\Delta_i + \alpha) \theta_t^{(n)}(x^1, \cdots, x^n)
$$

$$
+ \sum_{i \neq j, i, j = 1}^{n} r(x^i, x^j) \theta_t^{(n-2)}(x^1, \cdots, \check{x}^i, \cdots, \check{x}^j, \cdots, x^n), \tag{9.111}
$$

$$
\theta_0^{(n)}(x^1, \cdots, x^n) = \phi^{(n)}(x^1, \cdots, x^n),
$$

where $\check{x}^i, \check{x}^j$ mean that the variables x^i and x^j are deleted from the argument of $\psi_t^{(n-2)}$. Suppose that the covariance function $r(x, y)$ is bounded, continuous and $r(x, x)$ is integrable. Similar to the proof of Theorem 4-6.2 for moment functions, it can be shown that the solutions $\theta_t^{(n)}$ of the moment equations are bounded, continuous, and the series (9.109) for the characteristic functional $\Theta_t(\eta)$ converges uniformly on any bounded subset of $[0, T] \times H$.

Remarks: The Hopf equation was introduced by the namesake for the statistical solutions of the Navier-Stokes equations [36]. This subject was treated in detail in the book [26]. This type of functional equation has also been applied to other physical problems, such as in the quantum field theory [60]. The Hopf equation for the characteristic functional, if it exists, is a reasonable substitute for the Kolmogorov forward equation for the probability density function, which does not exist in infinite dimensions.

References

[1] Adams, R.A.: *Sobolev Spaces*, Academic Press, New York, 1975.

[2] Arnold, L.: *Stochastic Differential Equations: Theory and Applications*, John Wiley & Sons, New York, 1974.

[3] Baklan, V.V.: *On the existence of solutions of stochastic equations in Hilbert space*, Depov. Akad. Nauk. Ukr. USR., **10** (1963), 1299-1303.

[4] Bensoussan, A., Temam, R.: *Équations aux dérivées partielles stochastiques non linéaires*, Israel J. Math., **11** (1972), 95-129.

[5] Bensoussan, A., Temam, R.: *Équations stochastiques du type Navier-Stokes*, J. Funct. Analy., **13** (1973), 195-222.

[6] Bensoussan, A., Viot, M.: *Optimal control of stochastic linear distributed parameter systems*, SIAM J. Control Optim., **13** (1975), 904-926.

[7] Billinsley, P.: *Convergence of Probability Measures*, Wiley, New York, 1968.

[8] Chow, P.L.: *Stochastic partial differential equations in turbulence-related problems*, in Probabilistic Analysis and Related Topics, Vol. I, Academic Press, New York (1978), 1-43.

[9] Chow, P.L.: *Stability of nonlinear stochastic evolution equations*, J. Math. Analy. Applic., **89** (1982), 400-419.

[10] Chow, P.L.: *Large deviation problem for some parabolic Itô equations*, Comm. Pure and Appl. Math., **45** (1992), 97-120.

[11] Chow, P.L.: *Infinite-dimensional Kolmogorov equations in Gauss-Sobolev spaces*, Stoch. Analy. Applic., **14** (1996), 257-282.

[12] Chow, P.L.: *Stochastic wave equations with polynomial nonlineraity*, Ann. Appl. Probab., **12** (2002), 84-96.

[13] Chow, P.L., Menaldi, J.L.: *Exponential estimates in exit probability for some diffusion processes in Hilbert spaces*, Stochastics, **29** (1990), 377-393.

[14] Chow, P.L., Menaldi, J.L.: *Stochastic PDE for nonlinear vibration of elastic panels*, Diff. Integ. Equations, **12** (1999), 419-434.

274   *References*

[15] Chow, P.L., Khasminskii, R.Z.: *Stationary solutions of nonlinear stochastic evolution equations*, Stoch. Analy. Applic., **15** (1997), 671-699.

[16] Chung, K.L., Williams, R.J.: *Introduction to Stochastic Integration*, Birkhäuser, Boston, 1990.

[17] Courant, R., Hilbert, D.: *Methods of Mathematical Physics* II, Interscience, New York, 1962.

[18] Dalesskii, Yu. L.: *Differential equations with functional derivatives and stochastic equations for generalized processes*, Dokl. Akad. Nauk. SSSR, **166** (1966), 220-223.

[19] Da Prato, G., Zabczyk, J.: *Stochastic Equations in Infinite Dimensions*, Cambridge University Press, Cambridge, 1992.

[20] Da Prato, G., Zabczyk, J.: *Ergodicity for Infinite Dimensional Systems*, Cambridge Univ. Press, Cambridge, 1996.

[21] Da Prato, G., Zabczyk, J.: *Second Order Partial Differential Equations in Hilbert Spaces*, Cambridge Univ. Press, Cambridge, 2002.

[22] Dawson, D.A.: *Stochastic evolution equations*, Math. Biosci., **15** (1972), 287-316.

[23] Deuschel, J-D., Stroock, D.W.: *Large Deviations*, Academic Press, New York, 1989.

[24] Doob, J.L.: *Stochastic Processes*, John Wily, New York, 1953.

[25] Flandoli, F., Maslowski, B.: *Ergodicity of the 2-D Navier -Stokes equations under random perturbations*, Comm. Math. Phys., **171** (1995), 119-141.

[26] Fleming, W.H., Viot, M.: *Some measure-valued processes in population genetic theory*, Indiana Univ. Math. J., **28** (1979), 817-843.

[27] Folland, G.B.: *Introduction to Partial Differential Equations*, Princeton Univ. Press, Princeton, 1976.

[28] Freidlin, M.I, Wentzell, A.D: *Random Perturbations of Dynamical Systems*, Springer-Verlag, New York, 1984.

[29] Friedman, A.: *Partial Differential Equations of Parabolic Type*, Prentice Hall, Englewood Cliffs, N.J., 1964.

[30] Funaki, T., Woycznski, W.A.: Eds., *Nonlinear Stochastic PDEs: Hydrodynamic Limit and Burgers' Turbulence*, IMA Vol. 77, Springer-Verlag, New York, 1996.

[31] Garabedian, P.R.: *Partial Differential Equations*, John Wiley, New York, 1967.

[32] Gihman, I.I., Skorohod, A.V.: *Stochastic Differential Equations*, Springer-Verlag, New York, 1974.

[33] Gross, L.: *Abstract Wiener spaces*, Proc. 5th. Berkeley Symp. Math. Statist. and Probab., **2** (1967), 31-42.

[34] Hale, J.K.: *Ordinary Differential Equations*, Wiley-Interscience, New York, 1969.

[35] Hausenblas, E., Seidler, J.: *A note on maximal inequality for stochastic convolutions*, Chzech. Math. J., **51** (2001), 785-790.

[36] Hopf, E: *Statistical hydrodynamics and functional calculus*, J. Rational Mech. Analy., **1** (1952), 87-123.

[37] Hormander, L.: *Linear Partial Differential Operators*, Springer-Verlag, Berlin, 1963.

[38] Hutson, V., Pym, J.S.: *Applications of Functional Analysis and Operator Theory*, Academic Press, New York, 1980.

[39] Ikeda, N., Watanabe, S.: *Stochastic Differential Equations and Diffusion Processes*, North-Holland, Amsterdam, 1989.

[40] Itô, K.: *Stochastic integral*, Proc. Imper. Acad. Tokyo, **20** (1944), 519-524.

[41] Itô, K.: *On a stochastic integral equation*, Proc. Imper. Acad. Tokyo, **22** (1946), 32-35.

[42] Kahane, J-P.: *Some Random Series of Functions*, Cambridge Univ. Press, Cambridge, 1985.

[43] Khasminskii, R.Z.: *Stochastic Stability of Differential Equations*, Sijthoff & Noordhoff, Apphen aan den Rijn, The Netherlands, 1980.

[44] Keller, J.B.: *On stochastic equations and wave propagation in random media*, Proc. Symp. Appl. Math., **16** (1964), 145-170.

[45] Klyatskin, V.I. Tatarski, V.I.: *The parabolic equation approximation for wave propagation in a medium with random inhomogeneities*, Sov. Phys. JETP, **31** (1970), 335-339.

[46] Kotelenez, P.: *A maximal inequality for stochastic convolution integrals on Hilbert spaces and space-time regularity of linear stochastic partial differential equations*, Stochastics, **21** (1987), 245-358.

[47] Krylov, N.N.: *On L_p−theory of stochastic partial differential equations in the whole space*, SIAM J. Math. Anay., **27** (1996), 313-340.

[48] Krylov, N.N., Rozovskii, B.L.: *On the Cauchy problem for linear stochastic partial differential equations*, Izv. Akad. Nauk. SSSR., **41** (1977), 1267-1284.

[49] Krylov, N.N., Rozovskii, B.L.: *Stochastic evolution equations*, J. Soviet Math., **14** (1981), 1233-1277.

[50] Kunita, H.: *Stochastic flows and stochastic partial differential equations*, Proc. Intern. Congr. Math., Berkeley, (1986), 1021-1031.

[51] Kunita, H.: *Stochastic Flows and Stochastic Differential Equations*, Cambridge University Press, Cambridge, 1990.

[52] Kuo, H-H., Piech, M.A.: *Stochastic integrals and parabolic equations in abstract Wiener space*, Bull. Amer. Math. Soc., **79** (1973), 478-482.

[53] Kuo, H-H.: *Gaussian Measures in Banach Spaces*, Lecture Notes in Math. No.463, Springer-Verlag, New York, 1975.

[54] Kuo, H-H.: *White Noise Distribution Theory*, CRC Press, New York, 1996.

[55] Kuo, H-H.: *Introduction to Stochastic Integration*, Springer, New York, 2006.

[56] Landou, L., Lifschitz, E.: *Fluid Mechanics*, Addison-Wesley, New York, 1963.

[57] Lions, J.L. : *Équations Differentielles, Opérationelles et Problèmes aux Limites*, Springer-Verlag, New York, 1961.

[58] Lions, J.L., Magenes, E.: *Nonhomogeneous Boundary-value Problems and Applications*, Springer-Verlag, New York, 1972.

[59] Liu, K.: *Stability of Infinite Dimensional Stochastic Differential Equations with Applications*, Chapman-Hall/CRC, New York, 2006.

[60] Martin, T.M., Segal, I.: Eds., *Analysis in Function Spaces*, The MIT Press, Cambridge, Mass., 1964.

[61] McKean, H. P. Jr.: *Stochastic Integrals*, Academic Press, New York, 1969.

[62] Menaldi, J. L., Sritharan, S. S.: *Stochastic 2-D Navier-Stokes equations*, Appl. Math. and Opitim., **46** (2002), 31-53.

[63] Métevier, M., Pellaumail, J.: *Stochastic Integration*, Academic Press, New York, 1980.

[64] Métevier, M.: *Semimartigales: A Course in Stochastic Processes*, Walter de Gruyter & Co., Paris, 1982.

[65] Meyer, P.A.: *Notes sur les processus d'Ornstein-Uhlenbeck*, Séminaire de Prob. XVI, LNM bf 920, Springer-Verlag, (1982), 95-133.

[66] Mizohata, S.: *The Theory of Partial Differential Equations*, Cambridge University Press, Cambrige, UK, 1973. Academic Press, New York, 1980.

[67] Pardoux, E.: *Équations aux dérivees partielles stochastiques non linéaires monotones*, These, Université Paris, 1975.

[68] Pardoux, E.: *Stochastic partial differential equations and filtering of diffusion processes*, Stochastics, **3** (1979), 127-167.

[69] Pazy, A.: *Semigroups of Linear Operators and Applications to Partial Differential Equations*, Springer-Verlag, New York, 1983.

[70] Piech, M.A.: *Diffusion semigroups on abstract Wiener space*, Trans. Amer. Math. Soc., **166** (1972), 411-430.

[71] Piech, M.A.: *The Ornstein-Uhlenbeck semigroups in an infinite dimensional L^2 setting*, J. Funct. Analy. **18** (1975), 271-285.

[72] Revuz, D., Yor, M.: *Continuous Martingales and Brownian Motion*, Springer, Berlin, 1999.

[73] Rozovskii, B.L.: *Stochastic Evolution Equations: Linear Theory and Applications to Non-linear Filtering*, Kluwer Academic Pub., Boston, 1990.

[74] Sowers, R.: *Large deviations for the invarince measure for the reaction-diffusion equation with non-Gaussian perturbations*, Probab. Theory Relat. Fields, **92** (1992), 393-421.

[75] Stroock, D.W., Varadhan, S.R.S.: *Multidimensional Diffusion Processes*, Springer-Verlag, Berlin, 1979.

[76] Sz.-Nagy, B., Foias, C.: *Harmonic Analysis of Operators on Hilbert Spaces*, North-Holland, Amsterdam, 1970.

[77] Treves, F.: *Basic Linear Partial Differential Equations*, Academic Press, New York, 1975.

[78] Temam, R.: *Infinite-Dimensional Dynamical Systems in Mechanics and Physics*, Springer-Verlag, New York, 1988.

[79] Tubaro, L.: *An estimate of Burkholder type for stochastic processes defined by stochastic integrals*, Stoch. Analy. and Appl., **2** (1984), 197-192.

[80] Varadhan, S.R.S.: *Large Deviations and Applications*, SIAM Pub. Philidelphia, 1984.

[81] Varadhan, S.R.S.: *Stochastic Processes*, Cournat Institute Lecture Note, New York, 1968.

[82] Viot, M.: *Solutions faibles d'équations aux dérivees partielles stochastiques non linéaires*, These, Université Paris, 1976.

[83] Vishik, M.J., Fursikov, A.V.: *Mathematical Problems in Statistical Hydrodynamics*, Kluwer Acadedic Pub., Boston, 1988.

[84] Walsh, J.B.: *An introduction to stochastic partial differential equations*, Lect. Notes Math., **1180** (1984), Springer-Verlag, Berlin, 265-435.

[85] Watanabe, S.: *Stochastic Differential Equations and Malliavin Calculus*, Tata Inst. Fund. Research, Springer-Verlag, Berlin, 1984.

[86] Yosida, K.: *Functional Analysis*, Springer-Verlag, New York, 1968.

[87] Zakai, M.: *On the optimal filtering of diffusion processes*, Z. Wahrsch. Verw. Geb., **11** (1969), 230-243.

Index

Milton Keynes UK
Ingram Content Group UK Ltd.
UKHW040446071024
449327UK00020B/1028